P9-AGG-185

# THE SOLAR SYSTEM

Voyager at Neptune. PHOTO CREDIT: JPL/NASA

# The Solar System

## *A Practical Guide*

DAVID REIDY & KEN WALLACE

Allen & Unwin

**By the same authors:**
*The Southern Sky*: A practical guide to Astronomy

© David Reidy and Ken Wallace, 1991
This book is copyright under the Berne Convention. No reproduction without permission. All rights reserved.

First published in 1991
Allen & Unwin Australia Pty Ltd
8 Napier Street, North Sydney, NSW 2059 Australia

National Library of Australia
Cataloguing-in-Publication entry:

Reidy, David.
The solar system.

Bibliography.
Includes index.
ISBN 0 04 442288 1.
ISBN 0 04 442260 1 (pbk.).

1. Solar system—Observers' manuals. 2. Solar system—Popular works. I. Wallace, Ken. II. Title.

520

Set in 9/10 pt Times by Graphicraft Typesetters Limited, Hong Kong
Printed by South Wind Production Singapore Pte Ltd

# Contents

# Figures and tables

## FIGURES

## TABLES

# Introduction

If you go outside tonight and look at the sky, and it's clear, you will see the stars. You'll see the Milky Way, our galaxy. You may stand in wonder for a while, then you'll go back inside to watch television or read a book. Imagine for a moment that there is no inside to go to, that night after night you sit under the stars with your family. Then you will have the situation which existed for many millions of years when people lived simple existences as hunters and gatherers. In such a situation you would take a great interest in the stars and all else which happened above you.

Spending every night under the stars you would quickly recognise the patterns you saw. Maybe one pattern reminded you of a buffalo, or maybe a great hunter; you would make up stories about the patterns you saw. Over a lifetime these patterns would not change, becoming like old friends to those who observed them. From time to time other things would appear in the sky: stars which sped across the sky for a few moments only to disappear; great white cloud-like balls with hair streaming behind them which moved slowly through the patterns for some months and then were gone; and most intriguing, some stars which were not content to stay in one place but which wandered throughout the patterns, much as hunters wandered the Earth.

In the past 30 years we have seen the development of space flight, and crewed missions and robot probes have captured the imagination of many. The discoveries from these small machines have been incredible, revealing worlds as complex as our own, yet each different from Earth. Just before this introduction was written, Voyager 2 rendezvoused with Neptune, ending our first phase of planetary exploration.

The images the Voyagers, Mariners and Veneras have sent to Earth have given close-up views of planets which for thousands of years were nothing more than a point of light in the sky. For a few hundred years they have shown themselves as discs in telescopes, yet still they hold our fascination. To the amateur astronomer the observation of the planets and our Sun can offer years of enjoyment and the thrill of 'getting to know the neighbours'.

To the astronomer confined to the city, the great advantage of the planets is their brightness and the ease with which they can be found using even the smallest telescopes. From the Moon seen through binoculars to observations of the rings of Saturn or the moons of Neptune, the solar system can offer the first step on the road of amateur astronomy or be a journey in itself.

This book has been written to bring together not only the techniques needed by amateur astronomers to enjoy their views of the solar system or to make their work scientifically valuable, but also to reveal some of the history of solar system astronomy and the people who took the first tentative steps in our journey of exploration. We also try to present the latest information returned from spacecraft to explain what is happening on the planets we view.

This last task is dependent upon the knowledge we have now. As this book is being written Voyager has passed Neptune and the first analysis of the data it has returned is being undertaken. Magellan has left on its way to Venus, and Galileo on its way to Jupiter and the discoveries made by those two craft will tell us much more than we think we know now. If we were to wait until those craft completed their explorations before commencing this work, then we would never start. Missions such as Cassini to Saturn, Ulysses to the Sun and perhaps people to Mars mean our knowledge of the solar system will continue to increase for many years.

We hope that this book will give you as much enjoyment as the authors have had over the years observing the solar system. It's a worthwhile place to get to know.

A book such as this is not possible without the efforts of the many people whom we wish to thank. Firstly our appreciation goes to the amateur astronomers who gave their observations and photographs. They are named throughout the book along with their contributions. A few others also deserve special thanks. We gratefully acknowledge Benn Martin of the Jet Propulsion Laboratory without whose help the magnificent photographs from the United States' spacecraft would not be in this book.

We are indebted to Dr David Cooke and the staff at the Australian National Radio Astronomy Observatory, Parkes, for their assistance during the Voyager 2 encounter with Neptune. In particular we would like to thank Rick Twardy, director of the Visitors' Centre, for his help and support without which none of those wonderful four days in August would have been possible. Thanks also to Dr Frank Donivan of NASA and his staff for their assistance.

Thank you also to the photographic section of the Australian Antarctic Division for their help in obtaining the photographs of the Aurora Australis. Also thanks to Glenn Moore of the University of Wollongong for access to the university's collection of meteorites.

Finally, we would like to thank those who helped with the checking and rechecking of the manuscript for the book: Audrey Reidy, Lesley Wallace and Vincent Joseph.

# 1

# Astronomical terms and units

Astronomy is a science; in fact, it is the oldest of the sciences, and because it is a science it has developed its own special terms and measurements. Astronomers throughout the world have agreed to follow the conventions of the science as this makes communication between observers much simpler. As amateur astronomers, or just as interested observers, it is important to understand some of these conventions in order to make the most of the numerous books and articles published on the topic. This chapter is aimed at introducing some of the systems astronomers use to describe the heavens.

Because of its antiquity, astronomy has picked up much along the way which doesn't quite fit with our modern scientific approach to things. One of these, perhaps the best known aspect of astronomy, is the use of *constellations*. Constellations arose when people, looking up at the stars, began to relate what they saw to their surroundings. They imagined that they could see animals they used for food, hunters and other things which matched their life on Earth. Later, as they became more sophisticated, they saw pictures of their gods and mythological beings. The story of the hero and the captured princess, for example, gives us the constellations of Hydra, Perseus and Andromeda.

As European civilisation grew in the northern hemisphere, the early picture makers only had the northern stars to work with; there were, however, stars further south they had never seen. It wasn't until the fifteenth century that maps were made of the southern stars and constellation names applied. Although peoples like the Australian Aborigines had made their own constellations—the Emu, the Womera—European scientists proceeded to fill the southern skies with their own pictures, reflecting their times. Hence in the southern sky we have pictures such as Horologium, the hour glass or Microscopium, the microscope, along with more mythological creatures such as Centaurus, the centaur.

The boundaries of the older constellations were quite arbitrary, as only the brighter stars were used to make the pictures, fainter stars being associated with the nearest picture. Having such a set of pictures in the sky was very useful for early astronomers who were able to relate objects they saw to the nearest picture—for example, 'It's near the brightest star in the Hunter'. Later as the science of positional astronomy grew, the arbitrary nature of the constellation boundaries was not sufficiently accurate, so they were formalised and given distinct boundaries. It is these boundaries, still in use today, that give us the 88 constellations.

Constellations as they are used in modern astronomy have little to do with the pictures of old. They are regarded more as regions of sky, in the same way that countries are regions of the surface of the Earth. Naming stars and objects after the constellation in which they appear is still practised and there are a number of different systems. The official method is to give the stars' number from the *Smithsonian Astrophysical Catalog*, but as this gives stars names such as 233 765, it is rarely used in amateur astronomy. The more usual method is the *Bayer system*.

The Bayer system classifies the stars within a constellation according to their brightness, by labelling them with a Greek letter: alpha for the brightest star, beta for the next and so on to omega for the 24th brightest. The Greek letter is then attached to the genitive case of the Latin name for the constellation. So, for example, the brightest star in the constellation of Andromeda

**The Solar System data**

| Planet | Mass (kg) | Radii (km) Equatorial / Polar | Density (g.cm$^{-3}$) | Gravitational Acceleration (m.s$^{-2}$) | Escape Velocity (km.s$^{-1}$) | Rotational Period | Axial Tilt | Axes ($\times 10^6$km) semi major / semi minor | Period of revolution | Orbital velocity (km.s$^{-1}$) | Eccentricity | Inclination of orbit |
|---|---|---|---|---|---|---|---|---|---|---|---|---|
| Mercury | $3.3 \times 10^{23}$ | 2439 / 2439 | 5.42 | 3.8 | 4.3 | 58.65d | 0° | 69.7 / 45.9 | 87.969d | 47.89 | 0.2056 | 7.00° |
| Venus | $4.9 \times 10^{24}$ | 6050 / 6050 | 5.25 | 8.6 | 10.3 | 243.01d | −2° | 109 / 107.4 | 224.701d | 35.03 | 0.0068 | 3.39° |
| Earth | $6.0 \times 10^{24}$ | 6378 / 6356 | 5.52 | 9.8 | 11.2 | 23h56m | 23.44° | 152.1 / 147.1 | 365.256d | 29.79 | 0.0167 | — |
| Mars | $6.4 \times 10^{23}$ | 3397 / 3377 | 3.94 | 3.7 | 5.0 | 24h38m | 23.98° | 249.1 / 206.7 | 686.980d | 24.13 | 0.0934 | 1.85° |
| Jupiter | $1.9 \times 10^{27}$ | 71 398 / 66 850 | 1.314 | 22.9 | 59.5 | 9h51m | 3.08° | 815.7 / 740.9 | 11.86223y | 13.06 | 0.0485 | 1.30° |
| Saturn | $5.7 \times 10^{26}$ | 60 000 / 53 880 | 0.69 | 9.1 | 35.6 | 10h14m | 29° | 1507 / 1347 | 29.4577y | 9.64 | 0.0556 | 2.49° |
| Uranus | $8.7 \times 10^{25}$ | 26 145 / 25 500 | 1.19 | 7.8 | 21.2 | 17h12m | 97.92° | 3004 / 2735 | 84.0139y | 6.81 | 0.0472 | 0.77° |
| Neptune | $1.0 \times 10^{26}$ | 24 300 / 23 654 | 1.71 | 11 | 23.6 | 17h48m | 28.80° | 4537 / 4456 | 164.793y | 5.43 | 0.0086 | 1.77° |
| Pluto | $1.3 \times 10^{22}$ | 1142 / 1142 | 2.08 | 4.3 | 5.3 | 6.39d | 118° | 7375 / 4425 | 247.7y | 4.47 | 0.250 | 17.2° |

**1.1** The constellation of Orion is one of the most prominent in the night sky. The constellation is made of many thousands of stars, the brighter ones apparently forming the picture of a hunter. The internationally agreed-upon constellation of Orion has little to do with the ancient picture. This area of sky is called Orion to distinguish it from other regions. It is used by astronomers chiefly to name objects, such as the Orion nebula here seen in the lower centre of the picture. PHOTO CREDIT: KEN WALLACE

would be alpha Andromedæ, or α And. The sixth brightest star would be zeta Andromedæ, or ζ And. Of course, there are some cases where this falls down. For example, α Geminorum is fainter than β Geminorum—but such cases are rare. The Bayer system has one major limitation: there are only 24 letters in the Greek alphabet and all the constellations have many more than 24 stars. So other systems are used, some of them involving the use of Roman letters, others listing objects by their numbers in special catalogues. However, none is as widely used as the Bayer system.

For more modern concepts, astronomy uses the *metric system*: metres for length, kilograms for mass and seconds for time. As we have seen, much of the foundation of modern astronomy was laid before the invention of the metric system and, because of the strange nature of some things astronomers are interested in, astronomy also has quite an array of other units to measure many different quantities.

It is from the Babylonians that we inherit the method of expressing the brightness of a star. It is from the Babylonians also that we have 360° in a circle and 24 hours in a day. The Babylonians had a liking for the number six ($24 = 4 \times 6$, $360 = 6 \times 6 \times 10$) and so when it came to describing the brightness of stars they chose six classes. They placed the brightest stars in the first class, the next brightest in the next class, and so on until the faintest stars were in the sixth class. They called the classes *magnitudes*, and so do we. The brightest stars were magni-

**Table 1.1 The 88 constellations and their genitive cases**

| Name | Genitive | Abbreviation |
|---|---|---|
| Andromeda | Andromedae | And |
| Antlia | Antilae | Ant |
| Apus | Apodis | Aps |
| Aquarius | Aquarii | Aqr |
| Aquila | Aquilae | Aql |
| Ara | Arae | Ara |
| Aries | Arietis | Ari |
| Auriga | Aurigae | Aur |
| Bootes | Bootis | Boo |
| Caelum | Caeli | Cae |
| Camelopardelis | Camelopardelis | Cam |
| Cancer | Cancri | Cnc |
| Canes Venatici | Canum Venaticorum | CVn |
| Canis Major | Canis Majoris | CMa |
| Canis Minor | Canis Minoris | CMi |
| Capricornus | Capricorni | Cap |
| Carina | Carinae | Car |
| Cassiopeia | Cassiopeiae | Cas |
| Centaurus | Centauri | Cen |
| Cephus | Cephi | Cep |
| Cetus | Ceti | Cet |
| Chamaeleon | Chamaeleontis | Cha |
| Circinus | Circini | Cir |
| Columba | Columbae | Col |
| Coma Berenices | Comae Berenices | Com |
| Corona Austrinus | Coronae Austrini | CrA |
| Corona Borealis | Coronae Borealis | CrB |
| Corvus | Corvi | Crv |
| Crater | Crateris | Crt |
| Crux | Crucis | Cru |
| Cygnus | Cygni | Cyg |
| Delphinus | Delphini | Del |
| Dorado | Doradus | Dor |
| Draco | Draconis | Dra |
| Equuleus | Equulei | Equ |
| Eridanus | Eridani | Eri |
| Fornax | Fornacis | For |
| Gemini | Geminorum | Gem |
| Grus | Gruis | Gru |
| Hercules | Herculis | Her |
| Horologium | Horologii | Hor |
| Hydra | Hydrae | Hya |
| Hydrus | Hydri | Hyi |
| Indus | Indi | Ind |
| Lacerta | Lacertae | Lac |
| Leo | Leonis | Leo |
| Leo Minor | Leonis Minoris | LMi |
| Lepus | Leporis | Lep |
| Libra | Librae | Lib |
| Lupus | Lupi | Lup |
| Lynx | Lyncis | Lyn |
| Lyra | Lyrae | Lyr |
| Mensa | Mensae | Men |
| Microscopium | Microscopii | Mic |
| Monoceros | Monocerotis | Mon |
| Musca | Muscae | Mus |
| Norma | Normae | Nor |
| Octans | Octantis | Oct |
| Ophiuchus | Ophiuchi | Oph |
| Orion | Orionis | Ori |
| Pavo | Pavonis | Pav |
| Pegasus | Pegasi | Peg |
| Perseus | Persei | Per |
| Phoenix | Phoenicis | Phe |
| Pictor | Pictoris | Pic |
| Pisces | Piscium | Psc |
| Piscis Austrinus | Piscis Austrini | PsA |
| Puppis | Puppis | Pup |
| Pyxis | Pyxidis | Pyx |
| Reticulum | Reticuli | Ret |
| Sagitta | Sagittae | Sge |
| Sagittarius | Sagittarii | Sgr |
| Scorpius | Scorpii | Sco |
| Sculptor | Sculptoris | Scl |
| Scutum | Scuti | Sct |
| Serpens | Serpentis | Ser |
| Sextans | Sextantis | Sex |
| Taurus | Tauri | Tau |
| Telescopium | Telescopii | Tel |
| Triangulum | Triangui | Tri |
| Triangulum | Trianguli Australe | TrA |
| Tucana | Tucanae | Tuc |
| Ursa Major | Ursae Majoris | UMa |
| Ursa Minor | Ursae Minoris | UMi |
| Vela | Velorum | Vel |
| Virgo | Virginis | Vir |
| Volans | Volantis | Vol |
| Vulpecula | Vulpeculae | Vul |

tude 1, the next mag. 2 and so on to mag. 6. The catalogues of stars which the Babylonians made assigned all the 4000 or so stars they could see to these classes.

As astronomy became a more exact science, the need for a more exact system arose. After all, two observers quite often disagreed about which class a star should be in, so the system was formalised. Astronomers calculated that stars of magnitude 1 were one hundred times brighter than stars of magnitude 6 and so they could divide the classes accordingly. Instead of using a linear scale (mag. 2 is 20 times dimmer than mag. 1, mag. 3 is 40 times dimmer than mag. 1, etc.) they used a logarithmic scale in which each magnitude was 2.51 times fainter than the magnitude preceding it. Thus mag. 2 stars are 2.51 times fainter than mag. 1 stars, mag. 6 stars are 2.51 times fainter than mag. 5 stars and so on. Table 1.3 shows how this works.

Once astronomers started measuring their observations accurately they found that some of the magnitude 1 stars were too bright for that class, so some were given magnitude 0 and still

**Table 1.2 The upper- and lower-case Greek letters**

| | | | | | |
|---|---|---|---|---|---|
| Α | α | Alpha | Ν | ν | Nu |
| Β | β | Beta | Ξ | ζ | Xi |
| Γ | γ | Gamma | Ο | ο | Omicron |
| Δ | δ | Delta | Π | π | Pi |
| Ε | ε | Epsilon | Ρ | ρ | Rho |
| Ζ | ζ | Zeta | Σ | σ | Sigma |
| Η | η | Eta | Τ | τ | Tau |
| Θ | θ | Theta | Υ | υ | Upsilon |
| Ι | ι | Iota | Φ | φ | Phi |
| Κ | κ | Kappa | Χ | χ | Chi |
| Λ | λ | Lambda | Ψ | ψ | Psi |
| Μ | μ | Mu | Ω | ω | Omega |

**Table 1.3 A method for comparing the magnitude and brightness of stars**

| Difference in magnitude | Difference in brightness |
|---|---|
| 0 | 1.00 |
| 1 | 2.51 |
| 2 | 6.31 |
| 3 | 15.85 |
| 4 | 39.81 |
| 5 | 100.00 |
| 6 | 251.19 |
| 7 | 630.96 |
| 8 | 1584.89 |
| 9 | 3981.07 |
| 10 | 10 000.00 |
| 11 | 25 118.86 |
| 12 | 63 095.73 |
| 13 | 158 489.32 |
| 14 | 398 107.17 |
| 15 | 1 000 000.00 |

*Note*: To use this table:

1. Subtract the smaller magnitude from the larger.
2. Look for this value in the left column.
3. Read off the brightness in the right column.

others negative magnitudes. Venus, the brightest of the planets and the brightest object in the sky, except for the Sun and the Moon, can be brighter than mag. −4. As astronomers have built better telescopes and better electronics, the faintest objects which can be seen have become fainter and fainter. Past magnitude 6, small telescopes can see many stars down to magnitude 10, the world's best telescopes can detect stars of magnitude 26, a hundred million times fainter than the faintest visible to the naked eye.

The magnitude system has been increased to cover a much wider range of objects than the Babylonians ever considered, and it has been made even more accurate. Instead of just full steps of magnitudes, astronomers routinely use decimal magnitudes as well. So Venus may reach −4.4 at its brightest, or Uranus 5.9. In some cases, such as variable star observation, magnitudes are taken to two decimal places.

Magnitudes play an important part in amateur astronomy because they determine whether or not you will be able to observe an object or event. Under good conditions the human eye can see to around magnitude 6, with a small telescope this limit is magnitude 10. Under poor conditions, those found in all cities and many towns, the magnitude limit may be as low as 4. If, for example, you find the position of Neptune, you will be unable to see it with your eye as its brightness is generally mag. 7.6, but a small telescope will reveal it to you.

One of the first things to strike a beginning astronomer is the huge size of many of the numbers involved. The distance from the Earth to the Sun, for example, is 148 000 000 km and the mass of the Earth is a staggering 5 980 000 000 000 000 000 000 tonnes, the Sun being over 200 000 times more massive again. If astronomers were to write down numbers like this every time they needed to refer to them, they would very quickly grow tired, and the number of trees in the world would rapidly decrease. So scientists have developed a shorthand way of dealing with such numbers, called *scientific notation*.

Scientific notation lets the scientist ignore all those zeros in a number which don't do anything except say how big the number is. For instance, in the number 12 000, the 1 and 2 are important, the zeros just tell us that the number is thousands. One thousand has three zeros in it—it is equal to $10 \times 10 \times 10$, or $10^3$. Therefore 12 000 is twelve times one thousand, or $12 \times 10^3$. This is the basis of scientific notation. To further standardise the system, scientists take it one step further. Instead of putting 12 at the front, they always use a number between 1 and 10, so $12 \times 10^3$ becomes $1.2 \times 10^4$.

Astronomers also need to deal with very small numbers, so again scientific notation is used. A number such as 0.00047 can be written as $47 \div 100\,000$. or $4.7 \div 10\,000$. This could be written as $4.7 \div 10^4$, but as $10^{-4}$ is the same as $1 \div 10^4$, the number is written as $4.7 \times 10^{-4}$. Using scientific notation makes the very large and very small numbers of science much easier to handle and with a little use, quite easy to understand. As an example, the mass of the Earth can be written as $5.98 \times 10^{21}$ tonnes and the Earth–Sun distance as $1.48 \times 10^{11}$ m.

While on the subject of numbers, there is often confusion between the way in which the Americans treat large numbers and the way the English do. Everyone agrees that one million is 1 000 000, but for larger numbers there are two opinions. The number 1 000 000 000 ($10^9$) is one billion if an American sees it, but only one thousand million to the English. The English would call $10^{12}$ a billion, while the Americans would call it a trillion. In this book, we will use the word 'billion' to denote $10^{12}$ and 'trillion' for the number $10^{18}$. On the whole it is better to stick with scientific notation, however, as everyone agrees on that.

As stated earlier, length in the metric system is measured using metres, and metres are the standard unit of measurement used in astronomy. Metres are very useful on the Earth where it is impossible for anything to be more than 20 000 km from anywhere else, but in astronomy the unit is just too small. Just within our local area, immense distances are involved, the distance to Pluto being $6 \times 10^{12}$ m. If we look at the stars the problem is greater, the nearest star to the Earth (apart from the Sun) is Alpha Centauri it is $4.1 \times 10^{16}$ m, so astronomers use more convenient units.

Within the solar system, one obvious length to use as a standard is the average distance of the Earth from the Sun. Astronomers do indeed use this unit; they call it an *astronomical unit* (AU). This distance is $1.495979 \times 10^{11}$ m. Thus the distance of the Earth from the Sun is 1 AU, while the distance to Pluto is 40 AU.

Outside the solar system astronomers once again strike the difficulty of immense distances. The Sun's closest companion star, Alpha Centauri, is 272 000 AU away, the nearest large galaxy is $1.26 \times 10^{11}$ AU from us; clearly a larger unit is needed. Einstein's theory of relativity says that the speed of light is the same everywhere in the universe, so this constant velocity, 300 000 km per second, is used. If light travels in a straight line for one year, it will cover $9.46 \times 10^{12}$ km; this distance is known as a *light year* (ly). Using this unit of distance, Alpha Centauri is just 4.3 light years distant. The nearest large galaxy, however, is still two million light years away.

Other more esoteric units are used in some fields of astronomy. One is the parsec. Without going into how a parsec is defined, it is the same as 3.26 ly or $3 \times 10^{13}$ km. For the really immense distances at which modern astronomers work, a common unit is the megaparsec—one million parsecs.

One advantage of using light years as a unit of distance measurement is that it also tells you something about the age of the object. If tonight you were to look at Alpha Centauri, the light reaching your eyes would have taken 4.3 years to reach you, so you are not seeing Alpha Centauri as it is today, but as it was 4.3 years ago. If Alpha Centauri is destroyed as you are reading this then neither you nor all the world's astronomers would know about it for another 4.3 years. It is highly unlikely that Alpha Centauri would cease to exist now, but in 1987 while astronomers were observing a small companion galaxy to the Milky Way they saw a faint star die by blowing itself apart in a massive explosion. Because of the distance to this star, it really destroyed itself tens of thousands of years ago, but it wasn't until January 1987 that anyone on Earth knew.

Astronomers very quickly get used to the fact that the observations they are making today use light that may be many millions of years old. In most cases this can be ignored for the sake of simplicity, but with the frontiers of the universe being pushed further back all the time, it is not unusual for an astronomer to be looking at light which started its journey before the Earth was formed.

Within the solar system the distances are too short for us to worry about using light as a measuring rod. The Sun is 8.3 light minutes from the Earth and Pluto at its furthest is only 5.5 light hours from the Sun so, in general, time delays within the solar system are not important to professional or amateur astronomers. In a later chapter, however, we will see how even the short light travel times within the solar system have been used to good effect by some scientists.

One of the first jobs professional astronomers were given was to calculate the measurements we use for time. For many thousands of years astronomers would regulate the calendar so that crops could be planted, or important religious festivals observed. Indeed even today the date of Easter is still determined by astronomical phenomena. It is therefore not surprising that astronomers have their own way of dealing with time.

Our days are determined by the Sun: they start at sunrise and finish at sunset. Originally the time between sunrise and sunset was divided into twelve hours, each of 60 minutes, and this was accurate enough for many thousands of years. This system had the problem that hours in summer were longer than hours in winter, but this was not a great worry. Eventually the time system was extended to cover the hours of darkness to give us the situation we have today: 24 hours, all of the same length, and divided into smaller units for convenience.

Astronomers have a number of uses for time. Like everyone else they need to have a time system based on the Sun, known as solar time. One problem with solar time is that it is different at any given time in different places around the world—if it is 10 p.m. in Sydney, it is 12 p.m. in London, 7 a.m. in New York and midnight in Auckland. In order to communicate effectively, astronomers use a standard system of time measured at the zero of longitude, running through Greenwich, England, and called universal time (UT). When an astronomer makes an observation, it is recorded along with the universal time, so that it can be easily compared with observations made by other astronomers in other countries.

Astronomers also always use the 24-hour style of writing times rather than the a.m./p.m. system. Using a 24-hour clock makes calculations easier; it is simpler to work out the number of hours between 9:30 and 22:30 than between 9.30 a.m. and 10.30 p.m. Dates too can cause problems; in Europe and other places it is usual for 1/4/89 to mean the first of April, but in the United States and Canada this means the fourth of January. To overcome this difficulty it is best to write dates 1 April 1989, with the month in words (possibly abbreviated) to avoid misunderstandings.

A major problem which astronomers have is that their observations usually involve the stars or planets, and these keep a different time from the Sun. We know that each year the Earth rotates once around the Sun, and that it rotates once each day on its axis so, given that there are 365.25 days in a year, the Earth would spin on its axis 365.25 times. However, its one revolution about the Sun would make a total of 366.25 revolutions if the revolutions were counted with respect to the stars.

Because of this extra yearly rotation, the stars appear to rotate around the Earth slightly faster than the Sun does. To an observer this means that the stars appear to rise about four minutes earlier each night. Observations and calculations show that a star will complete one trip around the sky every 23 hours 56 minutes and 4 seconds, and that if a star rose at exactly 22:00:00 tonight, it would rise at 21:56:04 tomorrow night. To facilitate this, astronomers have a second time system, one based on the stars, called the *sidereal time* system. Sidereal time divides one sidereal day into 24 hours and minutes and seconds, the only difference being that sidereal time units are only 99.727 per cent as long as solar time units.

Sidereal time running nearly four minutes fast each day means that the solar and sidereal clocks of astronomers show different times for much of the year—indeed, they only show exactly the same time some time on 23 September each year. Sidereal time is important to amateur astronomers as well, as we will see when we look at right ascension and declination.

Calculating sidereal time is fairly simple with the help of a computer, but laborious if you have to do it by hand. To simplify this, Table 1.4 gives a listing of sidereal times throughout the year. A simple fact to remember is that sidereal time runs four minutes fast, so if sidereal time at 20:00 tonight is 4:15, then tomorrow night it will be 4:19. Thus over a short period, allowances are simple. Remember though that although it is only four minutes a night, over a month the difference is two hours.

The calendar is something we all take for granted; 365 days in a year, an extra day every four years, but of course the calendar is the work of astronomers too. The calendar which is used throughout the world today originated over 2000 years ago during the reign of Julius Cæsar. In the Roman calendar all twelve months had 30 days, with five extra days tacked on at the end of the year. The problem with this was that the calendar kept running fast by one day every four years, so occasionally a few extra days would be put in to balance it. Cæsar decided to formalise it, giving us the twelve months of different lengths and an extra day every four years.

This was a big improvement and the calendar was now right—or almost. The problem is that

**Table 1.4 Sidereal times throughout the year**

| Local Time | 18:00 | 19:00 | 20:00 | 21:00 | 22:00 | 23:00 | 0:00 | 1:00 | 2:00 | 3:00 | 4:00 | 5:00 | 6:00 |
|---|---|---|---|---|---|---|---|---|---|---|---|---|---|
| 6 Jan | 1:00 | 2:00 | 3:00 | 4:00 | 5:00 | 6:00 | 7:00 | 8:00 | 9:00 | 10:00 | 11:00 | 12:00 | 13:00 |
| 3 Jan | 1:30 | 2:30 | 3:30 | 4:30 | 5:30 | 6:30 | 7:30 | 8:30 | 9:30 | 10:30 | 11:30 | 12:30 | 13:30 |
| 21 Jan | 2:00 | 3:00 | 4:00 | 5:00 | 6:00 | 7:00 | 8:00 | 9:00 | 10:00 | 11:00 | 12:00 | 13:00 | 14:00 |
| 28 Jan | 2:30 | 3:30 | 4:30 | 5:30 | 6:30 | 7:30 | 8:30 | 9:30 | 10:30 | 11:30 | 12:30 | 13:30 | 14:30 |
| 5 Feb | 3:00 | 4:00 | 5:00 | 6:00 | 7:00 | 8:00 | 9:00 | 10:00 | 11:00 | 12:00 | 13:00 | 14:00 | 15:00 |
| 13 Feb | 3:30 | 4:30 | 5:30 | 6:30 | 7:30 | 8:30 | 9:30 | 10:30 | 11:30 | 12:30 | 13:30 | 14:30 | 15:30 |
| 20 Feb | 4:00 | 5:00 | 6:00 | 7:00 | 8:00 | 9:00 | 10:00 | 11:00 | 12:00 | 13:00 | 14:00 | 15:00 | 16:00 |
| 28 Feb | 4:30 | 5:30 | 6:30 | 7:30 | 8:30 | 9:30 | 10:30 | 11:30 | 12:30 | 13:30 | 14:30 | 15:30 | 16:30 |
| 7 Mar | 5:00 | 6:00 | 7:00 | 8:00 | 9:00 | 10:00 | 11:00 | 12:00 | 13:00 | 14:00 | 15:00 | 16:00 | 17:00 |
| 15 Mar | 5:30 | 6:30 | 7:30 | 8:30 | 9:30 | 10:30 | 11:30 | 12:30 | 13:30 | 14:30 | 15:30 | 16:30 | 17:30 |
| 23 Mar | 6:00 | 7:00 | 8:00 | 9:00 | 10:00 | 11:00 | 12:00 | 13:00 | 14:00 | 15:00 | 16:00 | 17:00 | 18:00 |
| 30 Mar | 6:30 | 7:30 | 8:30 | 9:30 | 10:30 | 11:30 | 12:30 | 13:30 | 14:30 | 15:30 | 16:30 | 17:30 | 18:30 |
| 7 Apr | 7:00 | 8:00 | 9:00 | 10:00 | 11:00 | 12:00 | 13:00 | 14:00 | 15:00 | 16:00 | 17:00 | 18:00 | 19:00 |
| 14 Apr | 7:30 | 8:30 | 9:30 | 10:30 | 11:30 | 12:30 | 13:30 | 14:30 | 15:30 | 16:30 | 17:30 | 18:30 | 19:30 |
| 22 Apr | 8:00 | 9:00 | 10:00 | 11:00 | 12:00 | 13:00 | 14:00 | 15:00 | 16:00 | 17:00 | 18:00 | 19:00 | 20:00 |
| 30 Apr | 8:30 | 9:30 | 10:30 | 11:30 | 12:30 | 13:30 | 14:30 | 15:30 | 16:30 | 17:30 | 18:30 | 19:30 | 20:30 |
| 7 May | 9:00 | 10:00 | 11:00 | 12:00 | 13:00 | 14:00 | 15:00 | 16:00 | 17:00 | 18:00 | 19:00 | 20:00 | 21:00 |
| 15 May | 9:30 | 10:30 | 11:30 | 12:30 | 13:30 | 14:30 | 15:30 | 16:30 | 17:30 | 18:30 | 19:30 | 20:30 | 21:30 |
| 22 May | 10:00 | 11:00 | 12:00 | 13:00 | 14:00 | 15:00 | 16:00 | 17:00 | 18:00 | 19:00 | 20:00 | 21:00 | 22:00 |
| 30 May | 10:30 | 11:30 | 12:30 | 13:30 | 14:30 | 15:30 | 16:30 | 17:30 | 18:30 | 19:30 | 20:30 | 21:30 | 22:30 |
| 7 Jun | 11:00 | 12:00 | 13:00 | 14:00 | 15:00 | 16:00 | 17:00 | 18:00 | 19:00 | 20:00 | 21:00 | 22:00 | 23:00 |
| 14 Jun | 11:30 | 12:30 | 13:30 | 14:30 | 15:30 | 16:30 | 17:30 | 18:30 | 19:30 | 20:30 | 21:30 | 22:30 | 23:30 |
| 22 Jun | 12:00 | 13:00 | 14:00 | 15:00 | 16:00 | 17:00 | 18:00 | 19:00 | 20:00 | 21:00 | 22:00 | 23:00 | 0:00 |
| 29 Jun | 12:30 | 13:30 | 14:30 | 15:30 | 16:30 | 17:30 | 18:30 | 19:30 | 20:30 | 21:30 | 22:30 | 23:30 | 0:30 |
| 7 Jul | 13:00 | 14:00 | 15:00 | 16:00 | 17:00 | 18:00 | 19:00 | 20:00 | 21:00 | 22:00 | 23:00 | 0:00 | 1:00 |
| 15 Jul | 13:30 | 14:30 | 15:30 | 16:30 | 17:30 | 18:30 | 19:30 | 20:30 | 21:30 | 22:30 | 23:30 | 0:30 | 1:30 |
| 22 Jul | 14:00 | 15:00 | 16:00 | 17:00 | 18:00 | 19:00 | 20:00 | 21:00 | 22:00 | 23:00 | 0:00 | 1:00 | 2:00 |
| 30 Jul | 14:30 | 15:30 | 16:30 | 17:30 | 18:30 | 19:30 | 20:30 | 21:30 | 22:30 | 23:30 | 0:30 | 1:30 | 2:30 |
| 6 Aug | 15:00 | 16:00 | 17:00 | 18:00 | 19:00 | 20:00 | 21:00 | 22:00 | 23:00 | 0:00 | 1:00 | 2:00 | 3:00 |
| 14 Aug | 15:30 | 16:30 | 17:30 | 18:30 | 19:30 | 20:30 | 21:30 | 22:30 | 23:30 | 0:30 | 1:30 | 2:30 | 3:30 |
| 22 Aug | 16:00 | 17:00 | 18:00 | 19:00 | 20:00 | 21:00 | 22:00 | 23:00 | 0:00 | 1:00 | 2:00 | 3:00 | 4:00 |
| 29 Aug | 16:30 | 17:30 | 18:30 | 19:30 | 20:30 | 21:30 | 22:30 | 23:30 | 0:30 | 1:30 | 2:30 | 3:30 | 4:30 |
| 6 Sep | 17:00 | 18:00 | 19:00 | 20:00 | 21:00 | 22:00 | 23:00 | 0:00 | 1:00 | 2:00 | 3:00 | 4:00 | 5:00 |
| 13 Sep | 17:30 | 18:30 | 19:30 | 20:30 | 21:30 | 22:30 | 23:30 | 0:30 | 1:30 | 2:30 | 3:30 | 4:30 | 5:30 |
| 21 Sep | 18:00 | 19:00 | 20:00 | 21:00 | 22:00 | 23:00 | 0:00 | 1:00 | 2:00 | 3:00 | 4:00 | 5:00 | 6:00 |
| 29 Sep | 18:30 | 19:30 | 20:30 | 21:30 | 22:30 | 23:30 | 0:30 | 1:30 | 2:30 | 3:30 | 4:30 | 5:30 | 6:30 |
| 6 Oct | 19:00 | 20:00 | 21:00 | 22:00 | 23:00 | 0:00 | 1:00 | 2:00 | 3:00 | 4:00 | 5:00 | 6:00 | 7:00 |
| 14 Oct | 19:30 | 20:30 | 21:30 | 22:30 | 23:30 | 0:30 | 1:30 | 2:30 | 3:30 | 4:30 | 5:30 | 6:30 | 7:30 |
| 21 Oct | 20:00 | 21:00 | 22:00 | 23:00 | 0:00 | 1:00 | 2:00 | 3:00 | 4:00 | 5:00 | 6:00 | 7:00 | 8:00 |
| 29 Oct | 20:30 | 21:30 | 22:30 | 23:30 | 0:30 | 1:30 | 2:30 | 3:30 | 4:30 | 5:30 | 6:30 | 7:30 | 8:30 |
| 6 Nov | 21:00 | 22:00 | 23:00 | 0:00 | 1:00 | 2:00 | 3:00 | 4:00 | 5:00 | 6:00 | 7:00 | 8:00 | 9:00 |
| 13 Nov | 21:30 | 22:30 | 23:30 | 0:30 | 1:30 | 2:30 | 3:30 | 4:30 | 5:30 | 6:30 | 7:30 | 8:30 | 9:30 |
| 21 Nov | 22:00 | 23:00 | 0:00 | 1:00 | 2:00 | 3:00 | 4:00 | 5:00 | 6:00 | 7:00 | 8:00 | 9:00 | 10:00 |
| 28 Nov | 22 30 | 23:30 | 0:30 | 1:30 | 2:30 | 3:30 | 4:30 | 5:30 | 6:30 | 7:30 | 8:30 | 9:30 | 10:30 |
| 6 Dec | 23:00 | 0:00 | 1:00 | 2:00 | 3:00 | 4:00 | 5:00 | 6:00 | 7:00 | 8:00 | 9:00 | 10:00 | 11:00 |
| 14 Dec | 23:30 | 0:30 | 1:30 | 2:30 | 3:30 | 4:30 | 5:30 | 6:30 | 7:30 | 8:30 | 9:30 | 10:30 | 11:30 |
| 21 Dec | 0:00 | 1:00 | 2:00 | 3:00 | 4:00 | 5:00 | 6:00 | 7:00 | 8:00 | 9:00 | 10:00 | 11:00 | 12:00 |
| 29 Dec | 0:30 | 1:30 | 2:30 | 3:30 | 4:30 | 5:30 | 6:30 | 7:30 | 8:30 | 9:30 | 10:30 | 11:30 | 12:30 |

To find the approximate local sidereal time, find the closest date and look up the sidereal time in the column corresponding to local time. Remember to allow for summer time if necessary.

**Table 1.5A Julian Date**

| Year | A | Year | A |
|---|---|---|---|
| 1990 | 2447891 | 2005 | 2453370 |
| 1991 | 2448256 | 2006 | 2453735 |
| 1992* | 2448621 | 2007 | 2454100 |
| 1993 | 2448987 | 2008* | 2454465 |
| 1994 | 2449352 | 2009 | 2454831 |
| 1995 | 2449717 | 2010 | 2455196 |
| 1996* | 2450082 | 2011 | 2455561 |
| 1997 | 2450448 | 2012* | 2455926 |
| 1998 | 2450813 | 2013 | 2456292 |
| 1999 | 2451178 | 2014 | 2456657 |
| 2000* | 2451543 | 2015 | 2457022 |
| 2001 | 2451909 | 2016* | 2457387 |
| 2002 | 2452274 | 2017 | 2457753 |
| 2003 | 2452639 | 2018 | 2458118 |
| 2004* | 2453004 | 2019 | 2458483 |

*Leap years

**Table 1.5B Julian Date**

| Month | Normal | Leap |
|---|---|---|
| January | 0 | 0 |
| February | 31 | 31 |
| March | 59 | 60 |
| April | 90 | 91 |
| May | 120 | 121 |
| June | 151 | 152 |
| July | 181 | 182 |
| August | 212 | 213 |
| September | 243 | 244 |
| October | 273 | 274 |
| November | 304 | 305 |
| December | 334 | 335 |

To use these tables, start by finding the number A, from table A for the year, add to this the number B from table B, for the month then perform the following calculation:

$$JD = A + B + \text{Day of the month}$$

This gives the Julian date for 12 noon UT on the date in question.

the Earth's year is not exactly 365.25 days long, but a fraction shorter. Using the Julian calendar meant that the years ran fast by about three-quarters of a day per century. Over the centuries these errors grew until by the 1500s they were noticeably large. At that time Pope Gregory decided to do something about it. He commissioned a calendar reform and it was found that removing the extra day from the leap years occurring in years ending in 00, when the first two digits were not divisible by four, would fix the problem. This idea was adopted, so the year 1700, which would under the Julian calendar have been a leap year, wasn't. Neither 1800 nor 1900 were leap years, but 2000 will be, as 20 is divisible by 4.

Having fixed the number of leap days, Gregory still had one problem; the calendar was still ten days in error. The solution was simple: the Pope decreed that the days from 5 October 1582 to 14 October 1582 simply wouldn't exist, the day following the fourth would be the fifteenth. There were riots throughout Europe protesting the loss of ten days of people's lives, but the reform went ahead anyway. To make things difficult, not everywhere followed the Pope's decree; England held out until 1751 and Russia didn't change until after the Revolution in 1918 when the discrepancy was thirteen days.

The calendar we use is quite good for the purpose for which it was designed, but for some astronomers it doesn't work quite so well. Some phenomena in astronomy take place over long periods of time, periods which can conveniently be measured in days. The problem is to determine how many days there are between 5 May 1989 and 6 August 1989. With our calendar there is no quick and easy method of working this out, so astronomers use a system known as *Julian date* (JD).

The Julian date gives each day a number. As I am writing, today is 2447744, that being the number of days since 1 January 4713 BC. Why that date? It was chosen to be far enough back in time for there to be little chance of records of observations before that coming to light. To find the number of days between dates, simply subtract their Julian dates; 5 May 1989 = 2447651; 6 August 1989 = 2447744; 2447744 − 2447651 = 93 days between the two dates. Tables 1.5A and 1.5B give a method of calculating the Julian date.

Angles too play an important part in astronomy. Although we know better, it is much simpler to treat the sky as a hollow sphere which revolves about us once a day, so angles are important in describing the positions of objects in the sky and also the size of objects. The Babylonians divided a circle into six parts each of 60 degrees, giving 360 degrees for a full circle. Each degree they divided into 60 smaller units, minutes, and each minute into 60 seconds, thus there are 3600 seconds in a degree. When writing angles we use the symbols °, ′, ″ for degrees, minutes and seconds respectively; for instance we might write 27°14′27″.

For most everyday applications we rarely need anything smaller than degrees to express measurements, but in astronomy the great distance of objects leads to the common use of minutes and seconds. The Moon appears from Earth to be 30′ of arc wide, just half a degree, the planet Jupiter about 1′ of arc. One minute of arc is the same size as a 1 cm diameter coin seen at a distance of 34 metres and about the smallest angle that can be seen with the unaided human eye. One second of arc would be the size of that 1 cm coin 1.8 km away. The same coin at a distance of one metre is the half degree size of the Moon.

Estimating angles is a useful skill for the amateur astronomer, and there are some simple shortcuts to make the task easier. The width of your hand (fingers spread) held at arm's length is about 15°; the width of a finger at the same distance about 1°. Smaller angles will only be needed when looking through a telescope, and then special instruments are used.

For a beginning astronomer, finding your way around the sky is perhaps one of the hardest tasks to master. To begin with, one part of the sky is much the same as another, so specifying exactly where something is can be tricky. If you've ever tried pointing out some object to a friend not acquainted with astronomy the following exchange will seem all too familiar:

*Astronomer*: You see that fuzzy patch just there?
*Friend*: Where?
*Astronomer*: Just there, between the two bright stars.
*Friend*: Which two bright stars, those ones there?
*Astronomer*: No, the sort of reddish one and the other white one.
*Friend*: That's orange not red.
*Astronomer*: OK, orange then. You see the fuzzy patch?
*Friend:* No.
*Astronomer*: Look! You see between the two stars...
*Friend*: Oh *those* two stars, I thought you meant those ones there.
*Astronomer*: Yes those; the red and white ones.
*Friend*: Which white one?
*Astronomer*: The one that makes sort of a triangle with the other white one.
*Friend*: Funny looking triangle.
*Astronomer*: So, you see the fuzzy looking patch?
*Friend*: Yeah.
*Astronomer*: Well that's the Orion nebula.
*Friend*: So?

Clearly one of the first things professional astronomers did was to come up with a sensible method of defining the position of things in the sky. A number of methods exist; the one which an amateur astronomer will use is largely determined by the sort of telescope available. The simplest method, and the one used with altazimuthly mounted telescopes such as Dobsonians, is the *altazimuth system*.

The altazimuth system combines two measurements, *altitude*—the angle which the object makes to the horizon—and *azimuth*—the direction of the object measured from north. Altitude is simply measured in degrees above the horizon, 0° being on the horizon and 90° being directly overhead. An interesting point to note is that an angle of 30° looks to be about halfway up the sky, whereas we would expect 45° to be halfway up. If you use the altazimuth system simply by eye then you must take this effect into account.

Azimuth is measured in degrees from north, around through east, so north is 0° azimuth, east is 90°, south is 180° and west is 270°. The system continues to 359°59′59″ just before north.

An altazimuth position is usually expressed in terms such as: altitude 27°, azimuth 210°. To use such a measurement is simple. With a compass or other method to find direction, turn to face 210° and look up into the sky 27°; the object will be where you are looking. Similarly, it is easy to make measurements of the position of an object using this system.

A major problem with the altazimuth system of expressing the positions of things is that they move, or more precisely the Earth rotates and the stars and planets appear to move. It is therefore necessary when using the altazimuth system

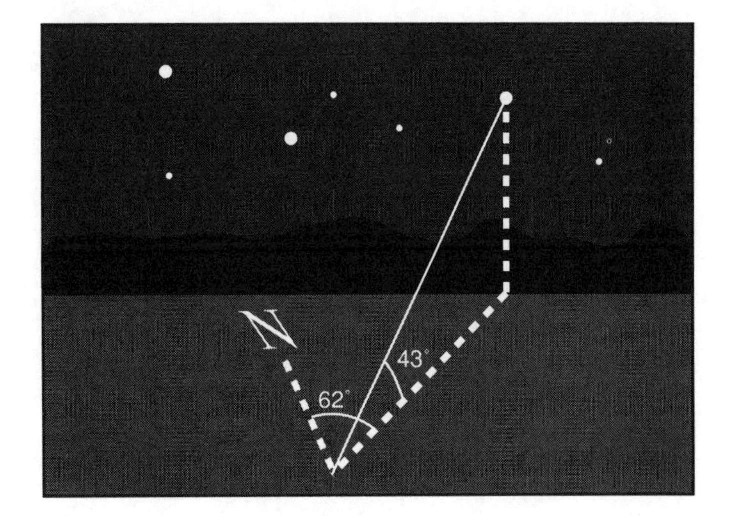


**1.2** Using altitude and azimuth it is possible to locate any object in the sky. Here a star is located by knowing that its altitude (elevation above the horizon) is 43° and that its azimuth (direction from north) is 62°. The disadvantage of this method of measuring position is that the altitude and azimuth of any object are constantly changing.

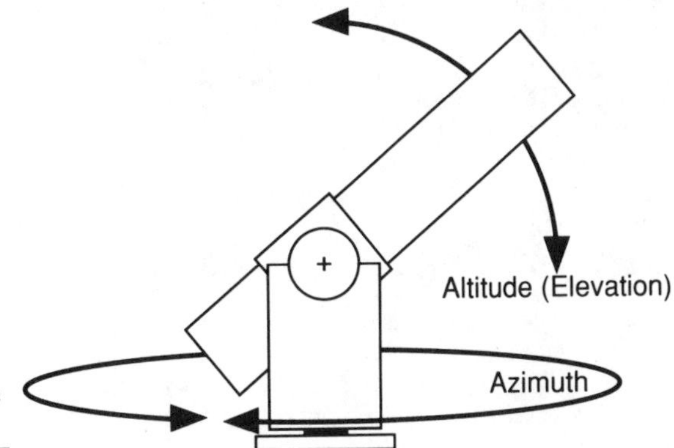


**1.3** The Dobsonian telescope mount is one example of an altazimuth mount. Here the telescope is free to turn in two directions, one parallel to the horizon (azimuth) the other up and down (elevation). This simple method allows quick accurate pointing of the telescope, but is unsuitable for photography and some other forms of observations.

to specify the time of the observation and the location of the observer, as both these factors alter the position of an object in the sky. Clearly it would be most inconvenient if star catalogues and maps all had to be drawn for specific locations and times, so astronomers invented another way of specifying the position of an object: the *equatorial system*.

The equatorial system relies upon a premise that astronomers proved was false many years ago, yet the idea is still useful. That premise is that the Earth is at the centre of everything and is surrounded by a sphere containing the stars, planets and other objects. This imaginary sphere is called the celestial sphere. Because of the Earth's rotation on its axis, the celestial sphere seems to rotate around the Earth about once a day carrying the stars and planets with it.

Astronomers draw lines on the celestial sphere in much the same way as geographers draw lines on the Earth. Directly above the Earth's south pole is a point on the sphere called the *south celestial pole* (SCP), directly above the north pole is the *north celestial pole* (NCP). Above the Earth's equator a line is drawn on the celestial sphere marking the celestial equator, and the correspondence to the lines on the Earth is taken further still.

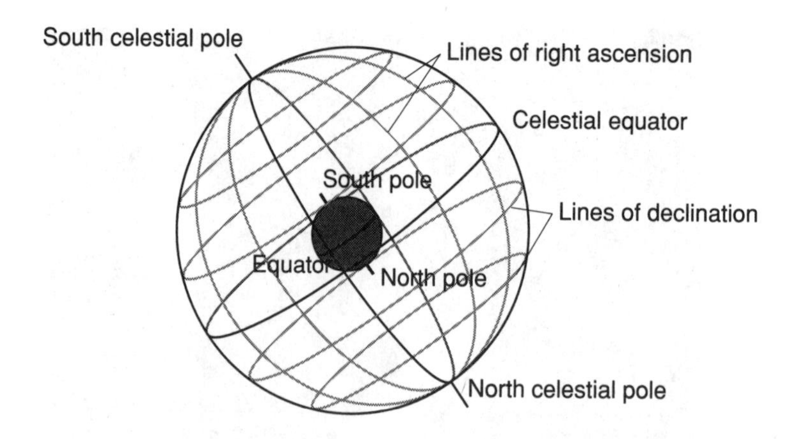


**1.4** The celestial sphere is the imaginary sphere astronomers construct around the Earth to make the measurement of positions easier. The sphere has two poles, the celestial poles above the geographic poles of the Earth and a celestial equator above the Earth's equator. Joining the poles are the lines of right ascension while parallel to the equator are the lines of declination.

**1.5** Looking towards the south, the lines of declination make circles around the south celestial pole while lines of right ascension curve towards the horizon.

Between the north and south poles on Earth there are lines of latitude; in the sky the lines are called lines of *declination*. Declination is measured in degrees from the celestial equator, positive degrees being towards north and negative degrees towards south. The north celestial pole is at +90° and the SCP is at −90°, while the celestial equator is 0°. Lines of declination are related quite closely to latitude. If you are at a latitude of 35°S then the declination directly above you is −35°, if you are at latitude 47°N, then the declination of the zenith is +47°.

Declination allows you to specify one half of an object's position, but you must have the equivalent of longitude to specify it completely. *Right ascension* (RA) is the other half of the equatorial co-ordinate system. Just as lines of longitude stretch from the north to the south pole on Earth, lines of RA go from the north to the south celestial poles. Unlike lines of longitude, however, right ascension is not measured in degrees, minutes and seconds, but in hours minutes and seconds, from 0 hours to 23 hours 59 minutes and 59 seconds.

A problem with the longitude system on Earth was the site for zero longitude; the British

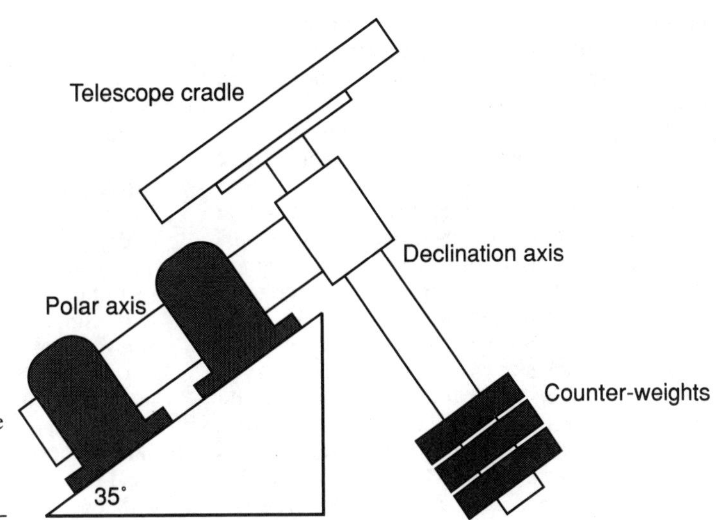


**1.6** A German equatorial mount achieves the goal of pointing the telescope in equatorial co-ordinates by mounting it on a plate on the declination axis. The axis of right ascension is pointed towards one of the celestial poles so that the telescope can follow any object in the sky by movement in that axis alone. The mount shown here is set for latitude 35° N or S.

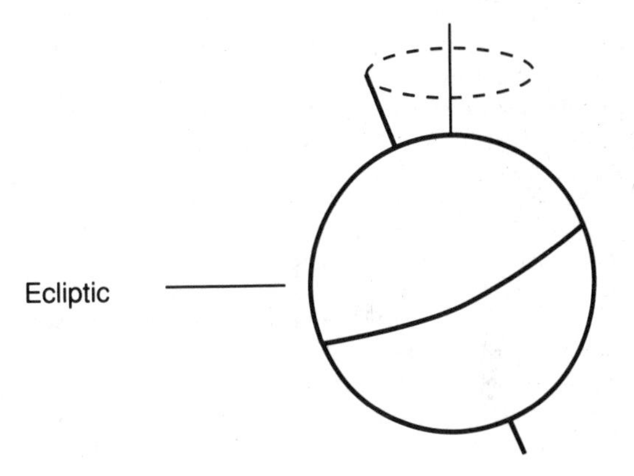


**1.7** The axis of the Earth is tilted at 23.5° to the plane of the ecliptic. Over 25 800 years the direction of the Earth's axis traces a circle 47° in diameter in the sky. This effect is called precession and it is responsible for the gradual change in the equatorial co-ordinates of the stars and other celestial objects.

used a line through the Royal Observatory at Greenwich, the French used a line through Paris and the Spanish through Cadiz. For many years these and other systems were used until 1884 when the nations agreed to use just one line, that through Greenwich. A similar problem of choosing a starting point arises with right ascension. Luckily there were no nationalistic considerations to take into account, so the decision was made on scientific lines. Each year, because of the Earth's orbit being tilted, the Sun appears to wander north and south. In December each year the Sun reaches its furthest point south, in June, its furthest point north. Twice a year it appears to cross the equator, in March and September. On 20 March (or sometimes 21 March, depending upon the time until the next leap year) the Sun passes from south of the equator to north. The point at which it does this is taken as 0 hours right ascension.

An important relationship exists between right ascension and sidereal time, the local sidereal time being equal to the right ascension at the zenith. Knowing this, it is easy to determine whether an object will be visible in the sky. If you know the sidereal time and the RA of the object, then, provided that it isn't too far north or south, if the RA is within three hours of the sidereal time, the object will be high enough to observe well.

The equatorial system of measuring the positions of objects was invented by the Greek scientist Hipparchus over 2000 years ago. At that time the point was in the constellation of Aries, and so was called the First Point of Aries. Because of the effect of *precession*, the point is no longer in Aries, it has moved into Pisces, but the name has remained the same.

Precession is an effect by which the positions of objects in the sky change slowly over a number of years. This isn't caused by anything to do with the stars themselves, but with the Earth. We all know that the Earth spins on its axis every 24 hours, regular as clockwork, although a better analogy might be a child's top. At some stage everyone will have played with a top; at the beginning it will stand upright, but after a while because it has wandered over a bump or something it starts to wobble. If you watch carefully, you will see that it wobbles with a set tilt and, apart from that tilt, is still quite stable. The tilt itself doesn't stay put, however, it rotates slowly, tracing a circle of its own. This effect is called precession.

The Earth's axis is tilted too, making an angle of 23.5° to the vertical, as we will see in Chapter 5. This tilt is responsible for our seasons. The tilt of the Earth wobbles, although it does so fairly slowly, taking 25 800 years to complete one revolution. One effect of this wobble is to change the First Point of Aries; thus the starting point of the right ascension system moves, and with it the co-ordinates which specify the positions of objects in the sky. Now 25 800 years is a long time, so the effect is not very large, but over a period of years it makes its effect felt. To the amateur astronomer the effect means little, but to the professional, with very precise instruments, it is necessary to correct for the effect.

Even for professional astronomers the effect is still small, so in their charts they only correct for it every 50 years or so. Charts are drawn to match a specific time, or epoch, with the system of RA and Dec. set to match that time, so charts may be referred to as epoch 1950.0, meaning that the co-ordinate system is set for 1 January 1950. Some charts now being drawn are for epoch 2000.0, but it doesn't matter which you use, as the stars are still in the same position. However, if you are mixing charts and catalogues it is important to ensure they are for the same epoch, or some lengthy calculations will be required.

Finally, as this book will deal with the planets, it is necessary to understand something about *orbits*: how they're described and how they work. If the orbits of the planets were circular and lay in the same plane, then the only thing you would need to describe the orbit would be its radius. Unfortunately, however, things aren't that simple. The orbits of the planets are *ellipses*, not circles, although an ellipse looks like a squashed circle, and indeed they are mathematically related to each other.

The shape of an ellipse can be described in a number of ways. An obvious one is to measure the long length of the ellipse and compare it with the shorter length. This is what is done, although half lengths are used. The longer length is called the major axis, and the shorter one the minor axis. The line from the centre of the ellipse to the longer end is called the semi-major axis and this length is one of the numbers we need to describe an orbit. Instead of measuring the semi-minor axis as well, the eccentricity of the orbit—that is, the amount of flattening—is measured. A circle, which has no flattening, has an eccentricity of 0; the planets of the solar system have eccentricities ranging from 0.0068 (Venus) to 0.250 (Pluto).

Because of the shape of the orbit, a planet will not stay a constant distance from the Sun. The closest point of the orbit is called perihelion, the furthest aphelion. In the case of the Earth, perihelion happens around 2 January each year when the Earth is 147.1 million km from the Sun and aphelion is around the 4 July when the Earth is 152.1 million km distant.

The orbits of the planets in the solar system do not all lie in the same plane, they are tilted by differing amounts with respect to each other. Because we live on it, we take the orbit of the Earth as the level and measure the *inclinations* of the other planet's orbits from that. We call the plane of the Earth's orbit the *plane of the ecliptic*. The tilts of the planets' orbits range from 0.77° for Uranus to 17.2° for Pluto. To the astronomer these tilts mean that the planets do not all follow the same path through the sky. The Sun's path traces out one line through the sky year after year, the *ecliptic*, while all the other planets wander a little to either side of this line only Pluto venturing far from it.

We will meet more of the terms used in astronomy in the chapters which follow. Like any new topic, it is difficult to come to grips with all of it immediately, but use brings familiarity and over time the concepts outlined here will become much clearer.

# 2

# Observing techniques

The stereotypical picture of an astronomer is an old man looking into the heavens through a large, ancient telescope. In reality, modern professional astronomers rarely look through a telescope at all; they have equipment much more sensitive than their eyes with which to observe. But amateur astronomers still use telescopes, and they still look through them—or do they? Actually it is not really necessary to have a telescope to be an amateur astronomer. Some parts of astronomy are still the preserve of the naked eye—the observation of meteors, for example. Other parts of astronomy need only binoculars; most comets are found by amateur astronomers using binoculars (and a lot of perseverance); and yes, many astronomers do have telescopes.

Whatever you decide to use, there are some techniques which will make your observing more enjoyable and your observations more worthwhile. Specific techniques for observing each of the objects in this book are included with each chapter, but there are some general hints worth noting now. Above all, remember that a little planning is always necessary before you observe.

The first thing to do is to choose a site from which to observe. Two things to look for are convenience of location and freedom from interfering light; unfortunately you can usually find one but not the other. Obviously the most convenient place to observe is close to home—outside the back door is ideal; you don't have far to carry your telescope and charts, you're close to shelter if the weather becomes inclement, and the time between deciding to observe and getting started may be only a few minutes. So if the backyard has all this going for it, why doesn't everyone observe from there?

The sad fact is that most people live in or near cities, and cities create problems for astronomy. The pollution in the air from cars and industry makes the air hazy, but even worse is the abundance of lights. Street lighting, illuminated advertising signs and security lighting on buildings and houses all contribute to the light shining up into space. This light reflects off the minute particles of pollution, making the sky near a city a muddy grey colour. Being in the suburbs is no better, because the light pollution from a large city such as Sydney, London or New York can clearly be seen from more than a hundred kilometres away. There is little that can be done about light pollution other than lobbying local governments for regulations about upward pointing lights. The trouble is that astronomy is a low profile hobby and councils and governments are unlikely to do very much, though some did turn street lights off for a period around the 1986 return of Halley's comet.

Regardless of what governments might do, there is still a problem with local lighting, but this is within your control. When looking for a site, wait until it is dark, as it is impossible to take into account all sources of lighting during the day. At night be critical in your search: that little bit of street light between the trees may not be a bother now, but it is sure to shine in your eye at some time during your observing. Even the worst locations usually have some dark areas from which to observe, and don't forget to turn out your own house lights. If it is impossible to find anywhere, then look for the place with the least problems and solve them. There is little you can do about street lights although if they are on a back street and you have an approachable local government you may be able

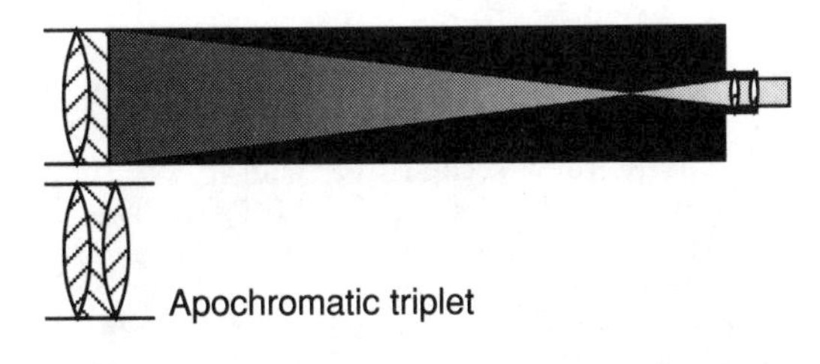


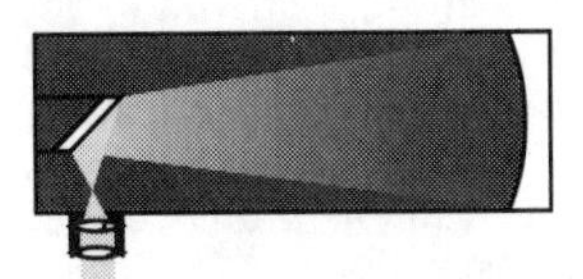

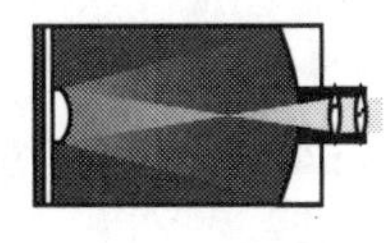

**2.1** An astronomical refracting telescope has a lens at the front of the tube which brings light to a focus near the lenses of the eyepiece, which make the image suitable for viewing. To combat the splitting of the light into colours by its passage through the lens, a doublet made from two different types of glass is used. To further enhance the colour correction of the objective lens, sometimes a triplet of lenses known as an apochromatic triplet is used.

A Newtonian reflecting telescope uses a paraboloidal primary mirror to reflect light to a focus. A small, flat mirror near the top of the tube set at 45° reflects the light through a small hole in the side of the tube to reach the eyepiece.

A Schmidt-Cassegrain telescope uses two curved mirrors to bring the light to a focus. The primary mirror reflects the light towards the front of the tube where a smaller secondary mirror reflects it through a hole in the primary to the eyepiece. A specially shaped sheet of glass at the front of the telescope provides corrections to the light to improve the image.

**2.2** This 200 mm aperture Newtonian reflecting telescope was made at home using prefabricated parts. The altazimuth mounting of the telescope is in the Dobsonian form, the telescope being held in a cradle in which it can rotate in elevation while the cradle turns in azimuth on the base. PHOTO CREDIT: KEN WALLACE

to have one or two of them turned off. It is not, however, a good idea to take the law into your own hands and turn them off yourself.

Lights on neighbours' houses are easier to deal with. Ask them to turn them off, but do it nicely. Better still, invite your neighbours around to have a look through the telescope and let them see for themselves the problem created by their lights. If the surrounding lighting can't be turned off, there are some other solutions. The first is a three-sided shelter made from a tarpaulin. This can be arranged to block a lot of light from reaching the telescope and it is also good at keeping light breezes away, though in strong winds it needs to be securely fastened to the ground. A smaller version of this is a hood to cover your head and the eyepiece of the telescope. Despite the fact that it may look a little silly, it can be quite effective. If only one or two lights are a problem, you can make small screens and place them at strategic places on fences or trees to block the light.

In addition to blocking extraneous lighting, you should check your telescope, as it can contribute to the problem. Dust or moisture on the lenses or mirror of a telescope will scatter any stray light, ruining the contrast of the image, so these must be kept very clean. On a Newtonian stray light entering the tube can reflect from the surface opposite the eyepiece hole, again reducing contrast. The inside of the tube is probably already painted black, but to reduce reflected light further, black cloth or velvet can be stuck to the inside surface of the tube. Sticking similar material to the area around the eyepiece will also reduce reflections into your eyes.

Of course, the best solution is to get away from the city and its problems, to a 'dark sky site'. The problem is that this normally involves a drive of a few hours each way, limiting excursions to weekends, and reducing the amount of time you can spend observing. On the positive side, though, planetary observing suffers least from the problem of light pollution because the objects observed are so bright: five of the planets are visible without a telescope and two more are easily seen using small instruments—even Pluto can be seen with a large telescope from the city. By making the most of the area near to home you will be able to observe on any clear night with little fuss.

Being organised is the key to enjoyable observing. Having to mess about for an hour getting your telescope and equipment ready before you begin is a sure way to limit the number of nights you will observe. On the other hand, if it takes only a few minutes to get everything set up you can take advantage of a spare hour or so much more often.

There are a number of little accessories for your telescope—ones which are not packaged with it—which will make your time much more enjoyable and productive. The first of these is a folding table. When you're observing there is always too much to hold, so invariably you resort to pockets or the ground to hold things. Pockets are the worst possible place in the case of spare eyepieces and such; they will collect dust and lint, knock together and become damaged at an alarming rate. You also need a place to keep charts and other books. A table makes these much easier to read, but don't forget some small paperweights to keep your place in a breeze.

You can't begin observing as soon as you walk outside: you would be able to see very little as your eyes won't have adjusted to the dark. The process of dark adaptation takes around fifteen minutes to complete. During this time the iris of your eye widens the pupil to allow more light in and the amount of visual purple in the retina of your eye increases. As a result of these processes, your eyes become much more sensitive to light. The widening of the iris takes two to three minutes, while the build-up of visual purple takes longer.

Once your pupils have widened you are ready to begin observing. Usually your eyes will have adjusted sufficiently by the time you've got the correct eyepiece in the telescope and have pointed it in the right direction. Although it takes a few minutes for your eyes to adjust, it takes only a few seconds for them to lose their sensitivity if they see a bright light. This is another reason for care when choosing your observing site; you shouldn't be able to accidently glimpse a bright light directly—headlights from passing cars are particularly bad. The Moon can also be a problem. If you're going to observe the Moon, it is quite bright enough to ruin your dark adaptation, so leave it until last if you want to observe other objects. Even viewing the Moon without a telescope is enough to impair your dark vision.

While observing you will need a light. Despite all the efforts you have probably made to be rid of extraneous light and let your eyes adapt to the dark, if there is no light you won't be able to see what you're doing. What you need is a very low-power torch. A penlight torch is ideal, though even with only one small battery it is much too bright. With your eyes dark adapted you will be surprised how little light you need to read a chart or book, or make your way around safely. To reduce the amount of light from the penlight, cover the end with three or four layers of red cellophane. Not only does the cellophane reduce the amount of illumination, but the red light has much less effect on your dark-adapted eyes than other colours.

There are two ways to observe, the 'Just for fun' method and the 'I'll do something useful and still have fun' method. If you want to observe just for the pleasure of looking around the solar system, then you have everything you're going to need. If, on the other hand, you want to make a record of what you see, then you need just a little bit more. To write or draw your observations you will need a clipboard of some sort and some form of lighting attached to it. The clipboard can be of any type, but you will need to add some clothes pegs to hold the bottom of the pages (both corners) so that they don't flap around in the breeze. To light your work you need some form of clip light which won't require a free hand. Clip-on reading lights are ideal; they are small, run off one or two batteries and are high enough off the page to provide good illumination. Once more, cover the light with red cellophane to preserve your night vision.

Once you begin observing seriously you will

probably want to keep a permanent record of your observations in some manner. Regardless of what method you choose, there is certain information you should record with your work. That information is:

1 the date and time of the observation in UT;
2 the magnification you are using;
3 the seeing conditions;
4 if you have more than one observing site, your location;
5 if you have more than one telescope, the aperture of the telescope.

Without this information, although your observations will still be valid they are of little value from a scientific point of view. Although you may not be part of any organised program, your observations could be quite important to someone else. Your observations of Mars and its polar caps could be quite valuable to someone researching the planet, but not if you haven't recorded at least the minimum of information outlined above.

Item 3, seeing conditions, needs a little explanation. If the air between us and space were perfectly still then the images of objects arriving at the telescope would be perfect. Unfortunately it isn't, so they're not. The currents of air disturb the passage of light, making the images in the telescope move around. On a night with perfect seeing you won't be able to see this movement, but on a very bad night the movement is so great that details on the object can't be made out. Most nights are between these two extremes. Judging the seeing conditions is something which comes with experience. The easiest way to record seeing is to use a scale such as that in Table 2.1. Seeing can then be written as 'III with periods of II' or the like.

**Table 2.1 Seeing index**

| Seeing Number | Name | Description |
|---|---|---|
| I | Perfect | Images steady, small details can be seen clearly |
| II | Near Perfect | Moments of calm lasting several seconds |
| III | Moderate | Occasional large air tremors |
| IV | Poor | Constant troublesome tremors |
| V | Very poor | Scarcely allowing the making of a rough sketch |

One way to ensure that you record this information with observations is to have a prepared observing sheet. By designing a simple header for each page, you can have them printed off at quite low cost and always have them ready. If you have the sheets hole punched, you then have a perfect way of keeping your observations in order by placing completed sheets in a ring binder.

Drawing is a common way to record observations of the planets. It doesn't matter if you're not an artist; time and patience will allow most people to become proficient at recording what they see in the form of a sketch. The main trick is, not to be too ambitious. Beginning by deciding to sketch the entire full moon is a good way to become discouraged, so begin with something simple, like the positions of Jupiter's satellites, or just one crater in a blank area of the Moon. You could begin by choosing one small area of the Moon and sketching it night after night—in no time you will be surprising yourself with the results you're achieving.

One thing all astronomers need is a set of charts of the sky. There is too much there to ever hope to learn where everything is, so a map of some kind is indispensable. For observations of objects in the solar system, with the exception of Pluto and some of the fainter minor planets, you do not need a very detailed map. Charts which show every star down to magnitude 13 in large scale are fine for work which needs them, but for planetary astronomy it is unnecessary.

To look at most planets, a chart down to magnitude 7 or 8, drawn on a reasonable scale (a dozen maps for the whole sky) is sufficient. You can mark the positions of the planets on the chart, then take it outside in an effort to find them. With the brighter planets, once you've found them on one night, it is easy to find them again on subsequent evenings. Uranus, Neptune and Mercury need to be marked though, so that they can be found amongst the nearby stars. As the planets move, their positions change, and so they will move on the maps. Publications from local astronomical societies, or the *Astronomical Ephemeris* from the United States Naval Observatory, and the Royal Greenwich Observatory, give the co-ordinates of the planets, their right ascension and declination, throughout the year so that positions can be marked.

Another option to the conventional star chart

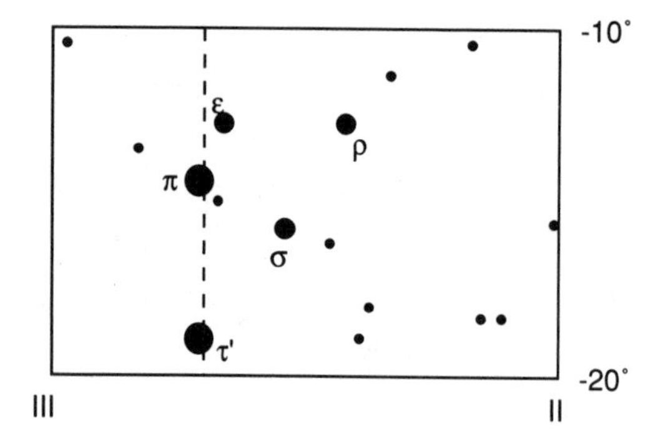


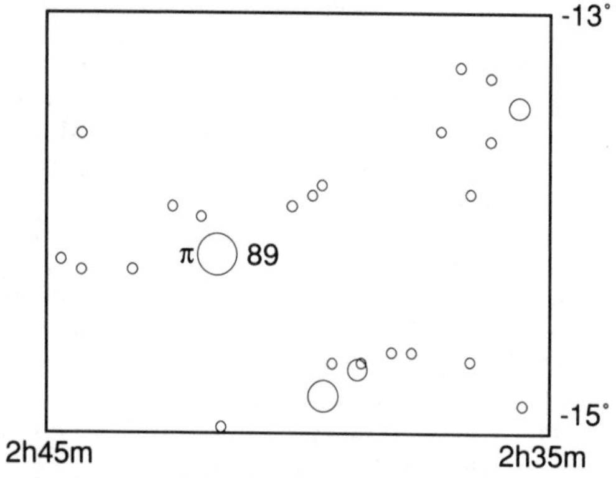


**2.3** An area of the sky as shown in two different star atlases. The first is a large-scale atlas which is suitable for general astronomical observations; it shows a region on the border between Cetus and Eridanus. Stars to magnitude 6 and selected deep-sky objects are shown. The second is a small segment of a chart from a much smaller scale atlas which shows only the region around the star $\pi$ Ceti.

is the planisphere. This is a small disk of cardboard or plastic designed to show the position of stars and objects in the sky from a given latitude on Earth. Using a planisphere is an ideal way to begin finding your way around the sky, or to find planets. Mark on the planisphere the positions of the planets you want to observe, take it outside, set it to the correct time and date, and you should have no difficulty finding the brighter planets.

Finding your way around the sky is a task all amateur astronomers need to master (Professional astronomers don't need to know—their computers do it all for them) and the only way is practice. Select an area of the sky with which you are familiar: the Southern Cross or the Big Dipper, Orion or Scorpius. Then, using a star chart, note the positions of some of the stars around the constellation and learn the names of the area. If you take this slowly and consistently, adding a new small piece of sky each evening, then in very little time you will know your way around. The key to learning the sky is to often look upwards. Letting even just a week go by will make it harder as the motion of the Earth changes the constellations' positions noticeably.

Above all, though, it is important to remember that for most of us astronomy is a hobby, and hobbies are meant to be fun. Forget the science if you like: to view the planets and stars just for the joy of seeing them is more than enough justification for going out at night.

The photographs in books like this and the impressive photographs taken by professional astronomers will eventually tempt many into trying astrophotography for themselves. Astrophotography is perhaps the hardest part of amateur astronomy to master, but the results can be stunning if you practise it.

Like normal photography, astrophotography involves the expenditure of money, but unlike conventional photography, the expenditure is minimal. The most important part of an astrocamera is the telescope, and if you already have that, the only additional expense is a connecting tube and something to hold the film flat and safe from stray light. To do this job, the rear of a single lens reflex camera is ideal. It doesn't have to have fancy features like light meters or programmed exposure modes, just a reasonable range of shutter speeds and a socket for a cable release—it doesn't even need a lens. An old, second-hand camera is all that is necessary, and they can often be bought quite inexpensively. Make sure that the camera has a B setting on the shutter speed. This setting leaves the shutter open for as long as the button is pressed, and is the only way to take exposures of longer than one second.

In addition to the camera you need some way of attaching it to the telescope. This is done using a camera adaptor available from any telescope supplier. The camera adaptor is a tube assembly which fits into the eyepiece tube of your telescope at one end, the other end attaching to a T-ring adaptor on your camera.

When buying an adaptor make sure that it is suitable for both prime focus and eyepiece projection photography as explained later in this chapter. Lastly you will need a cable release to activate the shutter without moving the camera. Get a release with a locking action so that you can use it with the camera's B setting to take long-exposure photographs.

Choosing the best film to use is as important to good results as any other part of the photographic process. A number of factors come to bear on your choice, the most important being who will process the results. If you depend upon a commercial processing firm to do your developing, then colour slide film is the best choice. Because the processing of slide film is standardised, your photographs will turn out dependent only upon your efforts, not those of the person exposing the final print. Slide film is also generally cheaper to have processed than prints.

One pitfall in using slide film is to forget to tell the processing company that there are astronomical photos on the film. To the untrained eye, a photo you may have spent a hour getting ready for and taking can look a lot like a blank frame of film, and thus may end up in the bin. Always ask for all slides to be mounted regardless of what they think. Another helpful hint is to take a photo of a 'conventional' object at the beginning of the film, and if possible once or twice along the film. A photo of a tree or house will give the person cutting up the slides a starting point so that your prize photo doesn't get cut down the middle by mistake.

If you have access to darkroom facilities then you can experiment with black and white photography, developing the film yourself and printing the results. This gives you much more control over your photographs, as often mistakes at the telescope can be corrected in the darkroom. It is also more enjoyable to be able to produce your astrophotographs yourself from beginning to end.

Whether you use black and white or colour slide film, choosing that film is going to have a bearing on the quality of your results. The quality of commercially available films is uniformly high. There is little to chose between different brands, and all manufacturers offer roughly the same range of films. What is important when choosing between films is the speed of the film, or how long it takes to build up an image. Film speeds are classified by a system from the International Standards Organisation (ISO), with films given numbers which increase in size as the film becomes faster.

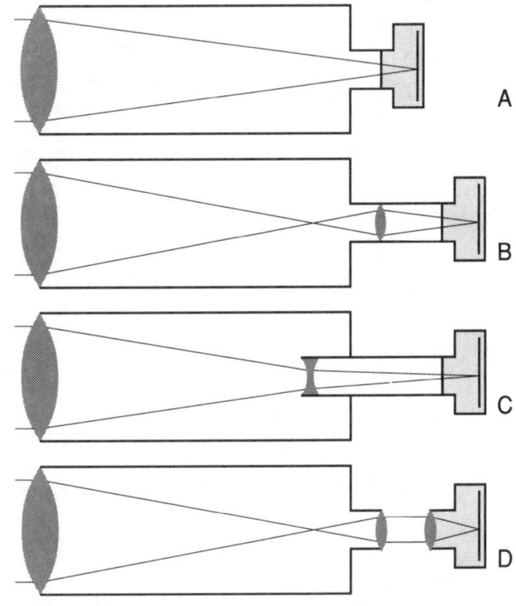


**2.4** The first three methods of projection photography rely on using the telescope as a camera lens to place a focused image on the film. Only in the last method is the camera used separately from the telescope.

a *Prime focus photography* uses the main mirror or lens of the telescope as the only image-forming element. This leads to bright photographs but small images.

b *Eyepiece projection* Here an eyepiece increases the magnification of the image from the objective. It gives a much larger image scale at the expense of a longer exposure.

c *Negative projection* Instead of an eyepiece being used to help form an image, a barlow lens is used. This gives similar results to those obtained using eyepiece projection.

d *Afocal projection* Afocal projection uses the telescope in the same way that your eye uses it. The image formed by the telescope is photographed in the normal way using a camera with its own lens attached.

The most popular film for taking snapshots has a speed of 100 ISO, but speeds from 32 ISO to 3200 ISO are all readily available. The faster a film is, the faster it builds up an image, so 400 ISO film will take the same picture in one quarter of the time it takes 100 ISO film. Unfortu-

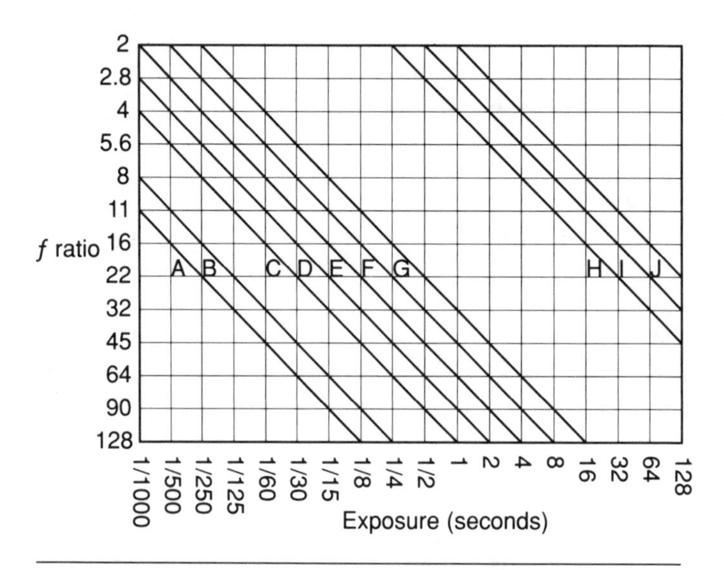


A Venus at greatest brightness
B Venus at greatest elongation
C Jupiter, full moon
D Gibbous moon
E First or last quarter moon
F Mars, crescent moon
G Saturn, thin crescent moon
H Solar corona
I Earthshine
J Total lunar eclipse

**2.5** This diagram gives the approximate exposures needed to take photographs of selected objects. It is a guide only. Use the bracketing technique described in the text for best results. To use the diagram, find the focal ratio of your optical system, then move to the diagonal line corresponding to the object you want to photograph and read off the exposure.

| Film speed (ISO) | Divide the time by | Film speed (ISO) | Multiply the time by |
|---|---|---|---|
| 400 | 4 | 32 | 4 |
| 200 | 2 | 64 | 2 |

The diagram is drawn for ISO 100 film. If you are using another speed film, adjust the exposure time as indicated.

nately this extra speed does have one drawback: photos taken on high-speed film are not as sharp as those taken on slower film, the reason being that the slower film has finer grains of chemicals in its emulsion. For photographs of the Sun and Moon, films of 100 ISO give good results but, because of the faintness of the planets their photos should be taken on 400 ISO film, at least as a rough guide.

The best introduction to solar system photography is the Moon. With a little effort you will be able to photograph craters, mountains and maria almost as easily as taking a photograph of a terrestrial landscape. The method used is prime focus or direct photography, where the camera body is attached to the focusing tube in place of the eyepiece. If the telescope is not equiped with a right ascension drive motor to counter the Earth's rotation, good results can still be obtained by using fast film and short exposures.

The Moon's size on the film is determined by the focal length of the telescope: the longer the focal length, the larger the image on the film; and the larger the image, the more detail on enlargements. It is important to know the size of the image in advance, because if it is too small the focal length must be increased. Methods of doing this are explained later. An easy way to find the image size of any celestial body is to multiply the effective focal length of your telescope, in millimetres, by the diameter of the object in seconds of arc, then divide the result by 202 265. The answer equals the image size in millimetres. Therefore, if you were to photograph the Moon which has a diameter of about half a degree, or 1800″, with only a camera and standard 50 mm lens you would get an image on the film a little less than half a millimetre in diameter ($50 \times 1800 \div 202\,265 = 0.44$ mm), an image so small that it would show no detail when enlarged. By using this simple formula, a

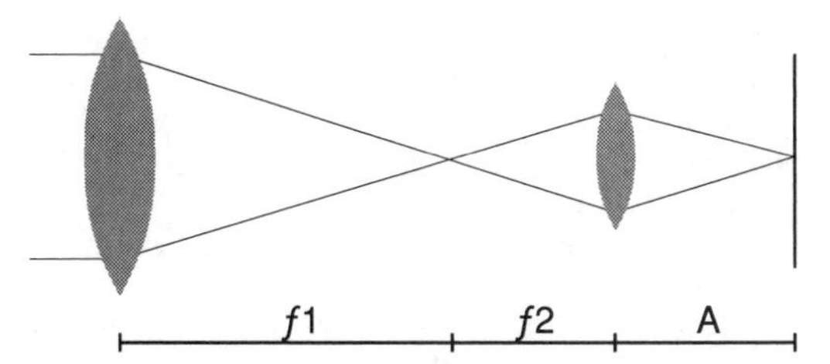


**2.6** To determine the magnification and focal ratio obtained using eyepiece projection, you need to know three distances: $f_1$, the focal length of the objective lens or mirror; $f_2$, the focal length of the eyepiece and A, the distance of the eyepiece to the film. Given these two distances
the magnification, M is $\frac{(A - f_2)}{f_2}$; and the focal ratio is given by $f_1 \times M$.

telescope of 700 mm focal length would give an image about 6.25 mm in diameter (700 × 1800 ÷ 202 265 = 6.23 mm), and by enlarging the image 15 times when printing, the diameter of the Moon on the finished print would be about 95 mm.

Some larger commercial amateur instruments like the popular 200 mm Schmidt–Cassegrain telescopes with their effective focal lengths of 2000 mm will present an image on the film 18 mm in diameter (2000 × 1800 ÷ 202 265 = 17.8 mm). A ten-times enlargement of this negative will give an image of the Moon almost 180 mm in diameter which will nicely fill a standard 200 × 250 sheet of photographic paper.

Having set up your equipment, found a dark site and pointed the camera and telescope at the Moon, the next question is, how long to make the exposure. Recommended exposure times for the lunar phases are given in Figure 2.5. For best results take a photo at the recommended speed, then follow up with two others, one at half the speed of the first and the other at twice the speed; for example, if the required exposure was 1/30 s the other two would be 1/15 s and 1/60 s. This simple technique, called *bracketing* the exposures, will ensure that all variables such as weather conditions, atmospheric clarity and other unknowns will provide at least one shot of the correct exposure.

Unless you have an exceptionally still night, many photographs will be slightly blurred because of atmospheric turbulence. If you process your own film, try shooting an entire roll each time you decide to photograph the Moon, then choose the sharpest negative for enlargement. The best procedure is to carefully focus the image, wait for a few seconds to allow any vibration to settle, then take a bracket of three exposures, letting vibration dampen between each. After three exposures, check that the Moon is still centred, refocus and take another bracket and so on until the roll is finished.

After some success with the prime focus technique, you may wish to try your hand at photographing the lunar surface in more detail. This can be achieved by inserting an eyepiece into the extension tube part of the camera adaptor, between the camera and telescope. This is called *eyepiece*, or positive, *projection*. The resulting image on the film is enlarged greatly over that given by prime focus photography. Instead of an eyepiece, some amateurs prefer to use a barlow lens. This system is known as *negative projection*.

The aim of using positive or negative projection is to increase the effective focal length of the system, and so increase the image size; this is where the problems begin. By increasing the focal length the exposure time must also be increased, and the longer the exposure time, the more prone the image is to atmospheric turbulence. Fast films will compensate somewhat, but will be grainier than the slower films.

By using a projection system you may be increasing the focal length of the telescope by two, three or more times—the amount depends on the eyepiece or barlow used and the distance of that lens from the film. Determining the length of exposure is not much trickier than for prime focus photography. You can use trial and error and shoot several exposures and determine the best from the developed film or if the image of the Moon fills most of the field in the viewfinder you can try the camera's own light meter. All results should then be recorded, so that future trial and error shots will not be necessary. However, you should still bracket the exposures.

A simple formula gives the magnification (M), the effective focal length (EFL) and the focal ratio ($f$) of any projection system. If $f_1$ is the focal length of the telescope objective, $f_2$ is the eyepiece focal length, and A is the distance from the eyepiece to the film then the magnification used is $M = (A - f_2) \div f_2$. Once M is established, the effective focal length is found by multiplying $f_1$ by M, and the focal ratio by multiplying the focal ratio of the telescope by M.

For example, if you had an 80 mm telescope of 800 mm focal length, and used a 25 mm eyepiece with the film spaced 100 mm behind the eyepiece, the projection magnification would be:

$$\begin{aligned} M &= (A - f_2) \div f_2 \\ &= (100 - 25) \div 25 \\ &= 75 \div 25 \\ &= 3 \end{aligned}$$

By multiplying the objective focal length of 800 mm by the projection magnification we obtain the EFL of 2400 mm. Since the telescope's focal ratio is $f$10 (800 ÷ 80), the focal ratio of our projection setup is $f$30 (10 × 3). For negative projection with a barlow the formula should be changed to $M = (A + f_2) \div f_2$.

A third type of projection system known as the *afocal method* has the advantage that you do not necessarily need a single lens reflex camera. The telescope with eyepiece is focused for visual observation and the camera, with lens focused at infinity, is placed directly behind the eyepiece. The camera may be mounted on a separate tripod behind the eyepiece or, preferably, on a bracket attached to the telescope tube. The EFL can be estimated by multiplying the magnification obtained with the eyepiece by the focal length of the camera lens. The focal ratio is then obtained by dividing the EFL by the objective diameter.

For example, an 80 mm telescope of 800 mm focal length will give a magnification of 32 with a 25 mm eyepiece (800 ÷ 25 = 32). If we then multiply the camera lens focal length, 50 mm, by 32 we arrive at an EFL of 1600, and by dividing the EFL by the objective diameter we get a focal ratio of $f$20 (1600 ÷ 80 = 20).

The camera and eyepiece separation is easily established by setting up a bright light source such as a 75 or 100 watt incandescent globe about 50 cm or so in front of the telescope in a darkened room. The tube should be covered by a piece of tracing paper stretched over the end to diffuse the light. With a single lens reflex camera the distance from the eyepiece is determined by noting whether the light is evenly distributed over the entire field of the view-finder; if darkness is seen at the edges, the camera is either too far or too close to the eyepiece, and should be moved until the light is uniform.

If the camera is not a single lens reflex type, it will be slightly more difficult to determine the camera and eyepiece distance. The camera and telescope should be set up as above with film removed, and a piece of tracing paper secured across the shutter opening. With the camera back open, observe any darkening at the edges and compensate by moving the camera until the light is even over the field.

Solar photography is also an easy exercise, but under no circumstances should any deviation be made from the safe observation methods described in Chapter 6. If a pre-filter or full aperture solar filter is used, the same methods as for lunar prime focus photography are used. Stray light can be a problem with the afocal method, but can be rectified by shielding the gap between camera and eyepiece with a cardboard tube painted black on the inside. If a solar projection screen is used, you should take a photograph of the projected image with the camera held as square to the screen as possible.

Exposure times for prime focus solar photography will vary depending on the properties of the filter used, but as a rough guide take a range of photographs from the recommended speed for the Moon as a thin crescent to the full moon for your particular $f$ ratio. For example, if the table calls for 1/8 s for a thin crescent and 1/125 s for the full moon, five exposures should be taken: 1/8, 1/15, 1/30, 1/60 and 1/125. An examination of the developed film will determine the best speed, and from then on exposures can be bracketed in sets of three, one at half the speed and one at twice the speed, similar to the procedure for the Moon.

Unlike that of the Sun and Moon, planetary photography is quite difficult. Many amateurs, after a few attempts at the planets, generally progress to deep-sky work. It is, however, fun to capture planets on film and it is the next logical step after experimenting with positive or negative projection of the Moon.

As you need to obtain the largest possible image on the negative, a long effective focal length is needed. However, the more magnification used, the longer the exposure needed, and it is not uncommon to use exposures ranging from one to ten seconds for the planets. Though short, this time is enough for factors like drive inaccuracies and atmospheric turbulence to blur the image.

Even the largest planet, Jupiter, whose average diameter is a respectable 40 arc seconds, will require a long effective focal ratio to give a good sized image on the film. If a telescope of 700 mm focal length were used with the prime focus method, the image on the film would be only

0.14 mm (700 × 40 ÷ 202 265 = 0.14 mm)—not much to work with; even twenty times enlargement will only give an image slightly less than 3 mm in diameter. It can easily be seen that projection is the only method that will give an image scale sufficiently large to show any detail.

The best planetary subjects, and the only ones worth attempting with the projection method, are Venus at or near greatest brilliancy, Mars at opposition, Jupiter and Saturn. The tiny, featureless discs of the outer planets, Uranus, Neptune and Pluto, are best captured with a standard camera and lens against a star field as described in their respective chapters. Recommended exposures for the planets are given in Figure 2.5, and it is necessary to calculate the focal ratio by the formulae given earlier for best results.

Astrophotography is an exciting part of amateur astronomy. With the low cost of equipment, the high quality of available films and the ease of having them processed, it is a part of the hobby available to all, and one in which a little patience and perseverance can achieve spectacular results.

# 3

# The Solar System

It is impossible to make a model of the system of planets we inhabit; the sizes of the planets and the distances which separate them make the task unworkable. It is therefore difficult to construct some mental concept of the solar system. If we were to image the solar system from the outside, no doubt we would think of a group of planets happily following their paths around the Sun in an orderly manner; it is difficult not to include the lines drawn on many diagrams in the picture too.

To gain some idea of the scale of the solar system, pace out a circle roughly 3 m in diameter, to represent the Earth's orbit. You have just compressed the size of the solar system by a factor of a hundred thousand million. If it took you twenty seconds to pace the circle, then you've compressed time too, but only by a factor of 30 million. Now let's build the solar system on this scale. If the Sun is at the centre of the circle, the orbit of Mercury is half a metre from it, flying around the Sun once every five seconds. Venus is next, a metre from the Sun orbiting once every twelve seconds.

Outside the room is the last of the small planets, Mars, 2.25 m from the Sun orbiting every 36 seconds. Jupiter, the largest planet, is nearly 8 m from the Sun, orbiting every four minutes. Saturn, Uranus and Neptune follow with Pluto 60 metres from the Sun, orbiting once every 84 minutes. On a scale such as this trying to represent the planets with anything is useless; the Earth is just a tenth of a millimetre in size, even Jupiter would be a ball just a millimetre and a half in diameter.

What would the solar system really look like from the outside? Would it be obvious to a visitor that planets did revolve about our Sun? A spacecraft arriving at the solar system would first encounter the Oort cloud of comets surrounding the Sun at a distance of around one light year. At that distance any planets orbiting the Sun would be lost in its glare. If the spacecraft were to travel on to Pluto, the situation would not improve much. The Sun would certainly be the brightest object in the sky, but few of the planets could be easily seen; only the gas giants would appear far enough from the Sun for them not to be lost in its glare. The four terrestrial type planets are much too close to be seen from Pluto, and would only be noticed as they occasionally transited the Sun's disk—even then they would only be seen with a telescope.

If we can ignore the difficulties for a moment, the view of the solar system from the outside can be quite instructive. It is important also to speed up time, as Pluto takes so long to orbit the Sun that its contribution to the picture only becomes clear after an extended period. At the centre of the solar system is the Sun, its gravity keeping all the planets and other bodies in orderly procession about it. The orbits of the planets, close together at first, become further apart as their distance from the Sun increases, but they stay in almost the same plane.

The planets of the solar system fall into two main groups: the four small rocky worlds near to the Sun and the four large gas worlds further from it. Another planet, a tiny, icy world with a huge moon, orbits the outer reaches occasionally journeying closer to the Sun. Add to this over 60 moons orbiting the planets, various other small bodies and comets around the Sun and the picture of the solar system is complete. However, just knowing what it looks like does not tell us about its origins. That has taken scientists

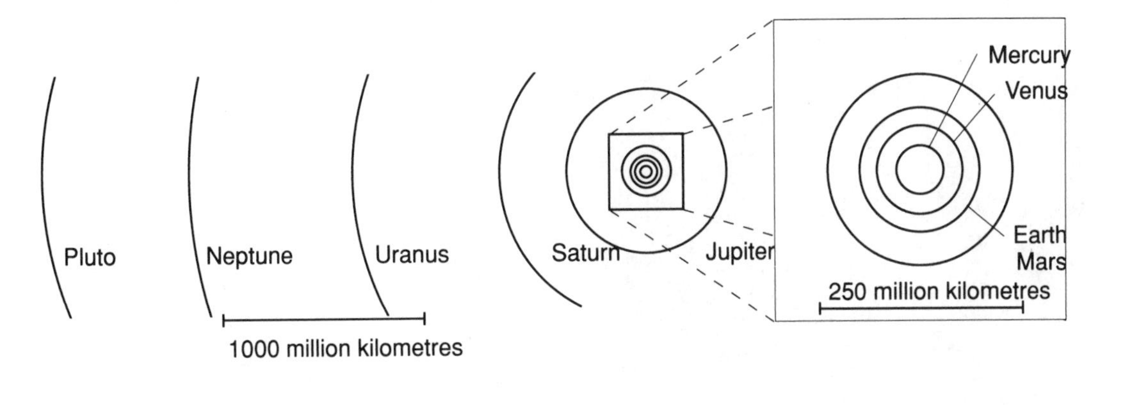


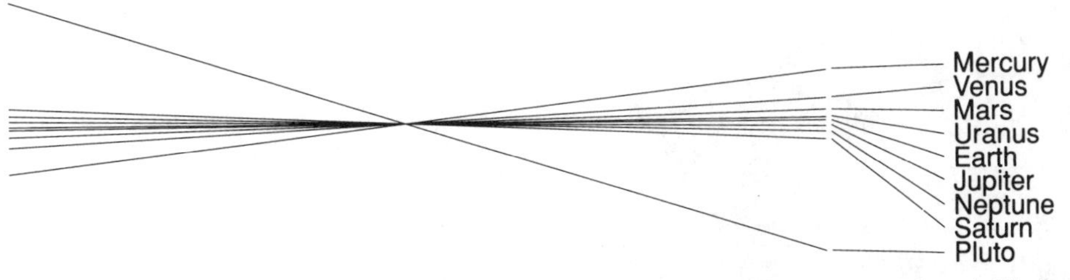


**3.1** The orbits of the planets about the Sun. The region closest to the Sun has been enlarged. The lower diagram shows the inclinations of the orbits of the planets.

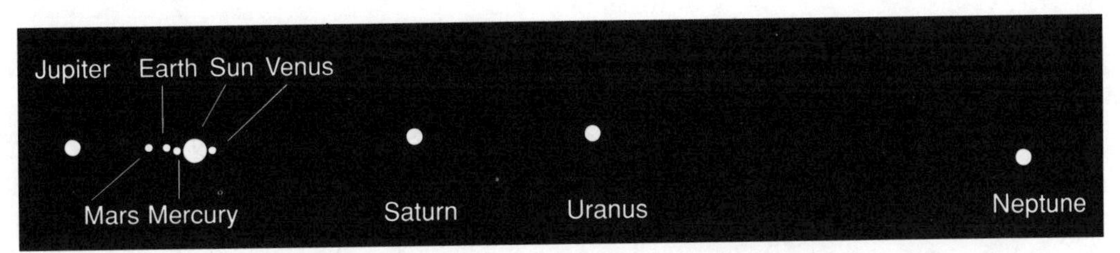


**3.2** If you were standing on Pluto during September of 1989 and looked towards the Sun, the planets of the solar system would have appeared stretched across about 45° of sky. Lost in the glare of the Sun, obvious as the brightest thing in the sky, would be the four small rocky worlds. Out further the gas giant planets would be more prominent, bright amongst the stars.

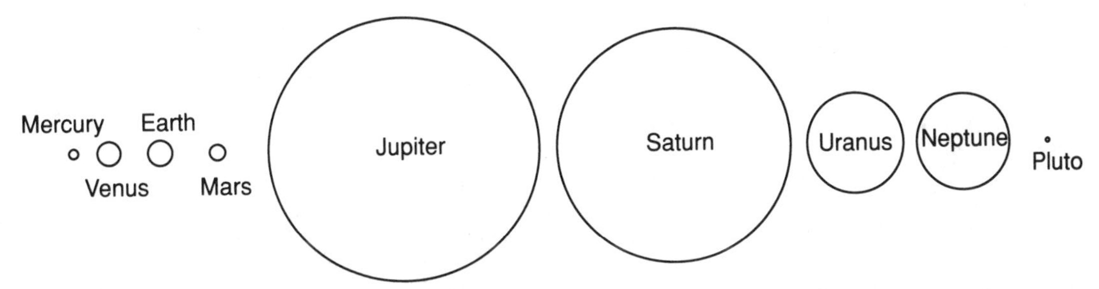


**3.3** The planets are here all drawn to the same scale; naturally the distances between them are not to scale.

**3.4** This photograph shows a typical region of our own galaxy, the Milky Way, in this case in the constellation of Sagittarius. The bright cloudy looking area is caused by thousands of faint stars, many too faint to be seen by themselves, contributing to the brightness of the region. The dark regions cutting through the bright band are not regions lacking in stars but clouds of dark gas and dust between the Earth and the light of the galaxy. The masses of stars here are typical of this region of our galaxy which is roughly in the direction of the galactic centre. Apart from the stars in this photograph, the planet Saturn can be seen as the brightest object just below the band of the Milky Way. Above Saturn is a short streak caused by a meteor passing into the atmosphere. Following the line of the meteor downwards about one centimetre is a bright object at the edge of a dark lane. This is the planet Neptune. PHOTO CREDIT: KEN WALLACE

many years of study, and still the answer is not clear.

The system of planets we inhabit takes its name from the star at its centre, so before looking at the origin of the planets and their moons we must first look at the origins of the Sun. Our Sun is a normal star, of around average mass and average brightness. The fact that the type of life we know on Earth exists depends upon the Sun being such a boring, average star. There are other stars in the universe which are very different from the Sun. Some are small, burning their nuclear fuel slowly, making it last over many thousands of millions of years. Any planets orbiting such stars would be cold places indeed. Other stars are the showoffs of the universe; they are large in size and burn through their fuel at an enormous rate, reaching the end after only a few hundred million years. Yet although it would be impossible for life to evolve in such a short time, our Sun and its planets are intimately connected with just such a star.

Stars are powered by the process of nuclear fusion in which four hydrogen atoms are turned into one helium atom and some energy is given off. This process, which is the same as that used in a hydrogen bomb, makes the star shine. The pressure it puts on the atoms of the star also tends to drive the star apart. Counteracting this radiation pressure is the force of gravity, the mass of the star holding it together. For a star to shine as our Sun does, there must be a delicate balancing act between these two forces. If a star is larger than our Sun, its gravitational field is larger, so it must use more hydrogen each second to counteract the pull; if the star is smaller than our Sun, it can use its fuel more slowly.

So what happens to a large star when its fuel runs out? Firstly, without the radiation pressure to hold it out, the star begins to collapse. In the process the centre of the star is heated to enormous temperatures, much higher than the ten million degrees normally found there. These new higher temperatures begin a new round of nuclear reactions, and the helium from earlier fusion is turned into other elements, mainly carbon. As these reactions peter out, the star contracts a little more, the temperature rises and the fusion of carbon to heavier elements com-

**3.5** The ring nebula in Lyra is an example of a planetary nebula. Planetary nebulae are so called, not because they have anything to do with planets, but because through the telescope their small size and circular shape make them look like a planet. At the centre of this nebula is a small, very hot star which formed the nebula by shedding its outer layers into space. This blowing off of material is common and is one of the ways in which material formed within a star is returned to space to form more stars and planets. PHOTO CREDIT: KEN WALLACE

mences. The fusion of carbon releases enormous amounts of energy, so much that the star is blown apart in an explosion called a *supernova*.

The energy released in a supernova explosion is more than the total amount of energy released during the star's life to that point. Indeed, supernovæ seen in distant galaxies are so bright that for a short time they outshine all the rest of the stars in the galaxy combined. Supernovæ are rare, the chance of any given star becoming a supernova in a given year is slim, but because of the large number of stars in a galaxy, any galaxy would expect a supernova every hundred years or so.

In our galaxy astronomers have been waiting for a supernova for a long time. The last supernova within our own galaxy occurred in 1604 and was viewed by Kepler and Galileo amongst others. It has now been nearly 400 years since that supernova; we are overdue for another. On the other hand we are unable to see a large part of our galaxy because of the intervening dust and gas, so it is possible that supernovæ have exploded within our galaxy but we have been in the wrong place to see them.

The next best thing to a local supernova happened on 23 February 1987. A star in one of the two small companion galaxies to the Milky Way, the Large Magellanic Cloud, became a supernova. The star was so bright that although individual stars in the LMC cannot be seen without a telescope, this star, which has been named SN1987A, was easily visible to the naked eye. The discovery of the supernova was made by a number of amateur astronomers around the world, as well as by professional astronomers and satellites. The study of SN1987A and the core of the star which was left behind has already caused a major rethinking of the theories which astronomers thought they understood before the explosion. Research on the star will keep many astronomers and theoretical physicists busy for many years.

The core of a star which has undergone a supernova explosion may form a *neutron star*, or even a *black hole*, but it is the gas that is thrown out into space which is more interesting. This gas is rich in the products of the star's final fusion; there is lots of hydrogen and helium from the outer reaches of the star and, more importantly, some carbon and heavier elements from the regions closer to the core. It is in these supernova explosions that the material which forms stars is returned to the universe.

When the universe was formed around twenty thousand million years ago, there was only hydrogen and a little helium. The first stars had only this from which to form. The exact process by which the stars and galaxies formed is still unknown, but it is must have involved condensation from large gas clouds. Within these clouds small eddies formed and some atoms got closer together than the average. In these regions the added mass, and hence increased gravity, would have drawn still more atoms in, the process becoming self sustaining. Eventually, when enough atoms congregated, the temperature at the centre of the ball of gas would have been high enough for the fusion reactions to start, and the star would have begun to shine.

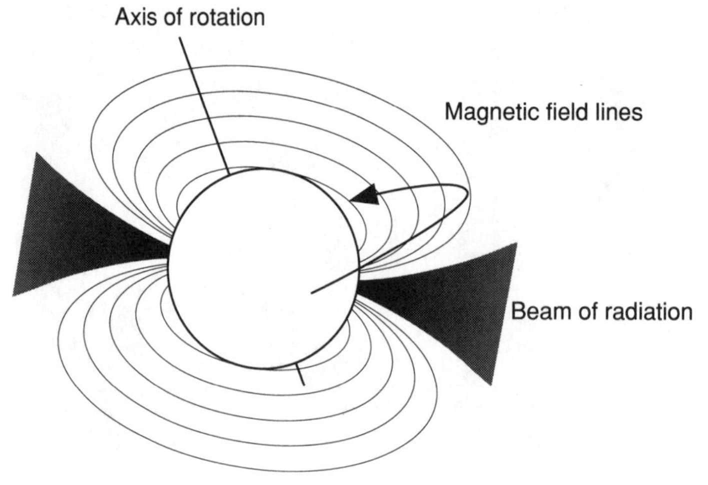


**3.6** A pulsar is what remains after a massive star blows off its outer layers in an explosion called a supernova. The core of the star collapses until the electrons in the atoms are compressed so much they combine with the nuclei of the atoms, forming neutrons. The ball of neutrons remaining spins very quickly because the star from which it formed was spinning. In addition, the magnetic field of the star is so strong that beams of charged particles and radiation stream out from the star's magnetic poles. The combination of these two searchlight-like beams and the rapid rotation of the star makes the star appear as if it is flashing on and off in a very regular manner, hence the name pulsar.

**3.7** The Orion nebula is one of the best known features of the sky; this huge cloud of gas and dust is 1600 light years from the Earth and the region seen in this photograph is around 30 light years across. Inside the nebula can be found a number of young stars which have begun to shine in the past few tens of thousands of years, and other stars visible only in infrared telescopes which have not yet attracted enough mass to begin shining in their own right. It is in a cloud such as this, and there are many scattered throughout the heavens, that the Sun and its planets began to form 5000 million years ago. PHOTO CREDIT: KEN WALLACE

Some of these early stars would have been the large ones mentioned earlier. They would have burned up their fuel relatively quickly and blasted new elements out into space. These remnants would have spread out and eventually the process of stellar formation would have begun again. We can see this process happening today, in regions such as the Orion nebula where a huge cloud of gas and dust from earlier stars has within it new stars burning brightly. The nebula also contains condensations which astronomers call *protostars*, stars still in the process of collecting gas and not yet undergoing fusion.

The gas from which these second-generation stars form is rich in the fusion products of the old generation. The cloud would be rotating slowly, as the star which formed it was rotating. As it condenses the speed of rotation increases due to the conservation of angular momentum. Because of collisions between particles, the outer parts of the cloud form a disc shape around the central condensation. The rotation of the outer reaches of the cloud will cause it to detach from the central body as gravity is not enough to hold it in place. These outer reaches of gas form rings which themselves have eddies which begin

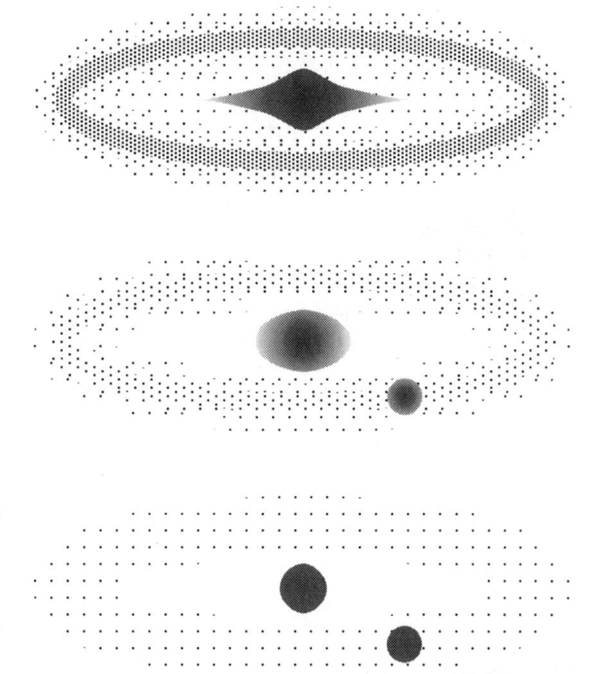

**3.8** As the disc of gas and dust which formed the early solar system began to condense the very centre of the disc collected most of the material and formed the Sun. Further out, however, bands of dust began to condense into smaller bodies. Eventually nearly all the material in the band was collected by the new bodies until planets were formed. When the Sun began shining, most of the remaining material not in the planet was blown away. Just a little still remains in the solar system; what there is can be seen as the Zodiacal light.

to condense to form larger bodies. As the central condensation builds up it will eventually begin nuclear fusion at its core. The energy released by such fusion will blow away any gas surrounding the new star and affect the materials in the disc according to their proximity to the new star.

Close to the new star the radiation pressure will be greatest so the lighter gases, hydrogen and helium will be blown away. Further from the star these gases will remain in the disc held there by the gravity of the eddies. In our solar system we see just such a range of planets, the four closest to the Sun being small rocky worlds, the next four being large gas giants which have held on to their hydrogen and helium.

In the eddies the planets begin to form. Grains of dust from the supernova remnant begin to clump together, which in turn causes more particles to clump and so on, until bodies perhaps 100 km in size are formed. It might be thought that all this would have taken a long time after the formation of the disc, but it didn't. Because the larger the clumps are the greater is their gravity, and as these increased in size the process accelerated. Some models suggest that it took 100 000 years for small (1 mm) grains to flatten out into the disc, but larger (1 cm) clumps would have settled in only ten years. After the formation of the disc the initial loose condensations of material should have formed 10 km bodies within 1000 years. This may seem an incredibly short time but the theory is borne out by the fact that the materials making the meteorites which we find today were formed in just the first 20 million years of the solar system.

Once these 10 km *planetesimals* had formed, the formation of complete planets began. As the planetesimals came together, their gravitational reach extended further into space and they were able to attract other planetesimals to form *proto-planets*. The collisions between the planetesimals making the proto-planet would have been enough to heat the rocks and minerals and begin the process of differentiation in the planet. Differentiation caused the heavier metal minerals to fall to the centre of the proto-planet while lighter material stayed near the top. Towards the end of this phase of formation, the collisions forming the planets would have been between very large objects, some as large as our Moon. Collisions of this magnitude may explain some of the anomalies which we see in the solar system: the strange rotations of Venus and Uranus and even the formation of Earth's Moon.

The legacy of the collisions which formed the planets can still be seen on their surfaces as craters. One look at our Moon through a telescope will reveal thousands upon thousands of craters pockmarking the surface. Other worlds, Mercury, Venus and Mars, and the satellites of the outer planets also show evidence of impacts during their history. Even on Earth such impact scars can be seen, but the effects of wind and water erase them in only a short time.

During the process of planetary formation, the largest objects would have come together first, followed by the smaller objects and so on. This is seen in the distribution of craters on the planets. The largest impact craters are often themselves marked by the impact of smaller bodies later on; it is rare to find the edge of a small crater interrupted by a larger one, indicating that the smaller ones were generally formed after the large craters stopped forming. By counting the number of craters in a given area on a body and looking at their size distribution, it is possible to estimate the age of the surface. This technique has been used extensively on the satellites of the gas giant planets.

It is the smallest objects in the solar system which are the oldest, they have not condensed into planets but have floated around since their formation in the early years of the solar nebula. The minor planets in particular (see Chapter 10) did not have the chance to get any larger, as the gravity of the forming Jupiter was sufficient to clean up most of the material in that region of space. It is also possible that Jupiter cleaned up most of the material that would have formed Mars. The comets, too, are remnants of the early period of the solar system. We will look more closely at them in Chapter 16.

It is fine to be able to quote the ages of events such as the formation of the solar system, but calculating such a date was not easy. Scientists guessed that the age of the Earth and the age of the solar system must be related; it is difficult to imagine a process by which the Earth might be created without the solar system, or that it might have suddenly appeared after the rest of the system. So if the age of the solar system is to be determined, the age of the Earth is a good place to start.

It has only been in the last 150 years that we have come to an understanding of the age of the Earth. Early methods of estimating the Earth's age were based on little more than guesswork. In 1654 John Lightfoot took Archbishop Ussher's famous calculation of the age of the Earth, derived by adding up the ages of everyone in *The Bible*, and came up with the idea that the Earth was created on 26 October 4004 BC at 9 o'clock in the morning, Mesopotamian time. Other peoples had their own reckoning; the Mayans of Mexico place the most recent creation of the Earth at 3114 BC, and other cultures had similar ideas. No one really had the faintest idea when the Earth had formed, but all agreed that it was a long time ago.

It wasn't until the middle of the eighteenth century that science began to make its first serious stabs at the Earth's age. In 1779 the Comte de Buffon calculated that if the Earth had cooled from a hot state, it was 75 000 years old. Buffon reached this result experimentally by making small models of the Earth, heating them and examining them as they cooled. Geologists then took the running on the Earth's age, proposing that the structures that they observed built up slowly over vast periods of time, perhaps millions of years, not as the result of catastrophic events every so often. This theory of uniformatism became the basis for many attempts to calculate the Earth's age by measuring the rate at which geological processes proceeded. Unfortunately, the data collected were so widely divergent that many different ages appeared, supported by different factions within the scientific community.

Geologists became so used to being entrusted with determining the Earth's age that they were taken aback when Glasgow physicist William Thomson determined in 1862 that the Earth was between 20 and 400 million years old. Disregarding uniformatism, Thomson said that the Earth had once been molten and had now cooled so that the surface was solid but the core was still liquid. He offered as one piece of evidence the observation that the deeper you go underground, the hotter it becomes.

To derive the Earth's age, Thomson calculated the amount of energy that would have been generated by gravitational contraction during formation, then allowed for the cooling of the Earth by radiating heat into space. By further refining his theory to allow for tidal heating, he narrowed the age range to 20 to 40 million years. This short period of time worried many scientists, mainly those proponents of

**3.9** When a radioactive substance is allowed to decay naturally, after a certain period of time only half the original amount of substance will remain. This time is called the half-life. What is important about the half-life is that it does not depend upon the amount of substance present. It could be 1000 tonnes or a milligram; after one half-life, only half of it will remain. The material which decays doesn't disappear, but remains as another type of atom: the daughter atom. By comparing the amount of parent atom material and daughter atom material in a given sample of radioactive substance, it is possible to calculate the amount of time which has passed since the material formed. This time will depend upon the type of daughter atom which forms and the environment in which the material is found.

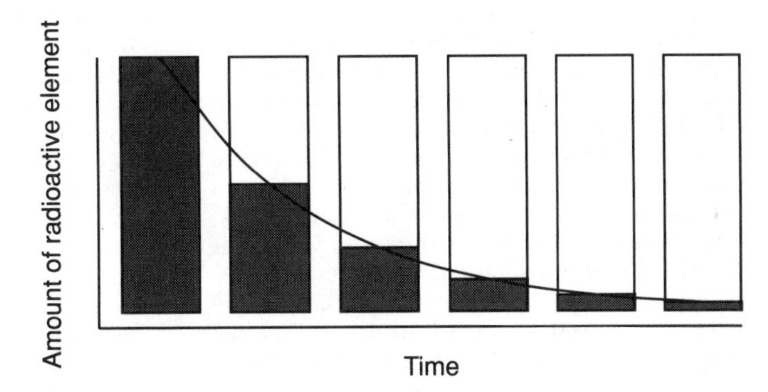


evolution who needed much more time for their process to occur. Thomas Huxley, Darwin's champion, attacked Thomson's* work, saying that much of his data was sketchy and that the 'accuracy of mathematical processes throw a wholly inadmissible appearance of authority over the results, pages of formulae will not get a definite result out of loose data'.

In 1899 it was John Jolly who devised the only new way of calculating the Earth's age. Jolly assumed that the oceans were originally of pure water and that all the salt in them was the result of chemicals dissolved out of the rocks by rivers and carried into the seas. Jolly measured the amount of salt in rivers, worked out the total amount of water flowing into the oceans and came up with a result of 80 to 90 million years. At the same time other geologists using older methods were coming up with similar results, so by the end of the nineteenth century geologists thought they had the problem solved. Different methods were coming up with around the same result: a hundred million years.

It was the discovery of radioactivity in 1896 that gave physicists the technique they needed to resume their study of the Earth's age. Henri Becquerel had discovered the phenomenon, and two years later Marie and Pierre Curie were the first to find the radioactive chemical elements polonium and radium. In 1902 Ernest Rutherford and others explained what the process was: the conversion of the atoms of one element into the atoms of another. In 1903 it was shown that radioactive elements such as radium gave out enormous quantities of heat as they decayed to lead; this heat would throw out Thomson's calculations of the Earth's age as he did not allow for this form of heating.

In the first decade of this century Rutherford and Frederick Soddy showed that radioactive elements turned into other elements in a very precise way. They found that the time taken for half a sample of a given radioactive material to turn into another element was always the same. For example, a 1 g sample of radium would turn into a 0.5 g sample in the same time as it took a 1 kg sample to turn into a 0.5 kg sample. This period of time was called a *half-life*. Half-lives of different elements were found to vary from a few thousandths of a second to thousands of millions of years.

The fact that the amount of radioactive material (parent atoms) and the number of decayed atoms (daughter atoms) in a sample could be

* For this, and other scientific work, Thomson was eventually elevated to the British peerage as Lord Kelvin.

determined gave scientists the system they needed to date rocks and other materials. By measuring the relative amounts of the parent and daughter atoms, the amount of time the sample had existed could be determined. If half the parent atoms had decayed to daughters, the sample must be one half-life old; if only a quarter of the parent atoms were left then the sample was two half-lives old, and so forth.

Rutherford and Yale University chemist Bertram Boltwood pioneered research into radioactive dating, as the technique became known. In 1905 Boltwood had dated 26 samples of minerals and found ages from 92 to 570 million years using the decay of uranium atoms to radium atoms as the basis of his calculations. Unfortunately most of the work was wrong, as the half-life of uranium-radium was uncertain.

In 1907 Boltwood repeated his work using the better researched uranium-lead decays, finding that for a given layer of rock, the uranium-lead ratio was constant, indicating that all parts of the layer had the same age. Boltwood then published results showing that two samples of minerals he had were 410 and 2200 million years old. Again the results were in error, but the fact that they indicated that the Earth was much older than any other estimates was something of a shock.

Other measurements of the age of rocks were carried out by more scientists, the accuracy of the method improving as more people experimented with it and associated techniques. It wasn't until 1921, at a meeting of the British Association for the Advancement of Science, that most groups of scientists, geologists, physicists, biologists and mathematicians were able to agree that the Earth was indeed several thousand million years old.

Over the ensuing years techniques have been refined and values of half-lives perfected until today radioactive dating is a normal part of any geological or paleontological investigation. The oldest rocks found so far, from the ancient parts of the crust in Australia and elsewhere, are 3800 million years old, pointing to the age of the Earth and the formation of the solar system some time before that. Present estimates, many made from measuring the age of meteorites using radioactive dating, put the formation of the solar system, the time when the Sun began shining, at 4500 million years ago.

# 4

# The Earth

It is human nature to wonder about our surroundings—what's over the next hill, how far it is to the next town. Even before people turned their attention towards the stars they explored their world. We live in a fortunate age because spacecraft have allowed us to see the Earth as it really is: a small, blue, water-covered world orbiting the Sun. Thousands of years ago, though, the answer was not so clear. Today we know that the world is spherical, but standing on its surface that is not a simple thing to prove. It took many years of observations and smart thinking before people came to the conclusion that the world was a ball. In fact various people in different parts of the world followed much the same reasoning and came to the same conclusion independently.

The method used, although it takes many years, is simple. Firstly you must understand lunar eclipses, and know that they are caused by the Moon passing into the shadow of the Earth. Once you've ascertained that lunar eclipses only happen at full moon when the Sun and Moon are opposite each other in the sky, this is not a big step. Then you need to notice the shape of the Earth's shadow. It is curved, thus the Earth has to be curved. It is possible for the Earth to be a flat, circular plate and still cast a curved shadow, but after a few eclipses have been observed, particularly those occurring near sunrise or sunset, you can deduce that the Earth must be spherical.

In Europe the belief that the Earth was flat was held by many until the boats of Magellan circumnavigated the globe. The reason was not the lack of scientific knowledge, but the fact that Aristotle had said that it was flat, and as Aristotle's work was the official basis of all scientific knowledge, despite being obviously wrong on many points, that was that.

Aristotle was not the only Greek thinker; they also had some very bright scientists. One of the burning questions two thousand years ago was, 'How big is the Earth?' Of course many disagreed with Aristotle and thought that the Earth was a sphere, but just how big was it?

It was the Greek scientist Eratosthenes who worked out the method for measuring the Earth's diameter and then performed the experiment. Eratosthenes noticed an important fact. At noon on the summer solstice (21 June) the Sun shone directly down a well in Syrene, Egypt. This meant that the Sun was directly overhead at that instant. Eratosthenes travelled to the city of Alexandria and at noon on 21 June measured the shadow cast by one of the obelisks there. He found that the shadow made an angle of 7°12′ with the obelisk. Therefore the Sun was 7°12′ away from being directly overhead.

Eratosthenes then assumed that the Sun was a long way from the Earth and that the rays from it would be parallel. The distance from Alexandria to Syrene was known: it was 5000 stadia (850 km). Knowing that this was 7°12′ or 1/50th of the Earth's circumference he calculated the Earth's radius as 25 000 stadia (6764 km), close to the correct value of 6370 km. Eratosthenes also attempted to calculate the distance to the Moon. The method he used was based on sound mathematics, but the instruments he had at his disposal were not accurate enough to produce reasonable results.

The nature of the orbits of the planets was one of the greatest scientific debates for over two thousand years. It was responsible for some of the greatest inventions in mathematics and

**4.1** Although the curved shadow of the Earth on the Moon during a lunar eclipse can be explained using a flat circular plate, this does not cover the case when the Moon is near the horizon during an eclipse. If the Earth were a flat plate, the shadow would be a straight line while the Moon was low in the sky, only becoming curved as it rose higher.

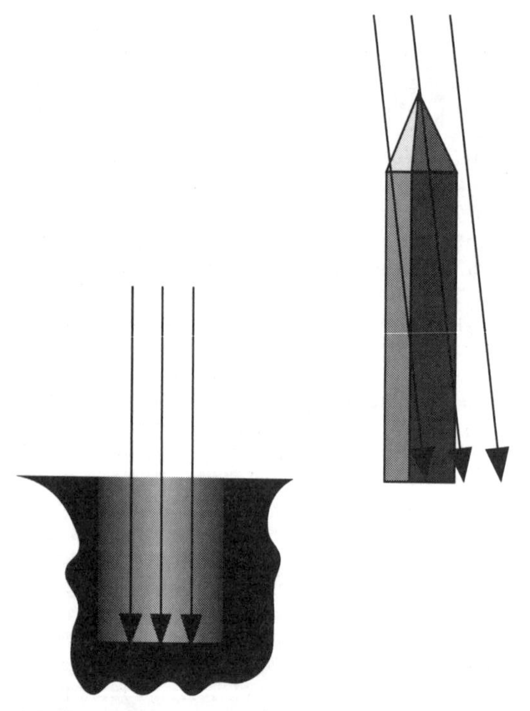

**4.2** Eratosthenes was able to calculate the curvature of the Earth and hence its radius. At Syrene he noticed that on the summer solstice the sunlight fell directly down a well, casting no shadows, while at the same time at Alexandria the obelisks and other structures did cast shadows. By calculating the angle of the Sun's rays at Alexandria and knowing the distance between the two cities, Eratosthenes was able to calculate the radius of the Earth.

science, and was also at least partially responsible for someone being burnt at the stake. So why all the fuss?

Look around you, and think of what you see and say every day. 'I was up before sunrise' is a common enough expression, but wait a moment: the Sun doesn't rise, the Earth turns and we come out of its shadow. But it doesn't look that way, does it? Get up early and you can *see* the Sun coming up. This was the problem: there is no simple way to prove to someone that the Earth is moving, so why believe that it is? Aristotle, the great philosopher, said that the Earth stood still and that the Sun, planets and stars orbited it in an orderly manner on crystal spheres, and there was no way of proving the great man wrong.

As the centuries passed, scientific knowledge became the domain of the few, kept in monasteries. Through the Dark Ages, the words of the great ancient minds were regarded as perfect, and none thought nor dared to challenge them. With the coming of the Renaissance and the invention of the printing press in the 1450s these works were taken as the height of learning; others expanded upon them and debated their merits, but basically accepted as true all that was written in them, without question.

The beginning of the end of these ideas came from the Catholic Church, although it was one of the main supporters of the theories of Aristotle and other philosophers; after all, it had kept the knowledge safe in its monasteries for over a thousand years. The problem was with the calendar: it didn't work. As we have seen, the system introduced by Julius Cæsar of having

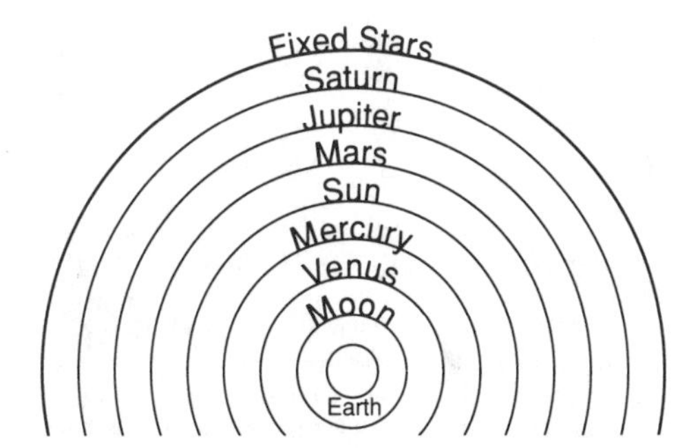


**4.3** Ptolomy's picture of the universe had the Earth at the centre of all with the Sun and planets orbiting it in perfect circles. The stars were all fixed to an outer sphere well beyond the planets. Ptolomy's system could not explain the motion of the planets completely; he had to rely on a series of epicycles and deferents to provide the retrograde motion exhibited by the outer planets.

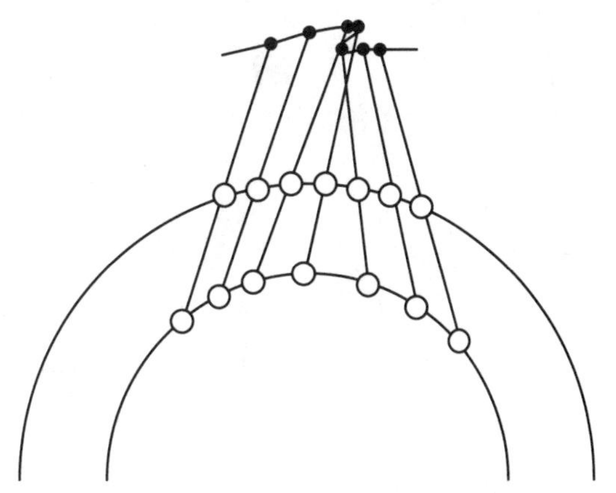

**4.4** The motion of the Earth and of a slower outer planet can give the appearance that the planet is moving in the opposite direction against the stars. As the Earth catches up with the planet, the planet appears to slow in the sky and travel west to east for a short while before regaining its normal motion. It was this phenomenon which lead Ptolomy to introduce deferents and epicycles into his explanation of the motion of the heavens.

leap years took much of the error out of the calendar, but there was still the problem of three-quarters of a day each century to worry about: by the 1500s this was an extra ten days. We saw Pope Gregory's solution to that problem in Chapter 1.

Gregory didn't just dream up the system for himself, in 1514 he set a mathematician, Nicklaus Kopernig, to work on it for him. Kopernig was a Canon in Frombork, Poland and had a doctorate in canon law from the university of Ferrara. Kopernig undertook the work requested of him by beginning with a complete reappraisal of the Aristotelian system. On 1 May 1514, under the latinised form of his name, Copernicus, Kopernig published *The Little Commentary* in which he proposed a Sun-centred system as a better description of the solar system than Aristotle's Earth-centred system.

Copernicus worked further on this theory but did not publish until close to death in 1543. In that year his full work *On the Revolution of Celestial Spheres* was printed. It is thought that Copernicus put off publishing until he was dying to keep from being persecuted for his statements. Indeed his work was roundly criticised by many, but not simply for religious reasons. Copernicus' system was not much better than the existing system of Aristotle which had been formalised by Ptolomy. Copernicus still used cycles and epicycles—in fact more of them than Ptolomy—and his theory left a number of important questions unanswered such as: if an object is thrown upwards on a spinning Earth, why doesn't it fall down further west?

Meeting in Trento, Italy, just two years after Copernicus' publication, the Catholic Church let the contents pass with little comment. The reason was simple: the system worked. As summed up by Dutch astronomer Gemma Fisius: 'It hardly matters to me whether he claims that the Earth moves or that it is immobile, so long as we get an absolutely exact knowledge of the movements of the stars and the periods of their movements, and so long as both are reduced to altogether exact calculation.'

So the reason for Copernicus not being condemned as a heretic was simple: no one believed that he really meant the Earth was moving; it was just convenient maths to make that assumption. After all, astronomers drew lines and points in the sky which weren't really there anyway, simply for mathematical convenience. Copernicus' work was not physical reality; it was seen as mathematical fiction. As a tool, the Church happily used it in the calendar reform of 1582.

The real problems for the heliocentric system

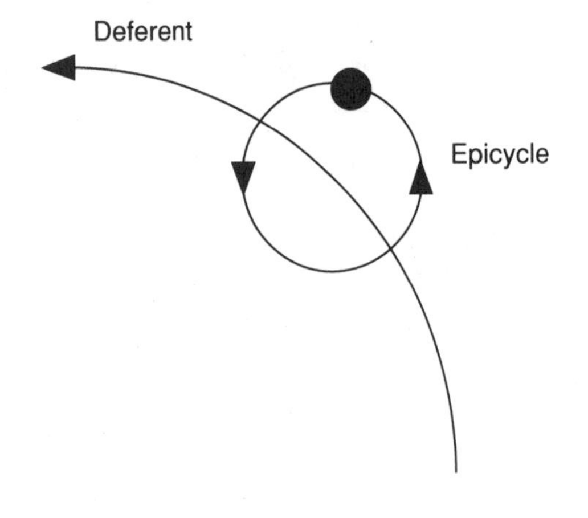


**4.5** To explain the phenomenon of retrograde motion and other irregularities in the orbits of the planets, Ptolomy used a system of epicycles and deferents. Instead of following a simple circle around the Earth, the planets orbited on smaller circles, epicycles, while the centre of the epicycle followed a larger circle, the deferent. In this way Ptolomy was able to keep the planets moving in perfect circles. Copernicus, too, used epicycles in his system of the universe to explain other orbital anomalies.

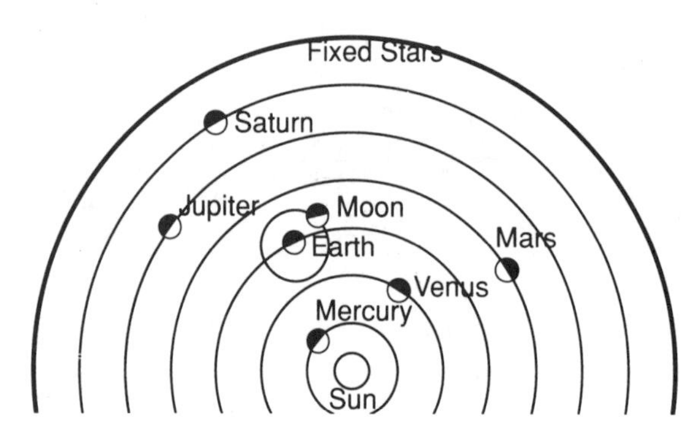


**4.6** Copernicus' view of the universe had the Sun at the centre and the Earth and all the other planets circling it. The Moon, however, kept its place circling the Earth. Though Copernicus made the break from Ptolomy by putting the Sun at the centre of the universe, he still held on to the idea of circular motion, for which he reintroduced epicycles, and the idea of a sphere of fixed stars outside the solar system.

did not begin until 1609 when Galileo Galilei took the newly invented telescope and pointed it upwards. Instead of the perfect sphere the Moon was meant to be, contaminated slightly by its proximity to the Earth, Galileo saw mountains and valleys like those of the Earth. When he looked at Jupiter he saw four dots shuttling back and forth around the planet, proving that not everything orbited the Earth.

Copernicus' work, though, was not the end of the matter. Although he was able to predict the motion of the planets very accurately, the system had little to do with scientific reality; after all, the planets still needed to move on deferents and epicycles. Many astronomers stuck to the old theories. One, Tycho Brahe, was perhaps the greatest recorder of the heavens in the time before the telescope. From his observatory on Hven, an island between Denmark and Sweden, Brahe observed the skies every clear night from 1573 until his death in 1602. Brahe's measurements of the comet he discovered in 1577 showed that the comet was outside of the atmosphere and passing near the planets, not within the atmosphere as Aristotle had said. Even with this evidence, however, Brahe was not moved from his belief that the Earth was the solar system's centre.

Brahe needed an assistant in his work, so in 1600 Johannus Kepler, a teacher from Graz, Austria, went to work with him. After Brahe's death in 1602 Kepler took over the mountain of observations from the previous twenty years. With this incredible detail at his fingertips, Kepler was in a unique position to begin the study of the orbits of the planets. He commenced his work with Mars.

Using Brahe's enormous collection of precise measurements, Kepler was able to draw the orbit of Mars around the Sun. Aristotle had said that the only perfect motion was motion in a circle, but Kepler found himself unable to make Mars' motion fit a circle. It took him four years and over 900 pages of calculations to realise that

**4.7** One of the problems besetting early astronomers was the method of calculating the area of an orbit, or the area under a graph. This had to be done by dividing the area under the graph into many small rectangles or triangles and then adding the areas of these up to get a total. This was tedious and no matter how many rectangles and triangles were used, the result was never exactly right. The invention of calculus by Newton and Leibnitz allowed the calculation to be made more accurately and more quickly, allowing great strides to be made in the study of orbits and in many other branches of science.

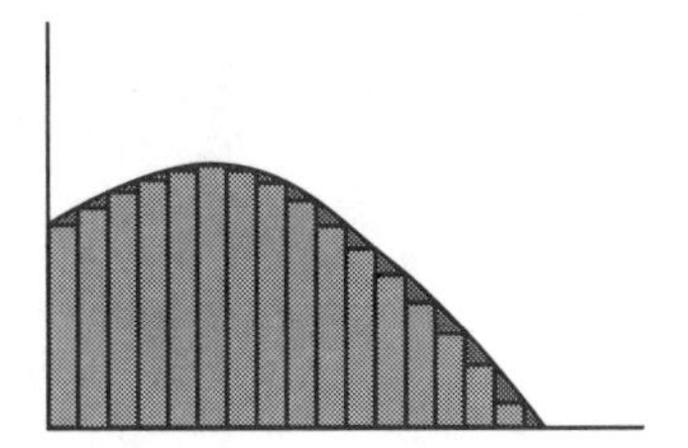

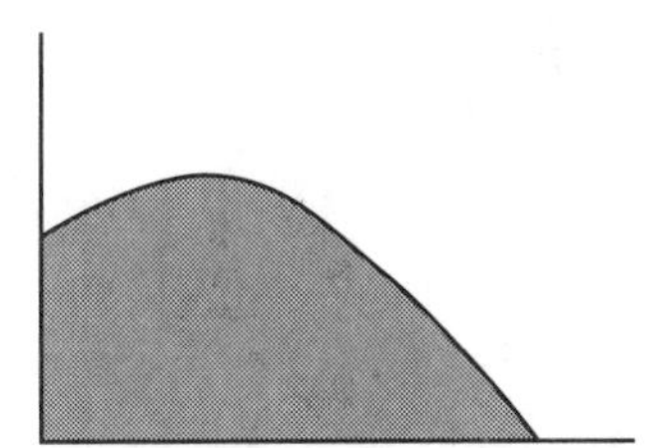

Mars' orbit was an ellipse rather than a circle: his first law of planetary motion.

Kepler was initially at a loss to explain how such an orbit could be stable. If there were a force holding the planet to the Sun, then that force must be weaker the further away the planet was. Kepler believed in William Gilbert's theory that the Sun was magnetic and held the planets by this force, so a solution was soon to hand. Kepler noticed that the further the planets were from the Sun, the slower they moved in their orbits and, combined with the extra distance, this made them move around their orbits over longer periods of time. This led him to his second law of planetary motion.

Further observations of Mars in its orbit led him to the conclusion that the planet moved faster when it was close to the Sun, and slower when it was far away. Using this fact, Kepler demonstrated that a line drawn from the planet to the Sun swept out equal areas of space in equal times: his third law of planetary motion.

In his calculations Kepler used a procedure known as the *method of infinitesimals* to calculate the areas of segments of ellipses. It wasn't until Leibnitz and Newton independently invented calculus that the calculations could be performed more easily. Indeed, it was Newton who finally put Kepler's work on a firm mathematical and scientific footing.

In 1665 Isaac Newton was 23. He had just finished his degree at Cambridge and was worried about his health. Newton had good reason to fear illness, because in 1665 the plague spread through England. To escape the risk of infection, Newton travelled to his family home in Woolsthorpe, Lincolnshire. In the two years he stayed there he worked out what held the universe together.

Being shy, it wasn't until 1685 that Newton made plans to publish his work, and then only at the urging of Edmund Halley. In 1687 when *Principia Mathematica* was published it took the scientific community by storm. Newton took the description of the universe out of the hands of the philosophers by asking not 'Why?' but 'How?'

The story of the apple falling from the tree and giving Newton the idea for gravity is probably apocryphal, but he did use an apple to explain his ideas. He proposed that gravity was the force which controlled the motion of all moving bodies. His calculations showed that gravity existed between all bodies, so that just as the Earth attracted an apple to it, the apple attracted the Earth, but the effect was much smaller in the latter case as the Earth is so much heavier.

Newton showed that this force had all the properties Kepler needed for his elliptical orbits to work. Using his calculus and his formula for universal gravitation, Newton was able to calculate the mass of the planets, show how the Moon caused tides on the Earth and demonstrate that it was a combination of the Earth's and Sun's gravity which determined the motion of the Moon. Thinking about the effect of gravity on light, Newton also predicted the existence of those objects we now call black holes.

Newton's work is still the basis of our calculations of planets' orbits and other solar system phenomena. In 1915 Albert Einstein's general

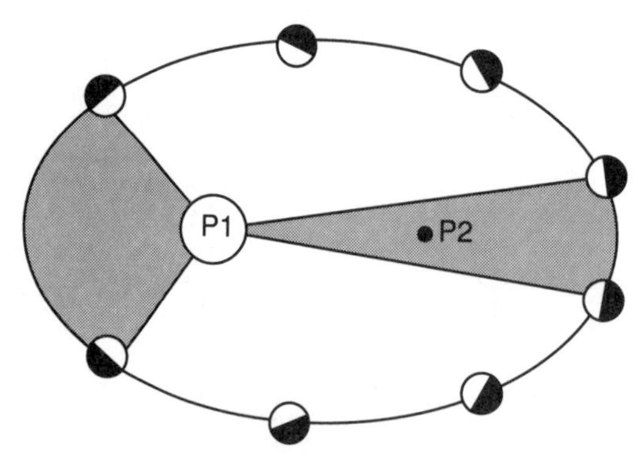


**4.8** Here eight positions of a planet in orbit around the Sun are shown, spaced equally in time. Kepler proposed that the orbit a planet followed around the Sun is an ellipse centred on two foci, P1 and P2, with the Sun situated at one focus (P1). Kepler's laws of planetary motion show that while the planet is far from the Sun it moves slower than when it is near the Sun. Kepler's laws also show that the area swept out in space by a planet in a given time is the same regardless of the planet's position in orbit.

theory of relativity gave a much better explanation of the causes of gravity and made it possible to make more accurate calculations. However, the mathematical complexity of the theory is such that Newton's work is preferred for all but the most exacting or exotic calculations.

Astronomy normally only deals with those things seen in the sky, but there is one part of the solar system which is much closer than all the other bits: the ground beneath our feet. When looking at the planets it is important to remember that many of the processes and features on the planets and their moons can be explained in terms of processes we see on the Earth.

The Earth is unique in the solar system: of all the planets it is the only one to have large amounts of water as solid, liquid and gas on the surface. It is this water which allows the abundance of life on the planet. With all its importance to those living on it, the surface of the Earth is but a minor part of the whole planet. Life is confined to the first few metres below the surface and a few kilometres into the air and below the seas, but beneath us are nearly 13 000 km of rocks and minerals.

The structure of the interior of the Earth is a puzzle still being pieced together by scientists. The layer on which we live is called the *crust*. It varies in depth from around 7 km thick under the oceans to more than 40 km under the continents. The deepest holes we have dug are only 6 km deep, barely scratching the surface. The first clue we have about the Earth's structure and composition is its density. The density of the Earth is 5.5 $g.cm^{-3}$, which is much higher than the 2.6 $g.cm^{-3}$ of the surface rocks. To balance the density, the material at the centre of the Earth must have a density of around 10 $g.cm^{-3}$, meaning that it is probably a metal.

The workers in deep mines are only too aware that the temperature of the Earth increases with depth. This, and the occurrence of volcanoes, shows that the inside of the Earth must be hot. But even given these facts, how is it possible to determine what is inside the Earth? Astronomers use waves of light and other radiation to bring them information about the universe. Geophysicists use waves also, but in their case it is the seismic waves generated by earthquakes.

When an earthquake happens, vibrations from it travel along the surface of the Earth, causing damage to buildings and other structures. These same vibrations also go down into the Earth. Given a large earthquake and sensitive equipment, the waves from an earthquake on one side of the Earth can be detected on the other side.

There are two types of wave generated by earthquakes, *pressure* (*P*) *waves* and *shear* (*S*) *waves*. P waves can travel through both solids and liquids, S waves only through solids. From looking at the paths these waves take away from an earthquake, geophysicists have shown that there is a liquid core within the Earth over 6500 km in diameter. Other techniques show that within the liquid core is a smaller (2400 km diameter) solid core.

Above the core is a thick layer of rock, the *mantle*, which extends from the outer core to the base of the crust, 40 km below our feet. We know something about the mantle because it regularly rises to the surface in volcanic eruptions. Some forms of eruption are caused by molten mantle material, *magma*, rising to the surface through cracks and fissures. When it reaches the surface the mantle material flows over the surface, depositing dark-coloured rocks such as basalts. The composition of these basalt

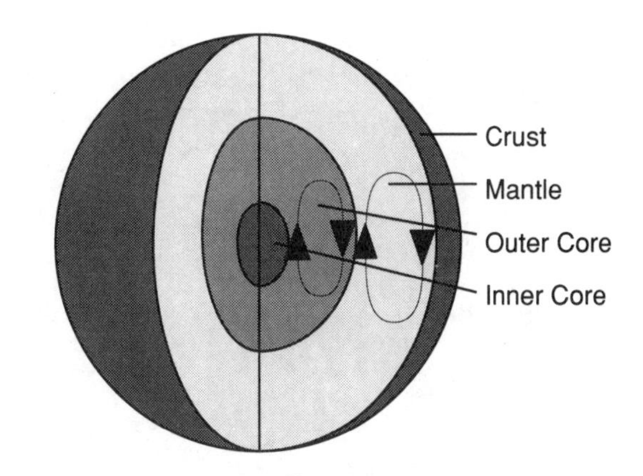


**4.9** The crust of the Earth which we live on is only a few tens of kilometres of hard rock; beneath it is the mantle, a solid though still mobile mass, 2900 km thick. Motion in the upper part of the mantle is responsible for the phenomenon of plate tectonics. Below the mantle is the outer core, a nickel-iron layer made liquid by the enormous temperatures found below the surface. It is the motion of this layer which makes the Earth's magnetic field by an as-yet not fully understood process. At the centre of the Earth is a ball of solid nickel-iron under tremendous pressure, which stops it too from becoming liquid.

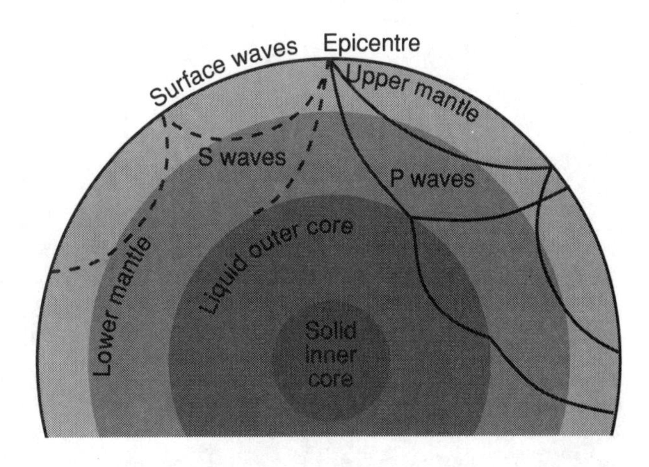


**4.10** When an earthquake happens three types of vibrational waves are sent out. Surface waves travel short distances along the Earth's surface. More important to the study of the Earth's structure are P and S waves which travel through the Earth and which can be detected by seismometers around the world. By looking at the intensity and arrival time of the waves at seismometers spread around the Earth's surface, scientists can reconstruct the path the waves took through the Earth and hence learn about the structure beneath the crust.

rocks tells us the composition of the material in the mantle, mainly dark-coloured magnesium and aluminium silicates.

The last piece of information we have about the Earth's interior comes from our knowledge of the Earth's magnetic field. The Earth's magnetic field has been known for centuries; its existence allowed people to explore the world using a compass as a directional instrument. We can think of the Earth's field as being caused by a large magnet within the Earth, inclined at 11.5° to the rotational axis. This means that the needle of a compass doesn't really point towards the north geographic pole, but to the north magnetic pole, presently in Canada. The centre of the magnetic field is quite close to the centre of the Earth—in fact, they differ by just 500 km. All these factors must be taken into account by anyone trying to explain the Earth's, or another planet's, magnetic field.

Although no one has yet completely explained the process by which the Earth's magnetic field is generated, some points are clear. Because the strength and direction of the magnetic field varies over time, the field cannot simply be due to a large lump of magnetic material within the Earth; it must be the result of some dynamic process. The best explanation is that there is a metallic, turbulent liquid within the Earth, a theory which fits well with seismic data. Putting all these facts together scientists have made a model of the interior of the Earth, as shown in Figure 4.9.

The surface of the Earth on which we live is a dynamic place. Things are constantly changing, and although these changes are not everyday

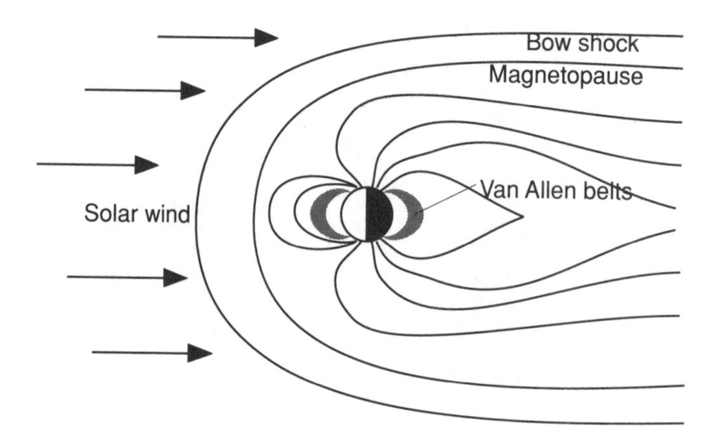


**4.11** The magnetic field of the Earth extends well past the atmosphere and out into space. It is affected by the solar wind, the stream of charged particles continuously given out by the Sun, giving the field a tear-drop shape. On the side towards the Sun the field is compressed as it deflects many of the particles away from the Earth, while past the Earth the field is drawn out in a long tail. Above the Earth, trapped in between the field lines of the magnetic field, are the charged particles of the Van Allen radiation belts.

**4.12** This view of the Grose Valley in the Blue Mountains of New South Wales is typical of scenes throughout the world. Layers of sedimentary rocks laid down over many millions of years are uplifted by the action of plate tectonics. Rivers cut their way through the rock forming valleys and exposing the layers to further erosion. Views such as these remind us of the ages which have passed, forming the landscapes we now see.
PHOTO CREDIT: DAVID REIDY

occurrences, evidence of them is abundant. The most dramatic evidence of the forces at work within the Earth comes from earthquakes and volcanoes, which in just a few minutes can create vast changes in the landscape. But by digging a bit deeper into the Earth's history, we can find much greater evidence of change.

We tend to think of the map of the world as fixed and unchanging, and indeed over the history of mankind it hasn't changed appreciably. The first serious challenge to this view came in 1915 from a German amateur geologist, Alfred Wegener. Like many before him, Wegener started with the fact that the east coast of South America and the west coast of Africa fit together remarkably well except that the Atlantic Ocean is between them. Wegener took the idea further, showing that the geological features of the two coastlines were also similar.

Working further north he showed three major geological features in Europe which continued on the North American continent. Wegener then proposed that the African and South American continents, and the European and North American continents, were once part of a larger landmass which had split apart. Wegener journeyed to Iceland in the North Atlantic Ocean, where he studied the active volcanoes and hypothesised that this is where the continents were splitting. Wegener's hypothesis, called continental drift, was scorned by all. Not only was Wegener not a real geologist, but his theories left one very important unanswered question: how could the lighter continents

**4.13** When magnetic maps were made of the ocean's floor, many scientists were surprised to see the same pattern of magnetic reversals on each side of the oceanic ridges. These matching reversals were used as support for the theory of plate tectonics which says that new oceanic crust is slowly created at the mid-oceanic ridges and that it spreads slowly outwards from there.

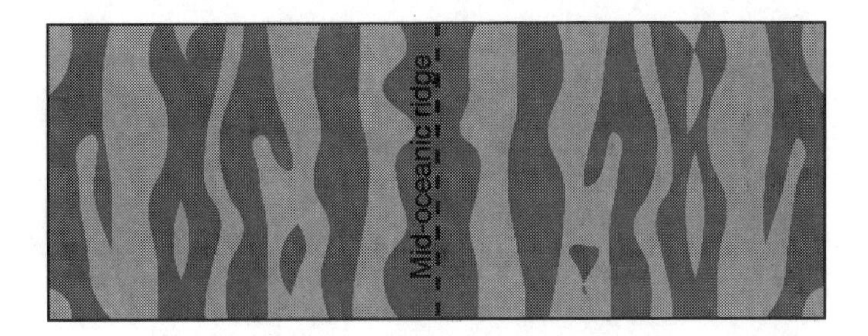


plough through the heavier basaltic rocks of the sea bed? There was no known force which could propel the land, and Wegener had been unable to come up with a convincing explanation.

In 1966 Walter Pitman was looking at maps of the magnetic field of the ocean floor. In the 1950s the invention of magnetometers had shown that the magnetic field of the ocean floor was not a constant north-south pattern as was expected; instead there were strips of floor where the pattern ran south-north. Pitman was looking at a region of floor near the Mid-Atlantic Ridge, a line of volcanoes running the entire length of the Atlantic Ocean underwater, except at Iceland where they poked above. Pitman noticed that the strips on one side of the ridge were identical in width and position to those on the other side of the ridge. Pitman had found the mechanism by which the continents moved apart. Instead of drifting through the oceanic crust, the continents rested on the crust like a raft and were carried along with it. At the oceanic ridge Pitman showed that new crust was being created as magma from the mantle slowly welled up to the surface. But if new crust was being created, what was happening to the old crust, and was the Earth getting bigger? Once the idea was in people's minds it was easy to explain: in other parts of the world, such as along the west coast of South America, old crust was being pushed back into the mantle to re-melt. These regions are called *subduction zones*. Thus for each metre of new crust created at a *spreading ridge*, a metre of old crust was consumed somewhere else.

The creation and destruction of crust explained the magnetic pattern on the sea floor. As the new crust was made at the ridge it acquired the magnetic field of the Earth at the time it cooled. The alternating direction of field was explained by the alternating direction of the Earth's field, firstly pointing north as is the case today, then south, then north and so on through history, each period leaving its signature in the rocks. Supporting evidence came from the amount of sediment on the sea bed. The further oceanographers went from the ridge the thicker the deposits of sand, silt and dead animals became. Near the ridge the deposits were very thin or non-existent, showing the youth of the sea floor.

The regions of continent and the attached oceanic floor were called *plates*, and the theory of the motion, *plate tectonics*. These plates are in constant motion, and the rate is quite measurable, given the right equipment. As an example, there is a ridge between Australia and Antarctica which is separating those two continents at around six centimetres per year, about the same rate that fingernails grow.

Plate tectonics only made its way into university and school texts in the early 1970s; by then the mechanism of the spreading was understood. Although nominally solid, the upper 700 km of the mantle is better considered to be more like plasticine or putty. The mantle material, because it is hotter at the bottom than at the top, has convection currents, like those which can be seen in soup bubbling on a stove. These convection currents rise, flow along the boundary with the crust for a way and then sink back to the depths. It is this motion beneath the crust which drags the crustal plates with it.

There are many places on the Earth where two plates meet. At some they move apart, at some together, and at others they simply move past each other. Where two plates move past each other a feature called a *transverse fault* develops. The San Andreas along the west coast of the United States is one such fault. Along such a fault, motion is often accompanied by violent earthquakes. The last major earthquake

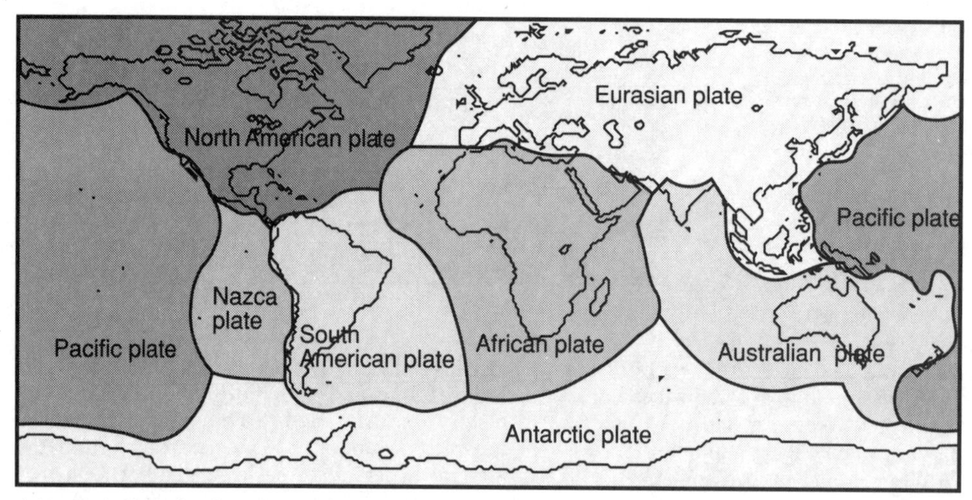


**4.14** By looking for the telltale geological features scientists have been able to define plates of crust on the Earth which move relative to each other very slowly and which explain many of the large-scale geological features of the Earth. The plates typically move 3 to 5 cm with respect to each other each year. This movement makes such features as the Himalayas where two plates collide with each other, or the chain of mountains beneath the Atlantic ocean where two plates move apart and new crust is created.

**4.15** Road cuttings are an ideal place to view the layers of rock otherwise hidden beneath our feet. Here on the road between Bathurst and Lithgow in New South Wales the orderly layers of sedimentary rock have been cut through by an intrusion of igneous rock. This intrusion of molten rock, a dyke, split the layers while they were deep beneath the Earth, slowly solidifying into the feature seen here. PHOTO CREDIT: DAVID REIDY

on the San Andreas fault was near San Francisco in 1989 and before that in 1906 but smaller tremors occur nearly all the time.

Faults modify the surface features of the Earth. In some cases the faults are near the plate boundaries, but smaller faults also occur away from the boundaries. The type of terrain caused by faults on Earth can often be seen on other planets and satellites of the solar system as well. Figure 4.18 shows the major types of terrestrial faults.

The atmosphere is the other great modifying element on the Earth's surface. It is because of the atmosphere that we have rain and winds, snow and ice, and the forces of erosion. Erosion eventually wears down all geological structures,

**4.16** Evidence of the movement of the Earth's crust is never hard to find. This section of sedimentary rocks shows bending due to movement of the crust. Because the bend is quite sharp, but the rocks have not fractured as a fault, it can be deduced that the bending took place while the rocks were well beneath the surface of the Earth. The area of New South Wales, in which this cutting and the one shown in Figure 4.15 are found, is near to Australia's Great Dividing Range, a mountain range formed by the collision of two tectonic plates over tens of millions of years. This join between the two plates is similar to the join now forming the Himalayas but is no longer active, having finished moving around twenty million years ago. PHOTO CREDIT: DAVID REIDY

**4.17** The evidence of past violent activity can be seen all around. Here a seam of quartz 100 mm wide cuts through layers of sedimentary rock. Once many kilometres below the Earth's surface the rock was split and liquid rock, magma, rich in quartz, squeezed into the gap. As the magma cooled, the quartz crystallised, making the vein seen here. Now exposed by uplift and erosion, the quartz and surrounding rock is weathering away at the surface. PHOTO CREDIT: DAVID REIDY

and in the process builds new ones from the particles worn away and deposited again elsewhere. Erosion and deposition have made features such as the Grand Canyon in the United States, the Nile delta in the Mediterranean Sea and much of the deserts which circle the globe. On other planets with atmospheres we can also see the effects of erosion.

Lying on the surface of the Earth are two layers which make this planet unique: the oceans and the atmosphere. It is the liquid water over 70 per cent of the Earth's surface and the oxygen-rich atmosphere which make life as we know it possible.

Although some of the water on the planet is in the form of ice caps at the poles and freshwater lakes and streams, the bulk of it is in the oceans. The oceans have a volume of $1.3 \times 10^9$ cubic kilometres, and a mass of over $10^{18}$ tonnes. The water in the oceans is salty; in fact 3.5 per cent of the mass of the oceans and seas is salt of one form or another. These salts come from the erosion of the land and the chemicals which become dissolved in rain water. The salts

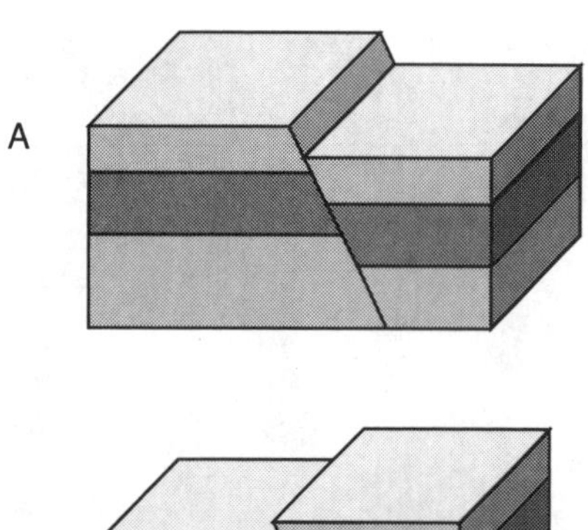


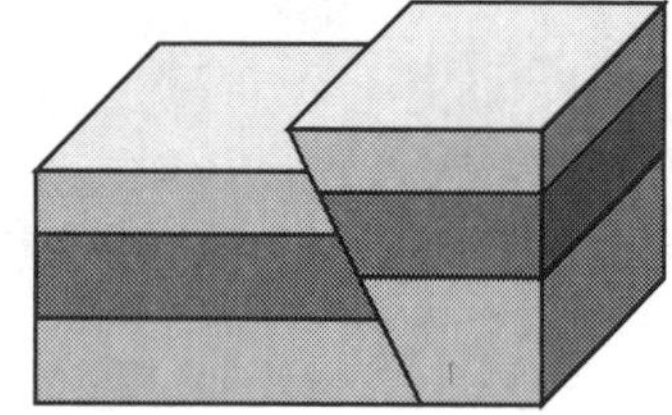


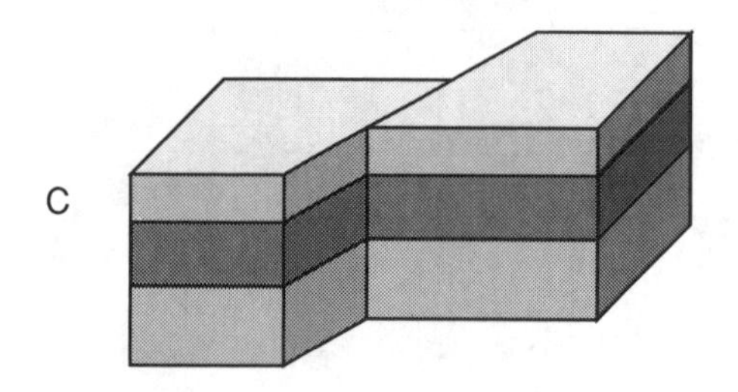


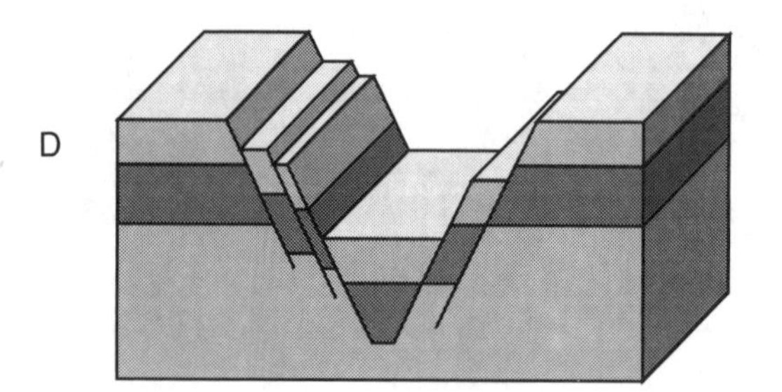


**4.18** By studying the types of faults found on Earth, many of the features on the outer planets can be explained.

A A *normal fault* is one in which two blocks move vertically with respect to each other. The fault is caused by the two blocks being pulled apart and one falling to fill the gap created.

B A *reverse fault* is similar to a normal fault, but is caused by the blocks being pushed together resulting in one moving upwards to overhang the other.

C In a *transverse fault* there is no vertical movement. The two blocks simply move past each other, distorting any surface features running across the boundary.

D *Horst and graben faulting* is caused when two blocks move apart and the material between collapses to fill the gap forming a valley (or graben) between lines of hills (horsts). This type of faulting can be seen on many satellites of the outer planets, and on Earth in such features as the Rift Valley in Africa.

**Table 4.1 The Earth's atmosphere**

| Gas | | Fraction |
|---|---|---|
| Nitrogen | $N_2$ | 78.1% |
| Oxygen | $O_2$ | 20.9% |
| Argon | Ar | 0.93% |
| Carbon Dioxide | $CO_2$ | 0.034% (increasing) |
| Neon | Ne | 0.0018% |
| Helium | He | 0.0005% |
| Methane | $CH_4$ | 0.0002% (increasing) |
| Water | $H_2O$ | 3%–0.1% |

in the ocean are not just common table salt, sodium chloride, but salts of calcium, magnesium and other metals. Even gold is found in sea water, but recovering it costs more than the gold is worth.

Gases as well as salts are dissolved in the oceans. Oxygen makes it possible for fish and other marine creatures to survive, and carbon dioxide also is dissolved in it. The amount of oxygen dissolved in the oceans is smaller than in the atmosphere, but there is 60 times as much carbon dioxide, making the oceans a major factor in the Earth's climate.

Of course, to those of us living on land, the atmosphere is a much more important factor in our existence. The blanket of gases which surrounds the Earth consists primarily of nitrogen (78 per cent) and oxygen (21 per cent) with the remaining 1 per cent being argon and carbon dioxide (0.034 per cent). Not included in this is the average 2 per cent water vapour. Gravity acts on the gases in the atmosphere, creating a sea level pressure of 101.3 kPa, or the equivalent of about ten tonnes per square metre.

The atmospheres of planets represent a delicate balancing act. The atoms and molecules in the atmosphere must be mobile enough to be gaseous, but not moving so fast that they can overcome the planet's gravitational attraction and escape into space. It is the temperature of the atmosphere which determines how fast the molecules in it are moving. The closer a planet is to the Sun, the higher the temperature of its atmosphere. On the Earth the gravity is strong enough to hold on to large molecules like oxygen and nitrogen, but not strong enough to hold on to fast-moving small molecules of hydrogen and atoms of helium, so these continually escape into space. Planets which have sufficient gravity to hold on to their light gases form the gas giant planets of the outer solar system.

The gases in the atmosphere all have very different properties. Nitrogen in the atmosphere has few important chemical properties and argon too is inert, taking part in no chemical reactions at all. Because the molecules of nitrogen and the atoms of argon are so massive, neither can escape from our atmosphere, so the amounts we have now represent the build-up since the formation of the Earth. Helium, the gas used in blimps, is present in the atmosphere, but has such light molecules that it can escape with little difficulty. So where does it come from? Helium is released by the decay of uranium and other radioactive substances. It is released into the atmosphere at a slow, but steady pace, just maintaining equilibrium with the amount that is lost.

The oxygen and carbon dioxide which are present in our air are both highly reactive elements. Although neither can escape because of the weight of the molecules, both are constantly being removed from the atmosphere by one process or another and being replaced by still other processes, mainly processes of life. At the base of every food chain* are plants. Using a complex series of chemical reactions called photosynthesis, plants take carbon dioxide from the air and water from the soil and, with the energy provided by sunlight, produce oxygen and sugars. The oxygen they release into the air, the sugars they use to make more plant material and so grow.

Further up the food chain are animals. They eat the plants, deriving energy from the stored sugars and starches. They combine these chemicals with the oxygen they breathe in to produce carbon dioxide and the energy they need to live, returning the $CO_2$ to the atmosphere. It is these two processes which maintain the oxygen and carbon dioxide levels in our atmosphere.

The structure of the atmosphere is in a delicate equilibrium with the ocean. If the temperature of the ocean increases, the amount of water and carbon dioxide in the air also increases, which in turn leads to a further increase and so on. The average temperature of the oceans is 3.9°C. If it were just 6° cooler the oceans would

* Recently small colonies of creatures have been found living around hot vents at the bottom of the ocean. These creatures derive their food directly from the mineral laden water issuing from the vents. These small colonies are an exception to this rule.

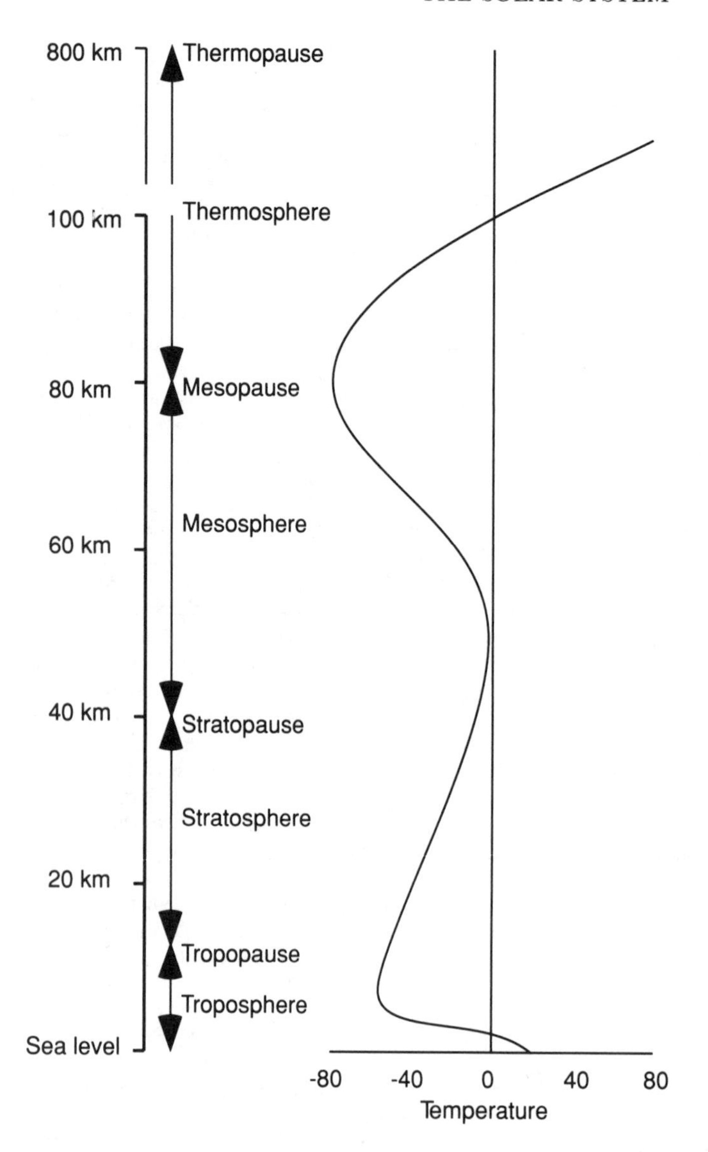


**4.19** This plot of the Earth's atmosphere shows the principal regions and the temperature of the air at altitude.

freeze; if it were just a few degrees warmer large amounts of $CO_2$ would be released into the atmosphere, disrupting the equilibrium there and causing a further increase in temperature.

This sea of air in which we live has a number of layers with differing properties. The lowest 12 km is called the *troposphere*. It contains 80 per cent of the atmosphere and 99 per cent of all the water. It is the region in which nearly all weather happens. The temperature of the air drops steadily through the troposphere, falling from 20°C near the surface to around −60° near the top.

Above the troposphere is the *stratosphere*. Here the temperature is a nearly uniform −60°C. In the upper part of the stratosphere a strange form of oxygen exists: $O_3$ or *ozone*, which is made from three oxygen atoms instead of the usual two. This layer from 20–50 km above the Earth absorbs much of the ultraviolet light from the Sun, shielding the plants and animals below from its damaging effects. In recent years a thinning of this layer around the poles has been noticed. This 'hole' as it has been called, is of concern, as chemical pollutants in the atmosphere appear to be causing the destruction of the ozone layer, which will affect all life on Earth.

Past the stratosphere is the *thermosphere*, so called because the temperature of the thin air here rises rapidly due to the absorption of sunlight. At this level some atoms lose an electron or two, producing a layer called the *ionosphere*. The ionosphere is important in radio communications as it is able to reflect radio waves, allowing a signal to bounce back to Earth over a much greater distance than could otherwise be achieved.

Apart from day and night it is the seasons which determine the rhythm of life on Earth. Even in the technological world in which we live, divorced from the concerns of planting and harvesting crops, everyone arranges their lives according to the seasons, even if it is just to organise their holidays so they can go skiing. The seasons of the Earth depend not upon the Earth's changing distance from the Sun, but upon the 23.5° tilt of its axis.

The tilt of the Earth's axis means that for some of the year the region around each pole will be turned away from the Sun and receive no sunlight at all. Right at the pole the sunless period is six months; as you journey closer to the equator the sunless period gets shorter, until at latitude 66.5° north or south the normal rhythm of day and night continues all year. It is in the regions closer to the equator than latitude 66.5° that most of the world's plants and animals (humans included) live, but still the seasons have their effects. Let us look at the southern hemisphere.

On 21 September each year the Sun is directly above the equator, moving south. On this day both day and night everywhere on Earth, even at the poles, are twelve hours long. This day also marks the last day of sunlight for the north pole. As the days pass the Sun moves further south and more of the northern hemisphere moves into its winter darkness. In the south the polar regions are seeing the Sun again. Because more of the southern hemisphere is being illuminated, the days in that part of the world are lengthening towards the summer solstice on 23 December. At noon on that day the Sun is directly overhead at latitude 23.5° S and the days in the southern hemisphere are at their longest. The south pole has been in constant sunshine for three months, and places south of 66.5° S have had at least a few days of uninterrupted sun. At the equator, though, the days are still twelve hours long, as they are on every day of the year.

After 23 December the Sun appears to move northwards again. On 2 January the Earth is at its closest to the Sun, though this has little effect on the climate. On 23 March the Sun is once again over the equator, this time heading north. At the south pole this day marks the end of six months of sunlight and the onset of six months of darkness. As the Sun moves north, the regions south of the equator receive less light each day, and that light is at a much steeper angle to the ground so it does not give as much warmth.

On 21 June the Sun has reached its northern limit. Places within 23.5° of the south pole are in continual night, and places closer to the equator have short, cold days. As the year passes the Sun moves southwards once more, bringing longer days, until on 21 September it is once again at the equator.

The seasons on the Earth are fairly mild; the tilt of the Earth's axis is fairly small; the periods of darkness over the inhabited parts of the globe are short; and the oceans and atmosphere cushion the planet from much of the effect. The layer of air surrounding the Earth carries heat from the warmer parts to the cooler ones, equalising the temperature to a large extent. The other thing the atmosphere does is trap heat. Without the blanketing effect of the atmosphere the Earth would have an average temperature of −15° C, instead of the temperate 10° we enjoy now.

Outside the atmosphere the astronomical effects of the Earth continue. The Earth's magnetic field extends well past the atmosphere and into space, with some quite spectacular consequences. The Sun, in addition to giving light and heat, constantly gives out showers of charged particles, mainly protons and electrons. Because these particles are charged they can be affected by the Earth's magnetic field.

As the particles head towards us, our magnetic field bends their paths towards the two magnetic poles. Here the particles spiral down into the atmosphere, heating the atoms in the

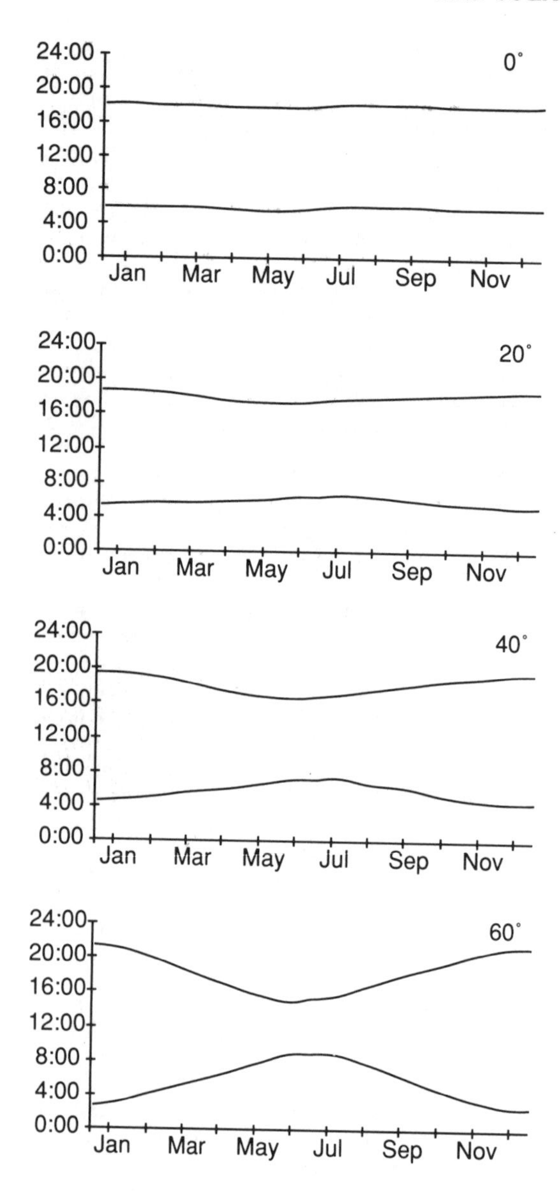


**4.20** The length of day and night throughout the year varies considerably depending upon your latitude. Graphs for latitudes from the equator to 60° south show this variation clearly.

upper regions and knocking electrons loose until the gas glows. This glowing gas is seen as the auroræ over the two poles. A second effect of the magnetic field is to trap bands of high-energy particles in orbit above the equator. These bands, known as the *Van Allen belts*, are filled by the Sun with fresh high-energy particles which spiral back and forth until they lose their energy and are able to escape the region.

The Van Allen belts were the first objects discovered by the space program. When the United States launched its first satellite, Explorer 1, on 31 January 1958, instruments on board registered high levels of radiation at certain heights above the Earth. The radiation was caused by the highly energetic charged particles which hit the metal in the spacecraft producing secondary x-rays and gamma-rays. It was James Van Allen

**4.21** This photograph of the Aurora Australis shows the rayed band form, a type of aurora which looks like moving curtains of light. This photograph was taken from Mawson Base in Antarctica. PHOTO CREDIT: AUSTRALIAN ANTARCTIC DIVISION, PHOTOGRAPH BY D. PARER

who had designed and built the instruments which detected this radiation, so his name was attached to the features.

There are hundreds of different phenomena triggered by the interaction of the atmosphere with various celestial bodies. The Sun and Moon are the chief instigators of most of these events. Some solar effects include beautiful sunsets, rainbows, many types of halos when thin cirrus cloud is present and occasional mock suns. The Moon frequently produces halos and occasionally moonbows, although the colour in these is not easy to see except with long-exposure photographs.

The full range of phenomena is beyond the scope of this book, and the interested reader should consult some of the works on atmospheric optical effects. The following section is a small collection of miscellaneous odds and ends that did not seem to fit anywhere else in this book, we hope you enjoy learning of them, and try to observe them.

If you regularly watch the sky, you may have noticed that the evening twilight sometimes seems reluctant to leave; the astute observer would note that this western sky brightening extends along the zodiac, and the same brightening is also visible prior to morning twilight in the east. For obvious reasons this phenomena is called the *Zodiacal light*.

The optimum time to view the Zodiacal light is when the ecliptic is close to being vertical to the horizon. In the southern hemisphere the best period is August-September in the evening,

**4.22** A photograph of the Aurora Australis taken from Davis Base in Antarctica. The lightshows of the Aurora are caused by the charged particles of the solar wind emitted by the Sun becoming trapped high in the Earth's magnetic field and shuttling back and forth. Near to the north and south poles the particles are able to enter the atmosphere, ionising the gases there and making them glow. PHOTO CREDIT: AUSTRALIAN ANTARCTIC DIVISION, PHOTOGRAPH BY M. ZAPPERT

and March-April in the morning. In the northern hemisphere, the best period is February-March in the evening and September-October in the morning. Tropical latitudes are favoured all year round.

The Zodiacal light is best seen away from city sky-glow, appearing as a tenuous conical shaft of light soon after twilight in the evening or before twilight in the morning. Being brighter than the Milky Way, it can provide an impressive naked eye spectacle, measuring up to 20° across at the base on the horizon, gradually tapering as its altitude increases. The glow is the result of sunlight reflecting on small particles of dust that lie within the plane of the solar system. The dust is believed to originate from comet 'fall out' and collisions between asteroids, providing a continuous supply of fresh material to the ecliptic.

Related to the Zodiacal light is the *Gegenschein* or *Counterglow*. This moves around the ecliptic 180° from the Sun (the anti-solar point). The Gegenschein is an elliptical patch that measures about 10° by 20° (slightly larger when seen from tropical latitudes). Being much fainter than the Zodiacal light, it needs a good clear, country night to be visible.

From the apex of the Zodiacal light cone, a faint parallel-sided projection known as the *Zodiacal band* extends to connect with the Gegenschein. This band can be difficult to see, being fainter than the Gegenschein.

The Earth's shadow, cast by the Sun, is visible as it passes across the lunar surface during total and partial eclipses. It is also possible to see the Earth's shadow at times other than during lunar eclipses, in fact on any clear, cloudless day you can see the shadow. As the Sun begins to set in the west, turn to the east. You will notice a dark band rising slowly from the eastern horizon expanding outwards to both the north and south. This is the Earth's shadow projected through our atmosphere and into space.

The colour of the shadow is quite striking: dark blue-grey with occasional purple and orange bands where it merges with the sunlit sky. After about three-quarters of an hour, as day gives way to night, the shadow begins to fade as it merges into the darkening sky. The shadow is most prominent in the east just after sunset, and in the west before sunrise. It should be observed where possible from a location with an unobstructed horizon; the effect is quite stunning from a mountain vantage point with a 360° view.

The *Aurora Australis* and *Borealis* (the southern and northern lights) are high-altitude light displays occurring near the Earth's poles. Residents of high southern and northern latitudes can expect to be treated to regular spectacular auroral displays, but the frequency for those living in temperate latitudes is much less. The authors, living in Sydney at latitude −35°, have on several occasions been fortunate enough to see splendid aurora south of the city under dark skies; hence it is not a prerequisite to live in a favoured region for auroral observation. Aurorae can be photographed easily, a standard SLR camera with a 50 mm or wide-angle lens is preferred. The camera should be mounted on a tripod and several exposures from a few seconds to one minute should be made, using a cable release. The fastest available colour slide or print film should be used.

If you live in a region where aurorae are common, you will find that local astronomical societies will have a section devoted to the study of aurorae. These groups supply predictions and welcome any observations and photographs.

# 5

# Our Moon

After the Sun, the Moon is the brightest object in the sky. It is also the only one on which details can be seen with the naked eye. Much of the world's culture has been formed around the Moon, the length of our months derive from the period of its orbit, and many natural cycles coincide with it. Before there were towns and cities, its light provided nocturnal illumination for many activities. Even today it plays a part in our culture, even if just for lovers to sit under or to provide poets with a rhyme for June.

The Moon has for a long time played a dominant part in astronomy. Even from the earliest of times its spherical shape has been evident by the changing illumination of its surface. The fact that it always keeps one unchanging face towards the Earth and that it really does orbit us must have bolstered the belief that other objects orbited us too. In the inner solar system our Moon is unique; neither Mercury nor Venus has a moon and the two moons of Mars are little more than captured minor planets a few tens of kilometres across. Yet our Moon is huge. Compared with the Earth it has one-eightieth the mass and one-quarter the diameter, making the Earth-Moon system more like a double planet than a planet and satellite. Consider that if Venus had a moon the same size as ours it would be clearly visible in our night sky, and easily seen orbiting that planet. How different the world might have been if it was obvious that the Earth wasn't the centre of everything.

The Moon and its motion through the sky have been studied for many thousands of years, sometimes for religious purposes, sometimes for agricultural, and at other times to save the skins of astronomers. Eratosthenes, the Greek scientist who first calculated the size of the Earth, was one of the first to make a serious attempt at calculating the distance to the Moon. Though his method was sound, his measurements were too coarse to allow the necessary calculations to be performed. We now know that the Moon is around 400 000 km from Earth and 3476 km in diameter.

Our Moon is large when compared with other satellites; only the gas giant planets have moons of comparable size, but when compared with the Earth the Moon is a giant. Jupiter's largest moon, Ganymede, although 5262 km in diameter, is only 0.0078 per cent of the mass of the planet. Our Moon, on the other hand, though smaller in diameter, is 1.23 per cent of the mass of the Earth; apart from Pluto and Charon, this is the greatest satellite to planet ratio in the solar system. The Moon's orbit is an ellipse with eccentricity of 0.054 inclined at 5°8' to the plane of the solar system.

The Moon's orbital period is not a simple matter, as it really depends upon how you want to measure it. There are four periods for the Moon. The time between successive full moons, known as the *synodic month*, is 29.53 days long. The period it takes for the Moon to rotate once with respect to the stars, the *sidereal month*, is 27.32 days. Because of the precession of the Moon's orbit around the Earth, there are also two other month types: the period between two successive perigees, the *anomalistic month*, is 27.55 days and the period between passages across the ecliptic, 27.21 days. To an amateur astronomer, only the synodic month is of real importance, as it determines the period between two identical phases.

The Moon also has another major influence on the Earth. It is because of the Moon that we have tides in the oceans and seas. People living near the sea will be familiar with the two high

**5.1** This photo of the Moon, taken one day after full moon, shows the ray system from the crater Tycho stretching across the lunar surface. The smaller ray system associated with the crater Copernicus in the Oceanus Procellarum can also be clearly seen. Apart from the eastern part of the Moon near to the terminator, it is easy to see why the time around full moon is not a good time to observe, the lack of shadows making even prominent features difficult to see. Full moon is a good time to trace the extent of the mare features on the Moon, from the Mare Crisium near the eastern limb to the Oceanus Procellarum in the west. In addition, dark features, such as the craters Grimaldi near the western limb and Plato at the north of the Mare Imbrium, stand out clearly. PHOTO CREDIT: KEN WALLACE

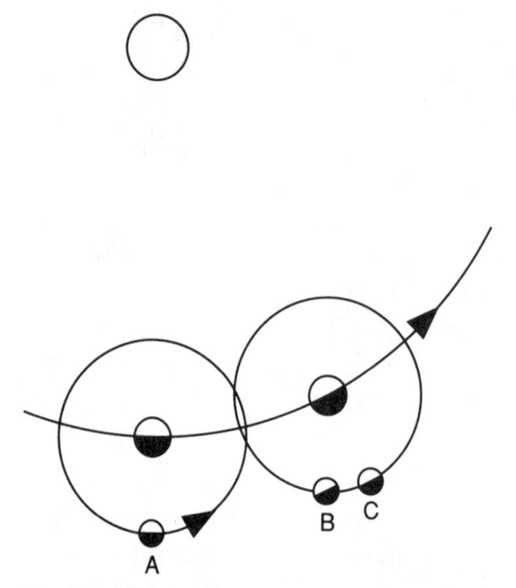


**5.2** The difference between the sidereal and synodic months is seen here. At position A we have the Moon at the start of a month; 27.32 days later the Earth and Moon are at position B, the Moon having rotated once on its axis with respect to the stars. It is not until 29.53 days have passed that observers on Earth again see a full moon (position C).

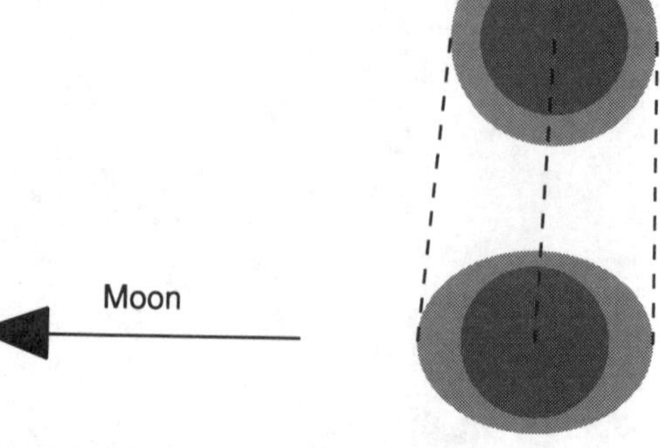


**5.3** The Moon's gravitational pull attracts both the Earth and the layer of water surrounding the Earth to form the tides. The water closest to the Moon experiences the greatest pull and forms a bulge beneath the Moon, the Earth itself experiences the next greatest pull and so the water on the far side of the Earth is left behind, causing a bulge there. This simple explanation would lead you to expect that high tide would always occur when the Moon was nearly overhead, but influences such as the drag of the seas against the land and the gravitational influence of the Sun vary the time of the tides.

tides and two low tides each day. In most places the difference between the height of the water between high and low tide is one to two metres but in some places, because of the arrangement of the land, tides can be as high as 15 m. In lakes and bodies of water not connected to the open sea, there are no measurable tides.

The tide is caused by the gravitational attraction between the Moon and the water in the oceans; this attraction makes the water pile up beneath the Moon. The Moon also attracts the Earth, moving it closer also. On the far side of the Earth, the water is largely shielded by the Earth and so it is left behind, causing another bulge. Figure 5.3 shows this. In between the two bulges is the area the water came from, so its depth is reduced. The bulges maintain their position below the Moon (or actually a little behind it, due to drag effects) while the Earth rotates beneath them. As the Moon circles the Earth each month, the bulges follow it around, so observers standing on the rotating Earth will pass through each of the two bulges and two shallows each 25 hours.

The Sun also has an effect on the tides, its gravitational effect being around one-third as strong as the Moon's. Without the Moon we would still have tides caused by the Sun, but it is the combined influences of the Sun and Moon which cause the tides we see. At the time near new moon, when the Sun and Moon are close to each other in the sky, their effects combine to produce large high tides and shallow low tides. At first or third quarter, when they are at 90° to each other, their effects cancel somewhat and high and low tide are much less different. A graph of the tides, high and low points (Figure 5.5) shows this well.

The gravity of the Moon does not only affect the Earth's water, the rocks of the Earth experience a tidal force as well. Each day the Earth's rocks are distorted slightly by the pull of the Moon, which heats the Earth a little and slows its rotation a tiny amount, making the days slowly longer. Similarly the effect slows the Moon making it move further from the Earth, in

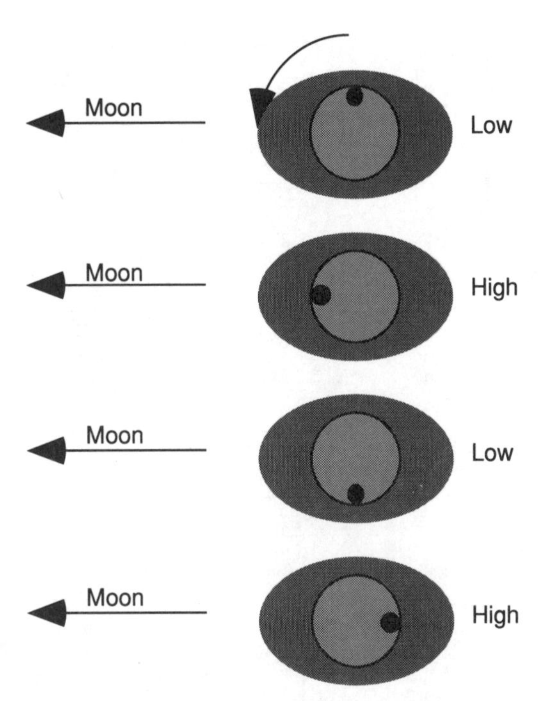


5.4 The rotation of the Earth moves an observer through two bulges of water each day, giving the appearance of two high and two low tides. The bulges themselves remain roughly aligned with the direction of the Moon.

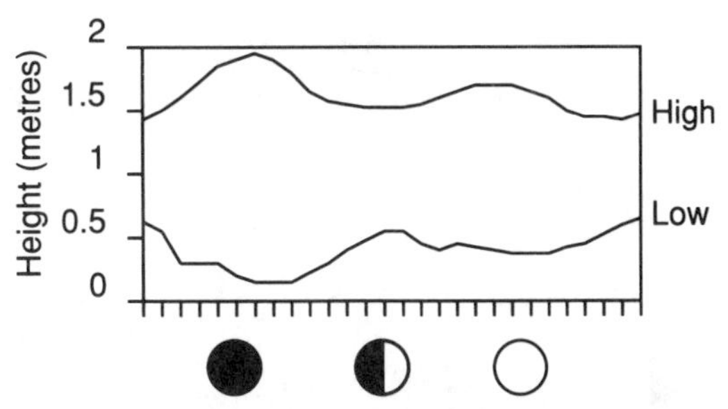


5.5 A graph of the high and low tide heights for Sydney Harbour over a typical month. It is clear that the greatest difference in heights is around the time of new moon, when the influences of the Sun and Moon are in the same direction, while the least difference is near first or third quarter, when they are at 90° to each other.

accordance with Kepler's laws. Tidal heating caused by the deformation, although not a major source of energy on Earth, has some spectacular effects on the moons of other planets of the solar system, as we will see in later chapters.

The Moon's orbit shows a feature which is exceedingly common amongst satellites in the solar system: it has synchronised rotation. That is, it rotates once on its axis in exactly the same time it takes to revolve around the Earth. This is no coincidence; it is gravity which forces a satellite to have a synchronised orbit about its primary planet. As a satellite forms it would have its own rotational period, probably much shorter than the period of its orbit. It is the gravitational effects of the primary planet which cause tides in the satellite, gradually slowing its rotation until it matches the period of the orbit and one face is permanently turned towards the primary planet. Our Moon exhibits this feature, always showing the same face.

Because of synchronised rotation, the Moon always keeps the same face pointed towards us and we should see only 50 per cent of the lunar surface. In reality, though, 59 per cent of the surface can be viewed by an effect called *libration*. Those regions seen due to libration are very close to the edge of the Moon and hence it is difficult to discern detail.

Libration is a complicated mix of three different effects. Firstly, the Moon suffers from a longitudinal wobble; as the Moon approaches perigee its motion through space speeds up, faster than its rotation, and so it does not turn fast enough to maintain the same face to us, giving us a view around its eastern limb. Then as the Moon moves slower, approaching apogee, it turns a little too much to keep facing us, giving a view beyond the western limb (Figure 5.7). The wobble in longitude is easily seen if you observe a lunar feature near the limb. Mare Crisium, for example, is a good gauge, swinging near to the limb and back. The east-west libration has an average period of 27.55 days.

The second type of libration is latitudinal and is caused by the slight tilt of the Moon's axis. We can see a little over the south pole during one half of each revolution with the north pole hidden; during the other half of its orbit we see over the northern limb while the south pole is hidden from view. The crater Plato, near the

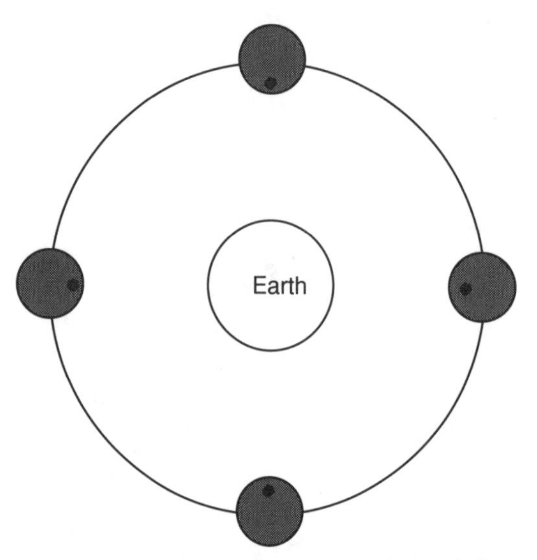


**5.6** Because the rotation period and orbital period are the same, the Moon always keeps the same face towards the Earth. This is captured, or synchronous, rotation. Synchronous rotation is common within the solar system; most of the moons of the planets exhibit it. In the Moon's case, the tidal effect of the Earth's gravity has slowed the Moon's rotation until it is synchronised.

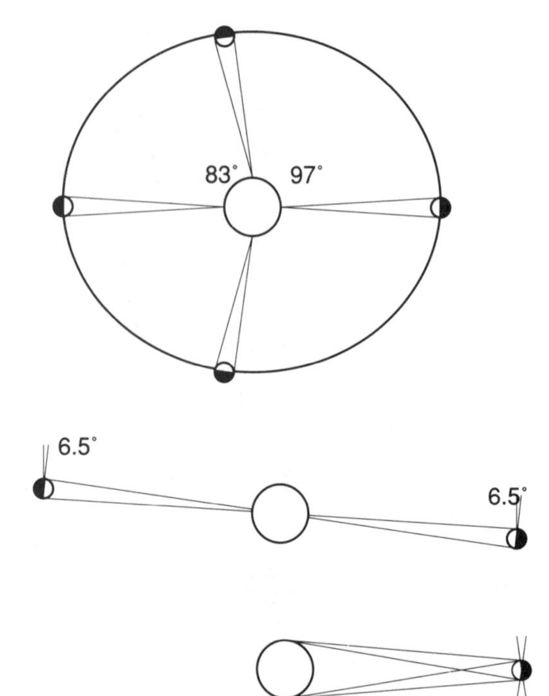


**5.7** Due to the combined effects of libration, it is possible to see more of the Moon's surface from Earth than the 50 per cent that would be expected from a synchronised orbit. The major effect is known as *longitudinal libration*. This comes from the Moon not moving at a constant speed throughout its orbit. The Moon moves faster when it is close to the Earth, and slower when further away, but it rotates at a constant rate. This allows a little more of the Moon to be seen around the sides.

The second effect is *latitudinal libration*, caused by the tilt of the Moon's axis so that some of the Moon can be seen around one pole during one half of the orbit and around the other during the other half of the orbit. The last effect is *diurnal libration*. This is caused by an observer being able to see a little further around the Moon's side when it is near to the horizon. All together these effects combine to allow 59 per cent of the Moon's surface to be studied from Earth.

northern limb, is a good reference point to judge libration in latitude. The north-south libration has an average period of 27.21 days.

The third libration is a minor one, but nonetheless it is real. It's termed *diurnal libration*. As the Moon is so close to the Earth, we see further over the limb when observing near the horizon than when the Moon is overhead. The sum total of these libration effects combines to give that extra 9 per cent visibility.

The 'man in the moon' is a familiar enough concept; some claim to be able to make out a picture in the light and dark areas of the lunar surface. This mottling of the surface of our neighbour caused great consternation amongst philosophers 2000 years ago; after all the Moon was part of the heavens and everything there was meant to be perfect, yet the surface of the Moon clearly wasn't. Although it had the 'perfect' spherical shape, philosophers decided that because of its proximity to the all too imperfect Earth, some of that imperfection had stained the surface.

To an amateur astronomer the Moon is far from imperfect; in fact, it is one of the best objects with which to begin observations. Its proximity to the Earth and its brightness mean that it presents views of its surface unlike those seen on any other astronomical object. Even with the naked eye, much can be seen on the

**5.8** The boundary between the Mare Imbrium and Mare Serenitatis eight days after new moon. The Apennine Mountains can be seen curving upwards from the lower left of the picture to meet the Caucasus Mountains from the north. To the north of the Mare Imbrium the Alpine Valley can be seen cutting through the mountains. The craters Plato, to the north, and Eratosthenes, to the south, lie on the terminator and the craters Archimedes, Aristillus and Autolycus can be seen in the Mare Imbrium. On the floor of the Mare Imbrium slight rises and falls can be seen due to the low angle of the Sun. PHOTO CREDIT: KEN WALLACE

lunar surface. At the distance it is, the 1′ resolution of the human eye can pick out features one-thirtieth of its diameter, or things around 100–200 km in size. At this scale the craters and mountains of the Moon are invisible, but the large dark patches are quite clearly seen.

The pretelescopic astronomers were unsure what to make of these large dark patches. Relating what they saw to the world about them, they decided that they must be seas like those of Earth. Armed with this knowledge they proceeded to name them after aquatic and meteorological phenomenon. The Latin word for sea is

**Table 5.1 The seas of the Moon**

| Latin Name | English Translation |
|---|---|
| Mare Australis | Southern Sea |
| Mare Crisium | Sea of Crises |
| Palus Epidemiarum | Marsh of Disease |
| Mare Foecunditatis | Sea of Fertility |
| Mare Frigoris | Sea of Cold |
| Mare Humboldtianum | Humboldt's Sea |
| Mare Humorum | Sea of Moistures |
| Mare Imbrium | Sea of Showers |
| Sinus Iridium | Bay of Rainbows |
| Mare Marginis | Marginal Sea |
| Sinus Medii | Central Bay |
| Lacus Mortis | Lake of Death |
| Palus Nebularum | Marsh of Mists |
| Mare Nectaris | Sea of Nectar |
| Mare Nubium | Sea of Clouds |
| Mare Orientale | Eastern Sea |
| Oceanus Procellarum | Ocean of Storms |
| Palus Putredinis | Marsh of Decay |
| Sinus Roris | Bay of Dews |
| Mare Serenitatis | Sea of Serenity |
| Sinus Aestuum | Seething Bay |
| Mare Smythii | Smyth's Sea |
| Palus Somnii | Marsh of Sleep |
| Lacus Somniorum | Lake of Sleepers |
| Mare Spumans | Foaming Sea |
| Mare Tranquillitatis | Sea of Tranquility |
| Mare Undarum | Sea of Waves |

*mare*, pronounced 'mar-ay', so Mare Tranquillitatis is the Sea of Tranquillity while Mare Imbrium is the Sea of Rains. Other features gained similar names: Oceanus Procellarum, the Ocean of Storms, and so forth. Because the plural of *mare* is *maria*, the dark regions of the Moon are known by that collective term. Table 5.1 gives a complete list of the 27 mare features of the Moon.

It is with the telescope that the Moon becomes a world in its own right. When Galileo turned his telescope towards it in 1610 he was confronted with a world of mountains and valleys and strange round holes, the craters. Modern optics are such that even a small pair of 7 × 50 binoculars offers a superior view of the lunar surface to that seen by Galileo. A good amateur telescope of 200 mm aperture will allow features as small as two kilometres to be seen on a clear, steady night. The craters of the Moon are also named, most of the work being done by the astronomers Grimaldi and Riccioli. It was they who decided that astronomers should be commemorated on the Moon, hence the craters of Archimedes, Ptolomaeus and Tycho. Copernicus is even remembered: a large crater in the Ocean of Storms.

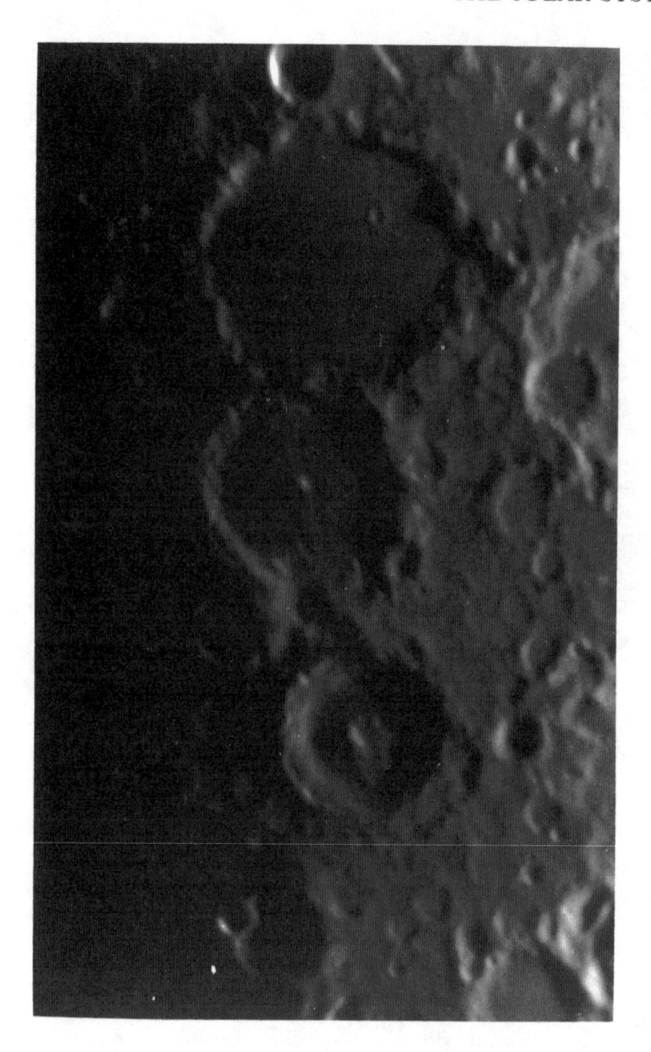

**5.9** These three large craters, sometimes called the 'big three' are almost on the Moon's central meridian, so seven days after new moon the sunlight is only just reaching their floors. The crater Ptolomeus at the top is almost bare except for one small crater, Ptolomeus A, towards the north-east. Alphonsus, in the middle, has a central peak and a number of faintly seen ridges running north-south across its floor. The lower crater, Arzachel, has a high central peak, the shadow of which can be seen falling westwards on the crater's floor. PHOTO CREDIT: KEN WALLACE

Even a cursory look will show that there are two main types of terrain on the Moon, the light-coloured regions and the dark-coloured regions. The lighter areas are populated by the mountains and valleys and are called highlands, whilst the dark areas are the lowlands. Closer examination of the highland regions shows that they are heavily cratered; in some places there are so many that individual craters become lost in the general jumble of walls and floors; the mare are much less densely cratered, with lots of uncratered areas visible from Earth. This leads to a simple conclusion: the mare must be younger than the highlands, and must have formed after the main impact period of planetary formation.

There must be other differences between the highlands and mare, apart from their ages. This is evident from their differing colour. As with rocks on Earth, darker rocks have a different chemical composition to lighter ones: they contain less quartz. On Earth this corresponds to rocks formed from mantle material such as the sea beds, whereas lighter rocks which have large quantities of quartz are part of the continents. We will see later how this tells us about the origin and history of the Moon.

Because it is the closest world to us, the Moon

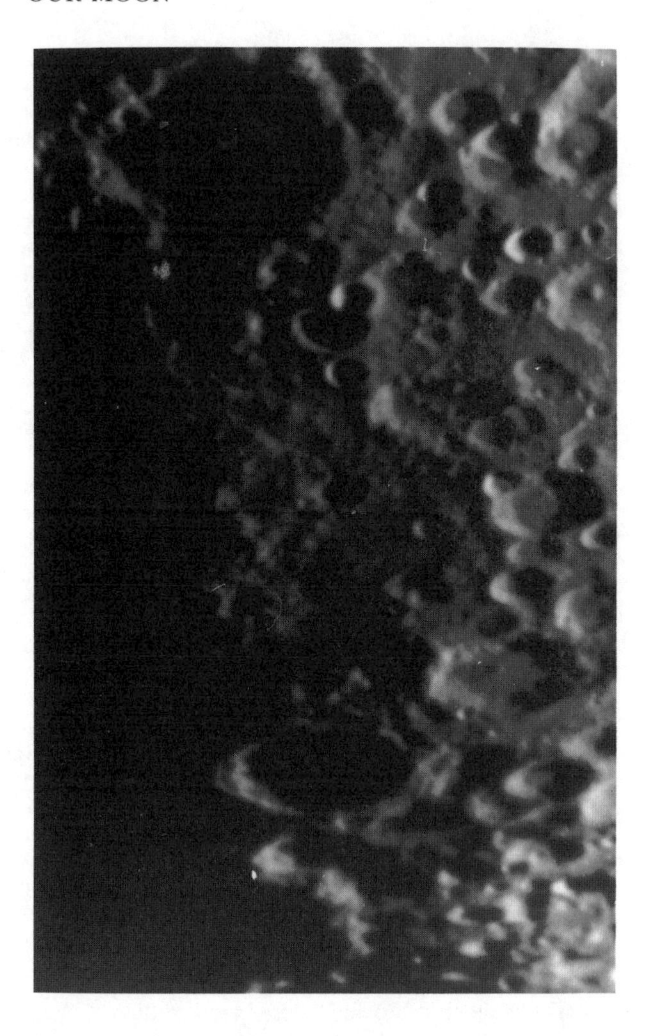

**5.10** In this heavily cratered area of the Moon's southern hemisphere it is difficult to find the beginning of one crater and the end of another. The large crater at the top of the photo is Maginus while in the centre of the photograph is the older crater Cysatus, almost completely destroyed and only just recognisable. The large crater near the bottom of the picture is Moretus, its central peak just catching the morning sunlight.
PHOTO CREDIT: KEN WALLACE

was an obvious target for the first spacecraft explorations. The exploration of space by machines began on 4 October 1957 when the Soviet Union launched Sputnik 1 into Earth orbit. Other satellites followed from both the Soviet Union and the United States. The first successful attempt to reach the Moon was made with the Soviet craft Luna 1 which passed within 5000 km of the Moon in 1959. Just nine months later the Soviets scored a hit with the Luna 2 probe crashing into the Moon in the Mare Imbrium.

In October 1959 the Soviets had a third success with Luna 3. This probe flew around the far side of the Moon, returning the first-ever pictures of that region. As telescopic observations show, the face of the Moon towards Earth is about 50 per cent highlands and 50 per cent lowlands. Scientists expected the same to be true of the far side, but as we have continued to see throughout the history of space exploration, what is expected and what is found are often very different things. The far side of the Moon turned out to be mainly highlands, with no major maria at all.

The highlight of the exploration of the Moon came with the manned landings in the late 1960s and early 1970s, but before these could go ahead much had to be found out about the Moon. Both the Soviet Union and the United States made plans to explore the Moon, the United

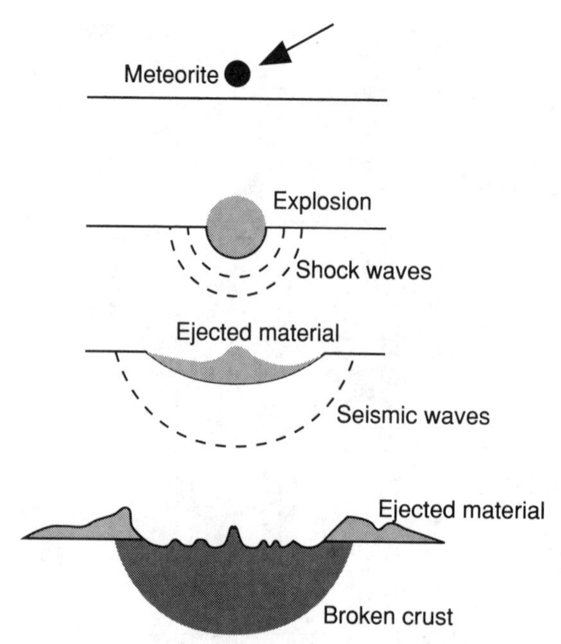


**5.11** A lunar crater is not formed simply by the impact of a small meteorite, but by the energy which is released when the meteorite vaporises on impact. This energy vaporises the lunar surface, making the crater and throwing material across the surrounding area and forming ramparts around the crater. Beneath the crater is an area of shattered crust caused by the shockwave from the impact.

States with its Ranger and Surveyor probes, the Soviet Union with its Luna and Lunokhod explorers. These craft provided close views of the lunar surface and provided information about possible landing sites for future missions.

It was United States president John F. Kennedy who sent the United States to the Moon by pledging the country, on 25 May 1961, to 'achieving the goal, before this decade is out, of placing a man on the Moon and returning him safely to the Earth'. This dream was realised on 20 July 1969 at 20h17m43s UT when the words 'Tranquillity Base here. The Eagle has landed.' were heard around the world. On that first trip, astronauts Neil Armstrong and Edwin Aldrin spent just two and a half hours outside on the surface; by the sixth and final trip in December 1972 astronauts spent more than 22 hours outside. Since that start, though, no more people have ventured to our satellite.

Though spectacular, the manned lunar landings were only a small part of the continuing research into the Moon. Now we know more about our companion in orbit, but still not everything. The most obvious thing about the Moon when viewed with a telescope is its craters. The name 'crater' comes from the Greek word for a shallow drinking cup or bowl, and indeed this is a good description of these features which cover the lunar landscape. Until extensive exploration of the Moon there was a debate as to the process of formation of the craters. Two opposing camps held sway, those who blamed the impact of meteorites and those who blamed volcanic activity. Both forms of crater can be seen on Earth. Holes such as the Wolf Creek meteorite crater in Australia show that the Earth has been hit by visitors from without, while craters such as those in the volcanoes of Hawaii show the effects of the Earth's internal processes.

Those who favoured the volcanic origin of craters pointed to the fact that the maria are obviously volcanic features and to the distribution of craters. It can be seen that very rarely do large craters break into smaller ones, indicating that as the volcanic activity declined the craters being formed decreased in size. Another factor is the linear arrangements of many craters, for example the 'big three' on the Moon's meridian, the craters Ptolomeus, Alphonsus and Arzachel, are almost in a straight line. The first real challenge to the idea of volcanic craters came from G.K. Gilbert who in the 1890s compared the craters of the Moon with the volcanic craters of the Earth and found many differences and few similarities. Gilbert was unable to explain, however, why lunar craters all appear to be round. Craters caused by meteoritic impact will be elongated unless the object struck from directly above, or so it was thought at the time.

The exploration of the Moon, and more importantly the discovery of other heavily cratered worlds throughout the solar system, put the idea of volcanic cratering to rest, though there were still some people supporting it into the 1970s. The similarities between the craters of Mercury, Jupiter's moons Callisto and Ganymede, Saturn's Mimas and Dione, Uranus' Titania and Miranda, Neptune's Triton and our Moon's are too great for volcanic processes on very different worlds to explain. The only credible explanation is impact cratering.

What Gilbert and many others failed to real-

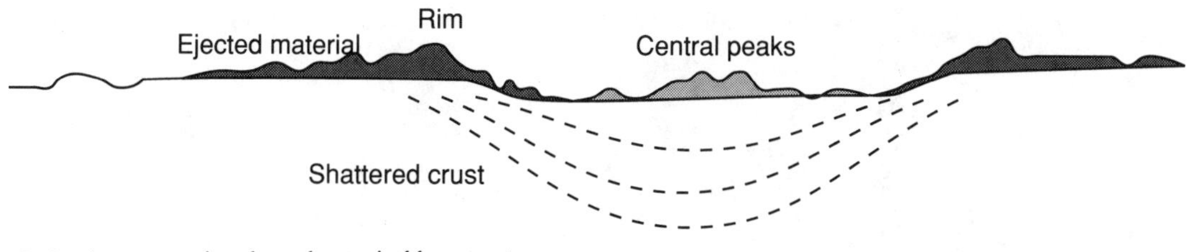


**5.12** A cross-section through a typical large crater.

ise is that the process of cratering is not like that of dropping pebbles into wet mud, but more like an explosion. The high speed with which a meteorite and a planet come into contact is such that the mechanical force of the impact is nothing compared with the explosion which results; it is this explosion which digs the crater. The minimum speed with which an object can hit the Moon is 2400 $m.s^{-1}$, because of the Moon's gravitational attraction for the object. Added to this speed must be any relative motion between the meteorite and the Moon. The kinetic energy gained by an object moving with a speed such as this is greater than the equivalent amount of TNT. Upon impact, this energy is released and the object explodes.

Imagine what happens when a rock moving at perhaps 3 km per second hits the Moon. Its speed is so great that it will penetrate a fair way into the surface of the Moon before coming to a halt, shattering the rocks below the crust and generating moonquakes. Most of the energy of the impact is converted into heat, turning the meteorite and the surrounding rock to vapour. This hot expanding gas then shatters the surface and blasts the crater, throwing rock and debris over the surface. The crater Tycho in the southern highland region of the Moon demonstrates the effect: spreading out from the crater over many thousands of kilometres is the lighter coloured rock from below the surface thrown across the Moon by the explosion which formed the crater.

The energy released by a meteorite in this type of collision is sufficient to excavate a crater around ten times the meteorite's diameter to a depth of one-fifth of the crater's diameter. Much of the material thrown up lands near the crater's rim, building it up above the level of the surrounding land. If the meteorite which forms the crater is more than about a kilometre across, the resulting hole will be unable to be supported by the lunar crust, and part of the walls will collapse and fill the basin leaving terraced walls around the crater's rim. Near the centre of the crater, where the greatest amount of material has been removed, the crust may rebound, forming a central peak. Sometimes molten rock from below the surface will run into a crater, filling its floor and leaving a smooth surface.

**Table 5.2 Lunar crater densities**

| Region | Crater Densities 10 km craters per million square km | Age $10^9$ years |
|---|---|---|
| Highlands | 1000 | more than 4.0 |
| Fra Mauro | 130 | 3.9* |
| Apennine | 95 | 3.9* |
| Tranquillitatis | 50 | 3.7* |
| Fecunditatis | 20 | 3.4* |
| Putredinis | 25 | 3.3* |
| Procellarum | 20 | 3.3* |
| Copernicus | 10 | 0.9* |
| Tycho | 2 | 0.2* |

* These sites were dated using samples returned by the United States' and Soviet Union's sample return missions.

Apart from providing spectacular views for astronomers and making it difficult to land a spacecraft safely, craters are very useful in dating the age of the surface of a planet or moon. The cratered surfaces of Mercury and those of the Jovian satellites reveal much about the origins and history of those bodies. On a planet such as Earth, the action of weather quickly erases any evidence of meteoritic impacts, those still visible being recent events. On an airless body like the Moon there are few processes which can erase the evidence, the only one being further impacts and their consequences.

On these airless worlds we can count the number of craters of a given size in a given area and calculate a crater density. In planetary studies the usual density is the number of craters per

**5.13** This photograph of the lunar landscape was taken by one of the United States' surveyor spacecraft in the highland region just to the north of the crater Tycho. The fine powdery surface is strewn with small rocks which have lain untouched for thousands of millions of years. In the distance a low line of hills can be seen against the horizon. PHOTO CREDIT: NASA

million square kilometres, about the size of the Mare Imbrium on the Moon. A study of the Moon with a telescope will reveal much about cratering. The first thing you will notice is that there is a much larger number of small craters than large ones. The reason is simple: there are more small meteorites than large ones. You will also notice what the volcanic theory people noticed, that smaller craters interrupt larger ones, not the other way around; this means that later cratering was caused by smaller objects after the large ones had all been swept up.

Scientists have spent a lot of time and effort counting the craters on the Moon. Typically they look at the number of craters 10 km or larger in a given area; you can repeat these observations yourself as 10 km craters are 5″ across and thus within the range of most amateur telescopes. The crater densities of selected parts of the Moon are given in Table 5.2. Once the cratering densities were known it was possible to begin calculating the age of the lunar surface. Obviously the heavily cratered regions are older than the relatively clean areas, as they have been there longer to be hit. Continuing this line of reasoning, the large craters which are in the maria, such as Copernicus and Archimedes, must be younger than the maria. The crater Tycho, with its rays of ejected material scattered across most of the lunar surface, must be still younger or the rays would have been covered.

If higher crater densities mean an older surface, then the question becomes: how much older? If the cratering in the highlands is twenty times that of the maria, does that make the highlands twenty times older? If the highlands were the same age as the Earth, $4.5 \times 10^9$ years, the maria would only be 200 million years old, but this doesn't work as the cratering densities of the maria suggest that they must be around $3.5 \times 10^9$ years old. The samples returned by Apollo 11 were eagerly awaited so that radioactive dating could be carried out on them. They confirmed the suspicions of many by being $3.7 \times 10^9$ years old. Subsequent samples placed all the maria between 3.9 and 3.2 thousand million years old.

This apparent discrepancy led scientists to a better understanding of the early solar system. They found that the system had experienced a relatively heavy bombardment by meteorites until $3.8 \times 10^9$ years ago, but after that cratering quickly died down to its present low level. These conclusions are borne out by the studies of craters on other worlds.

The last great craters on the Moon are the Imbrium and Orientale basins. These features are called basins because at around 1000 km in width they are much larger than the features we normally call craters. Both features occurred towards the end of major cratering, around 3.9 thousand million years ago, the Orientale basin being slightly younger as material from it is found on top of the Mare Imbrium. These features are both impressive, but only the Mare Imbrium can be seen well from Earth; the Mare Orientale can be seen only during favourable librations. At the outer edge of the Mare Imbrium the Carpathian Mountains rise 9 km above the surrounding region. Inner mountain rings once existed too, and their remnants can be made out in the floor of the basin, a floor which was flooded by the next event in lunar history.

Even as the last meteorites were falling on the Moon, the major period of lunar volcanism was beginning. The story of lunar volcanism is essentially the story of the maria. It is about the maria that we know the most. The cautious Apollo planners selected flat sites on the maria for five of the six landings. The maria are made of basalts very similar to those of the Earth's oceans, and it is believed that they originated in the same way, the slow release of mantle material. These lunar rocks do show one major difference from those of the Earth, they have much less water in their crystals. The samples returned to Earth from the Mare Serenitatis and Oceanus Procellarum give the age range of the mare features: $3.9 \times 10^9$ for Serenitatis to $3.2 \times 10^9$ for Procellarum.

Unlike the volcanic eruptions with which we are familiar on Earth, the lunar eruptions were much quieter and instead of issuing from one central vent, the lava escaped from long cracks flowing for many hundreds of kilometres before solidifying. The maria were not formed from just one eruption, however; they are the result of many small flows each producing a few tens of metres thickness of lava. A few hundred eruptions of this type are necessary to account for the 5 km depth of the maria.

It is interesting that the maria are only on the Earth-facing hemisphere of the Moon. It is conjectured that it is the Earth which is responsible for the features. Just as the gravity of the Moon creates tides in the oceans and land of the Earth, so the Earth makes tides in the rocks of the Moon. Since the Moon has one side permanently facing the Earth, the stress now induced by the tides is small, but as the Moon slowed to its present rotational period, the forces were enough to crack the lunar surface while it was still forming. It is through these cracks that the lava for the mare flowed.

After the eruption of the mare material, some settling took place. As is usual, the lava contracted as it cooled, causing shrinkage. Marks on surrounding features show that in some places the lava was 100 m higher than today's level. Near the edges of the maria cracks can be seen, caused by the contracting lava. In other places wrinkles have formed due to the stresses in the surface.

Many other features occur on and around the maria. Some fascinating ones when seen near to the terminator are the *rilles*. These are long meandering channels which led early selenologists to suspect that rivers had once flowed on the Moon. Closer examination has shown that the rivers were of lava which carved out the depressions, or perhaps underground lava tubes which collapsed. Other evidence of the Moon's past activity can also be seen. Faults in the surface are common and can be seen best by the shadows they cast on the surrounding terrain. One well-known feature is the Straight Wall in the Mare Nubium, a fault over 100 km long and 300 m tall in places.

The first Moon landings found that the surface of the Moon is covered with a thin layer of dust a few centimetres deep. This dust is the remnants of impact craters across the lunar surface. The fine powdered rock is thrown up by the impact and, because of the lack of atmosphere, nothing has disturbed it for millions of years. It has been calculated that the regolith, as this layer is known, accumulates at around a metre per 500 million years; when this is compared with the rate of sedimentary build-up on Earth, it is incredibly slow. Over the years the fine dust packs down to form the lunar equivalent of sedimentary rock, although the top few millimetres is kept fine by the occasional impact of tiny micrometeorites.

Many amateurs intent on photography or deep-sky observation would prefer we had no moon, or at best a trivial little minor planet-sized body that did not interfere with their observations. There are, however, many advantages to the Moon's size and brightness. It can be observed quite well from the most light-polluted cities of the world. Its large angular diameter and brightness are suited to any small instrument—an 80 mm refractor can see craters on the lunar surface as small as ten kilometres in diameter. It is also a prime target for the beginning astronomical photographer.

Amateurs with a broad astronomical interest should plan their observing month around the lunar phases. Take advantage of moonlit nights for lunar and planetary viewing from the city, where light pollution does not detract from these bright objects, and reserve those moonless weekends for trips to dark-sky sites, to view faint comets, galaxies, clusters and the like.

The most obvious feature of the Moon is the fact that it undergoes phases, similar to those of Mercury and Venus. As the Moon is illuminated by the Sun, only half of its sphere is lit up at any time. How much of this lighted part is visible to us depends on the relative positions of the Earth, Moon and Sun.

With the Moon situated between the Earth and the Sun, it is said to be new moon, and the Moon is invisible as its dark side is facing us. A day or so later, on its eastward journey, a thin crescent appears on the Moon's western side, which is visible in the early evening sky. It con-

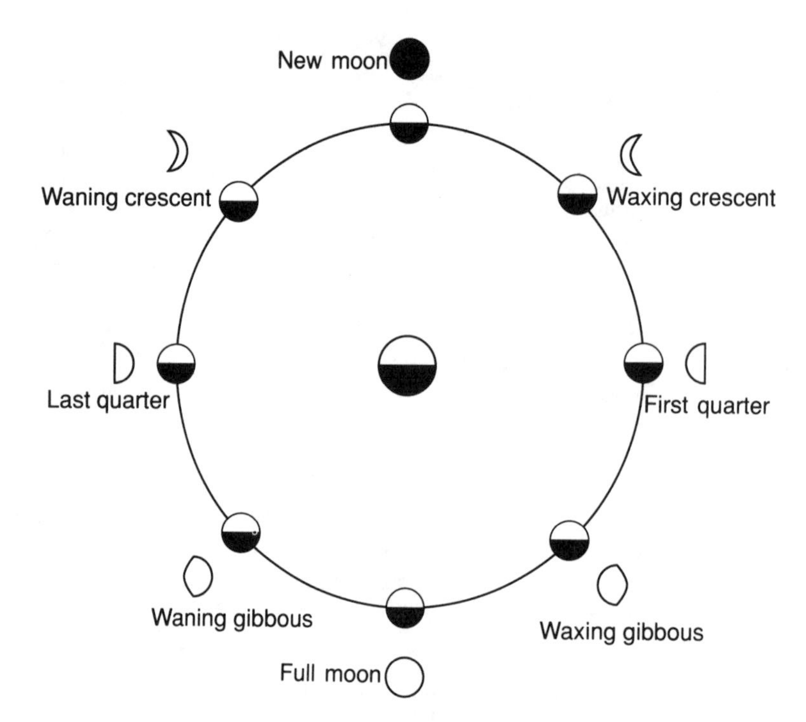


**5.14** As the Moon orbits the Earth the amount of surface which can be seen as illuminated from Earth changes. From new moon the phase grows to first quarter then on to full moon; past full the amount of illuminated surface decreases until once again the phase is new. The appearance of the phases here is as seen by a southern hemisphere observer. Northern observers see the phase shapes reversed.

tinues to wax, until first quarter is reached seven days after new moon. At first quarter one-half of the disc is visible and the Moon will be seen 90° from the Sun. Now the phase turns gibbous, growing until full moon, when the entire disc is completely illuminated. Full moon occurs fourteen days after new moon when the Moon is 180° from the Sun.

The process is then reversed. As the Moon wanes, less of the western half is illuminated. After seven days, one half of the disc is again visible, and the phase is termed last quarter. Fourteen days after full moon the Moon is again new and situated between us and the Sun. At new and full moon, the Moon is at *syzygy*; in line with the Earth and Sun. At first and last quarter, 90° to the Earth-Sun line, the Moon is at *quadrature*. The phase of the Moon at any given time will be exactly reversed two weeks later.

As the Moon follows an elliptical path about the Earth, its apparent size will vary slightly. At apogee its distance is 396 300 km, and at perigee it's only 346 300 km. In angular size, this means an increase from 29′22″ to 33′31″ of arc. Or, put another way, when most distant from the Earth it will be about seven-eighths the size it appears when nearest. This variation in size can be measured by building a simple sighting apparatus as shown in Figure 5.16. You may be surprised to learn how small an angle the Moon subtends against the sky; most people tend to think the Moon is at least the size of a large coin held at arm's length.

You may have noticed that the Moon rising or setting just above the horizon looks much larger than when seen overhead: this is the famous 'Moon Illusion'. Try using the sighting apparatus to see if there is any actual change in size. A first sighting should be made with the Moon on the horizon, then about three hours later and again when overhead. You'll find that the size remains constant, proving that the apparent size change is only illusory. When the Moon is near the horizon, it looks large as the brain compares it with trees and houses. High in the sky, with no familiar objects nearby, it appears small as the eye tends to accept a wide angle view. To be accurate, it must be admitted that when the Moon is seen on the horizon it is slightly smaller, as it is about 6400 kilometres further from the observer than when viewed six hours later at the zenith.

One of the most beautiful sights in the sky is

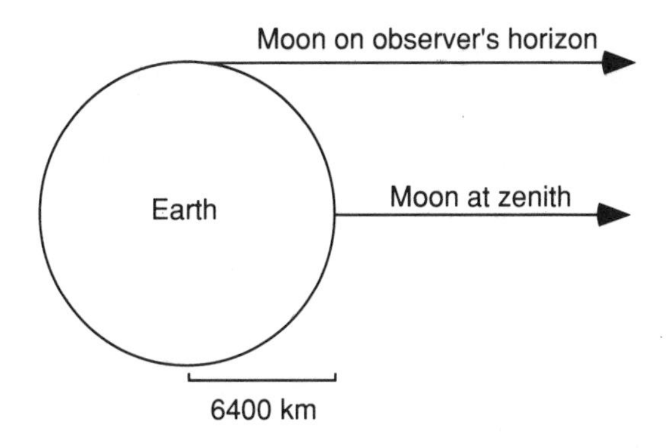


**5.15** When the Moon appears on the horizon it is 6400 km further from the observer than when it is overhead. This amounts to a 1.6 per cent reduction in the Moon's size when it is low in the sky, but the effect is hardly noticeable.

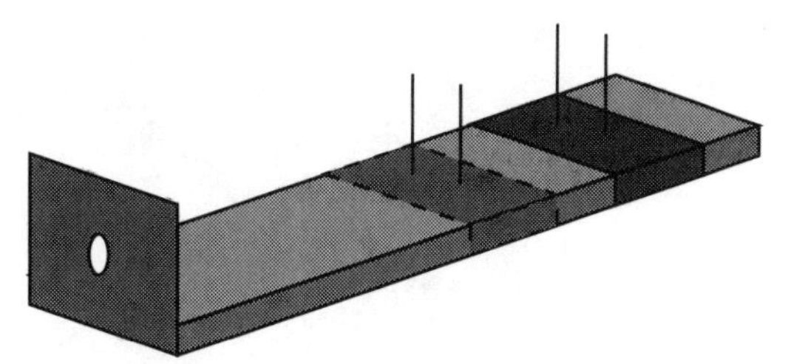

**5.16** A simple piece of apparatus can be used to measure the angular size of an object. By moving the sliding portion with two pins affixed until the pins exactly frame the object the angular size can be calculated from

$$\theta = \tan^{-1}\left(\frac{s}{d}\right)$$

where: s is the distance between the pins
d is the distance along the stick from the eyehole
$\theta$ is the separation in degrees

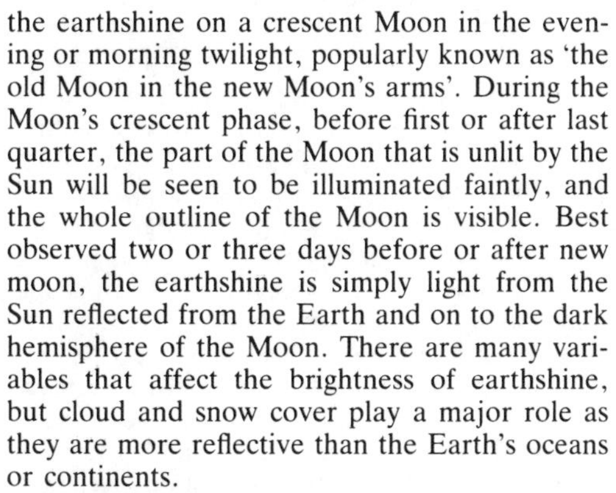

the earthshine on a crescent Moon in the evening or morning twilight, popularly known as 'the old Moon in the new Moon's arms'. During the Moon's crescent phase, before first or after last quarter, the part of the Moon that is unlit by the Sun will be seen to be illuminated faintly, and the whole outline of the Moon is visible. Best observed two or three days before or after new moon, the earthshine is simply light from the Sun reflected from the Earth and on to the dark hemisphere of the Moon. There are many variables that affect the brightness of earthshine, but cloud and snow cover play a major role as they are more reflective than the Earth's oceans or continents.

As the Moon reaches first quarter, the earthshine becomes too faint to see, not only from the Moon's increasing brightness, but because of the Earth's phase as seen from the Moon, which is always the opposite of the Moon's phase seen from Earth. The opposite change can be seen from last quarter to new moon: the earthshine gradually brightens, eventually becoming lost in twilight as the Moon nears the Sun as a thin crescent. A lunar inhabitant would see the Earth as a beautiful blue-white globe covering almost two degrees of sky, close to four times the diameter of the Moon from Earth. Since the Earth is a better reflector of sunlight than the Moon, it would appear about a hundred times as bright.

The Moon, viewed through a small telescope, is positively the most spectacular of all solar system objects. For sheer beauty, first prize must go to Saturn, but the Moon takes a lot of beating, especially for detail that will keep the average amateur occupied for many years. The Moon's brightness, even in small telescopes, can be overwhelming. To prevent eye fatigue, a blue or neutral density filter can be used. An increase in magnification, atmosphere permitting, will also dull the image.

Sketching lunar features is a rewarding aspect of astronomy, though you can't expect to make any new discoveries as the lunar surface has been surveyed by many spacecraft with better clarity than any Earth-based telescope could hope to match. You will, however, have the satisfaction of recording more detail than can be obtained in any of the best amateur photographs. Motor drive inadequacies and the Earth's atmosphere frequently hamper the photographer taking close-up shots using eyepiece projection, resulting in slightly blurred photographs, giving the visual observer the edge.

To draw lunar features, a soft lead pencil, blank paper and clipboard are necessary. It is a good idea to make yourself comfortable in a chair, as your drawings will take some time. Owners of Newtonian telescopes may have some difficulty here with the eyepiece some distance from the ground. First, draw a rough outline

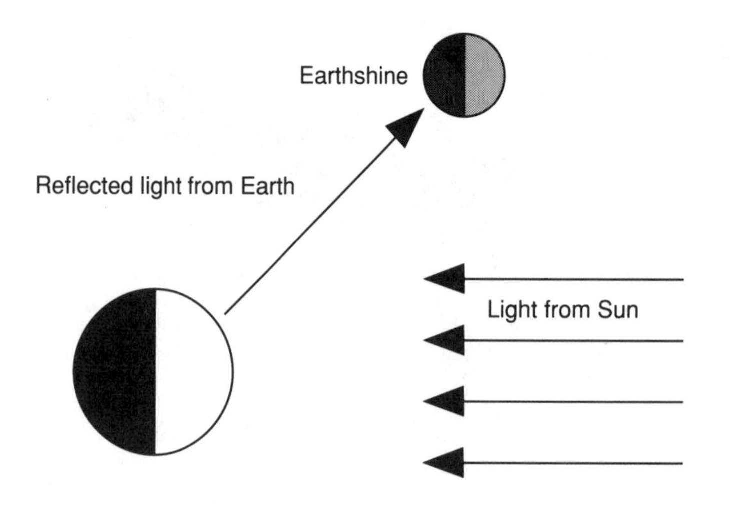


**5.17** Earthshine on the Moon is caused by light reflected from the Earth's sunlit hemisphere reaching the Moon and reflecting back to Earth again. It is seen as a faint illumination of the dark part of the Moon during crescent phases.

and then begin to fill in detail, shading can be outlined and filled in later. Charcoal is a good medium for this. You should not be too ambitious at first; choose a crater away from the terminator (the dividing line between the illuminated and dark portions of the disc) without much detail. If you work on features on the terminator, where the Sun is low, you should work quickly, as the shadows change rapidly.

There is far too much detail on the Moon to describe in this short chapter, and we will leave it to the reader to obtain a good Moon map and explore at leisure. We will, however, describe the principal features and follow with brief descriptions of some of our favourites. A word of caution on lunar atlases: before the Apollo missions the convention of east-west was reversed, so that an astronaut facing the rising Sun would be looking east. If you have a map that shows Mare Crisium near the western limb you should change it to read east. Remember that most features are best viewed at or near the terminator, the line between light and dark, where shadow relief is best. At full moon the telescopic view can be disappointing; shadows are eliminated as the Sun shines from directly above. On the other hand, the ray systems radiating from some craters are best viewed now.

The maria are the most obvious feature to the unaided eye and can be classified into two types: roughly circular regions surrounded by mountains (Mare Imbrium), and irregular outlined areas without bordering mountains (Mare Tranquillitatis). With the exception of Mare Crisium, the maria are connected, although their boundaries are fairly easy to see.

The maria, although not heavily cratered like the highland regions contain some of the most interesting of the lunar features; most craters on the maria were formed late in the lunar bombardment history, and are not defaced by later impacts, Copernicus being a fine example. The general appearance of these lava-filled regions, is that of smooth greenish-grey plains.

Undoubtedly, the most striking of lunar features are the craters. Craters come in all shapes and sizes; because of foreshortening, most craters, except those near the centre regions, appear elongated. Craters have been formerly classified into several groups: the largest are the Walled Plains (such as Clavius in the southern highlands), then the Ring Mountains (Romer), and Ring Plains (Archimedes). Craters follow and these are subdivided into several classes, from Crater Plains down to Crater Pits. In this chapter, for convenience, all classes regardless of size will be referred to as craters.

There is no end to the variety. Some have smooth regular floors, others massive, central mountain peaks or ranges. Walls can be terraced or sharply defined, and some exhibit extensive ray systems. An endless supply of detail awaits the amateur's scrutiny. Nearside craters have been named after famous (deceased) people for over 300 years; today on the farside there are memorials to the crews from the Apollo 1 and Challenger disasters, and Soviet cosmonauts, Gagarin, Komarov and Belyayev. In a break in

**5.18** This photo, taken a few days after new moon, clearly shows the features along the eastern limb of the Moon. The round Mare Crisium stands out near the limb, as does the diamond-shaped Mare Fecunditatis just below. To the north of Mare Crisium a small part of the Lacus Somniorum can be seen with the craters Hercules and Atlas just a little further north. In the south a heavily cratered part of the highlands can be seen. Along the southern limb of the Moon run the Leibnitz ranges, a peak of which is catching the dawning sunlight well into the otherwise dark hemisphere of the Moon. PHOTO CREDIT: KEN WALLACE

tradition, there are also several farside craters named after living American astronauts and Soviet cosmonauts.

The ray systems radiating from some craters are particularly striking, and are best observed at full moon, the prime example being Tycho. Under the low lunar gravity, material ejected from the formation of an impact crater will travel several hundreds or thousands of kilometres. The most prominent of rays originate from the most recent of craters, where not enough time has passed for them to darken by natural processes or micrometeorite bombardment.

Mountain ranges on the Moon are relatively higher than those on the Earth. The altitude of Earth's mountains is measured from sea level, but on the Moon, without seas, their height is compared with the average surface level. We cannot draw many similarities to Earth's lofty peaks, as the formation of the lunar mountain ranges was caused by catastrophic events and not plate tectonics or crustal folding. The mountain ranges mostly form borders around the maria. Valleys sometimes slice though the ranges like the spectacular Alpine Valley in the Alps Mountains.

Traversing some areas of the surface are the rilles and clefts, which can be traced on crater floors, and the maria. They can attain lengths of several hundreds of kilometres and widths of about 5 km. It seems that they are lava channels or huge collapsed lava tubes. They often link crater chains and frequently show little regard for objects in their path like the Hyginus Cleft which splits the crater Hyginus in two.

At all times during the cycle of phases there is plenty to see on the Moon. The three- to five-day old Moon shows detail that will keep the observer busy for hours. Try drawing some of the easier craters to begin with. The magnificent crater Petavius (160 km in diameter) near the eastern limb has complex walls and a large, central mountain group; the floor is cut by a rill that extends from the central mountains to the south-west wall. Close study of Mare Foecunditatis will show two small (13 km) craters: Messier and Messier 'A' (Pickering). Two rays extend westward from the 'A' twin giving the impression of a comet. On the north-east edge of Mare Serenitatis lies the exquisite crater Posidonius (100 km), many clefts cross its interior. Atlas (90 km) and Hercules (70 km) make an impressive pair near the north-east limb.

**5.19** This view of the Alpine Valley to the north of the Mare Imbrium was taken seven days after new moon. The valley stretches for 130 km through the mountains and is in places over 10 km wide and 3.5 km deep. The crater Cassini and two of its internal craters can be seen towards the right side of the photo and the crater Aristillus and its central peak can be seen at the bottom of the photo. PHOTO CREDIT: KEN WALLACE

The southern region of Mare Nectaris extends into the crater Fracastorius (95 km) forming a prominent bay. Just west of Mare Nectaris is a line of three superb craters with much detail. Theophilus (105 km) is a deep crater with a complex, central mountain region; as dawn reveals all of Mare Nectaris a long ridge is visible crossing the mare from Theophilus to a smaller crater named Beaumont. The walls of Theophilus intrude upon the much-damaged square-shaped crater Cyrillus which contains central peaks and a conspicuous small crater. A wide valley connects Cyrillus to the southernmost of the trio, Catharina (110 km), which consists of the remnants of a large ring crater with many smaller craters littering the floor.

From first quarter onwards, many fine sights are visible. The 'big three', Ptolemaeus, Alphonsus and Arzachel, dominate the low southern latitudes. The largest of the three, Ptolemaeus (145 km), has many small craterlets on its floor, and the deep crater Lyot. Alphonsus (110 km), the final resting place of the Ranger 9 spacecraft which transmitted spectacular pictures to Earth before crash landing in 1965, has a small central peak. Arzachel (95 km), with 4000 m high walls, has a large, central mountain and conspicuous inner crater. The finest mountain range on the Moon, the Apennine Mountains (1000 km in length) borders the south-east part of Mare Imbrium, and at this time some of its 3000 peaks will be sunlit; the highest, Huygens rises about 5500 m. The north-east sector of Mare Imbrium is rimmed by the Alps, an obvious feature of which is the Alpine Valley (110 km in length).

The giant crater Clavius (240 km) lies in the heavily cratered southern highlands. Peaks on its walls rise 5 km above the interior, several craters break the wall, and there is a string of craters crossing the floor. Tycho (85 km), with high mountainous walls and central peaks, is visible north of Clavius, the ray system beginning to show across the south-eastern highlands. The 130 km long Straight Wall, a fault in the Moon's surface, is easily detectable in the smallest of instruments on the south-eastern edge of Mare Nubium. Several ghost craters are visible on Mare Nubium, among them Fau Mauro, Parry, Bonpland, and Guericke; these lava-flooded craters were filled during the formation of the mare. To the west of Fau Mauro are the Riphaen Mountains, not particularly large as lunar ranges go, but a beautiful sight as the terminator passes them.

On the eastern shores of Oceanus Procellarum lies the magnificent crater Copernicus (90 km); in an isolated position, it is considered by many to be the finest crater on the Moon. Copernicus has something for everybody: a ray system, slumped walls 5 km above the crater floor and a central mountainous area with three main peaks. Nearby to the east are crater chains, and to the north the Carpathian Mountains, separating Copernicus from Mare Im-

**Table 5.3 Lunar landing sites**

| Mission | Region | Date |
|---|---|---|
| Apollo XI | Mare Tranquilitatus | July 1969 |
| Apollo XII | Oceanus Procellarum | November 1969 |
| Apollo XIV | Fra Mauro | February 1971 |
| Apollo XV | Hadley Apennines | July/August 1971 |
| Apollo XVI | Descartes Highlands | April 1972 |
| Apollo XVII | Taurus Littrow Valley | December 1972 |

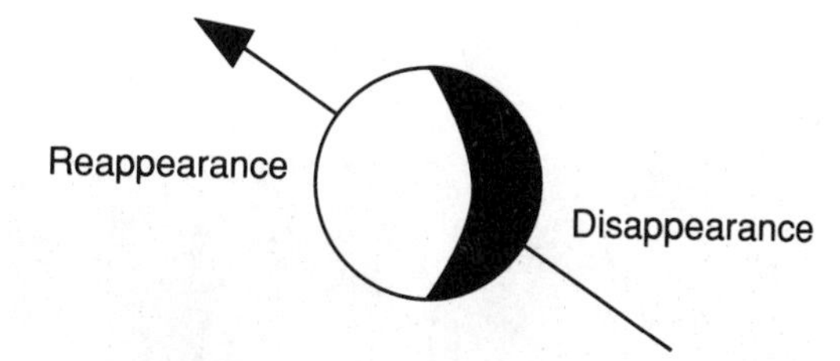


**5.20** The apparent motion of a star behind the Moon during an occultation. In reality the Moon is moving through the sky from west to east.

brium. The northern sector of Mare Imbrium contains the equally famed Plato (95 km), much elongated because of the closeness to the southern limb. Plato's dark smooth floor contains many small craterlets and light spots visible in large amateur instruments. To the west of Plato, the solitary Straight Range (65 km long) juts out of the mare. A little further west is Sinus Iridum, a large crater-like bay flooded by material from Mare Imbrium; only the northern and western walls are still visible.

Gassendi (90 km) is a very interesting crater at the entrance to Mare Humorum. Its floor contains an intricate cleft system and central peak, the northern wall has been infringed by the crater Gassendi A. Aristarchus (46 km), with walls rising over 2 km from the crater floor, is the brightest of the lunar features, being visible on the dark western side of a young moon, illuminated only by earthshine. Two or three dark bands can be seen extending from the central peak and up the inner western wall of Aristarchus when the Sun is low; they seem to darken appreciably as the lunar day progresses. Aristarchus with its attendant ray system, lies within Oceanus Procellarum; nearby is the darker crater Herodotus (35 km), and the most famous of the lunar clefts, Schröter's Valley.

As full moon nears, the beautiful ray systems reach their best. Tycho's are the most extensive and they dominate the disc; separated from the crater by a dark halo, some individual rays extend over 1500 km. The ray systems of Copernicus and Kepler exhibit much fine structure. Other obvious ray systems emanate from the craters Proculus, Anaxagoras and Olbers. Near the south-west limb lies Wargentin (90 km), an unusual crater filled to the brim with lava, forming a plateau elevated above the surrounding moonscape.

The above sampling is only a very small glimpse of the magical world of the Moon. The objects are all visible in telescopes of 60 to 80 mm in aperture. Although requiring late nights or early mornings, it is well worth viewing the lunar surface with the Sun illuminating features from the opposite direction. The perspective seen as the Sun sets over craters will make interesting comparisons with your pre-full moon drawings.

There are other places of interest on the Moon which are not natural features. Perhaps the most popular are the sites where people first walked upon that world. In the years around the Moon landings there was much interest in observing the sites of the Apollo landings. It still gives the authors a thrill to see those places. The landing sites of the six Apollo missions are given in Table 5.3; with a lunar atlas you should have little difficulty finding them. There is nothing to see there; the artefacts left behind are much too small to be seen from Earth, but it's worth the view anyway.

The passage of one astronomical body in front of another is known as an *occultation*. As the Moon passes in front of many stars, and the occasional planet or asteroid, during its monthly orbit of the Earth, it will hide them from view. As many variables affect the lunar orbit, occultations are the only sure method that astronomers can use to monitor that orbit. As there is not enough telescope time available among the world's professional observatories to carry out these routine tasks, the timing of these events are one facet of planetary work where the amateur can make an active contribution to our understanding and knowledge of celestial mechanics.

The standard type of occultation has two events: a disappearance when the Moon first occults the star, and a reappearance as the Moon moves away from the star. If the Moon is waning from full to new, it will approach the

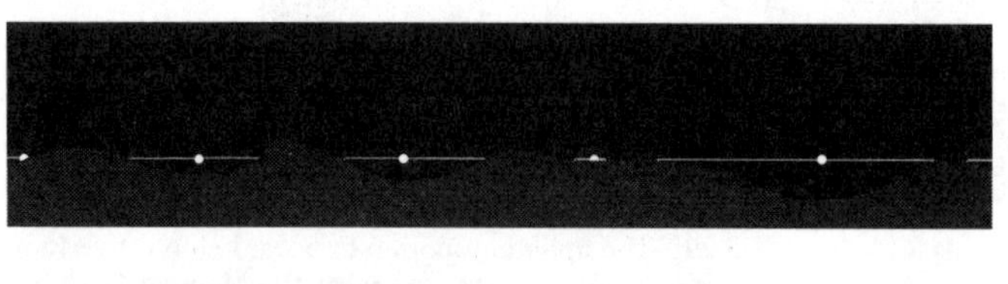

**5.21** During a grazing occultation, provided you are observing from the correct location, the star being occulted is seen to flicker in and out between the mountains of the Moon. By timing these disappearances and reappearances maps of the edge of the Moon and accurate calculations of the lunar size can be made.

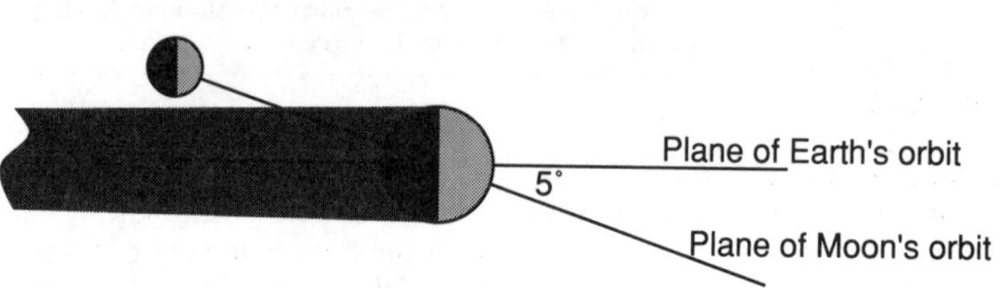


**5.22** From the side it is even easier to see why a lunar eclipse can only occur near a node. The 5° tilt of the Moon's orbit carries it well above the Earth's shadow at other times.

star with its bright limb, and if waxing from new to full the dark limb will make first contact. More spectacular are the *grazing occultations*. As the name implies, a star just grazes the lunar limb, blinking in and out as it passes behind mountains and valleys, as shown in Figure 5.21. Occultations, especially those of the graze type, require a combined effort by as many individuals as possible, and it is recommended that you join an astronomical society that co-ordinates observations of these events.

Any small telescope may be used for occultation timings, but it should be borne in mind that even a bright star near the Moon's limb will appear faint. It is generally easier to time occultations before full moon as the stars disappear on the dark limb. Reappearances on the other hand, are a little more difficult as the observer does not know exactly where the star may suddenly burst forth. The only other equipment needed is a digital watch with a stop-watch facility; the watch should be set against a telephone time service or international short-wave radio time signal just prior to the observation.

Allowances must be made for the time lag between the actual event and your reaction time. This is known as your 'personal equation' and can be established by pressing the stop button when the dial reaches a certain number, say 10 seconds; the time elapsed after 10 measured in 1/100ths of a second should be noted. Follow this procedure several times and take the average: this is your personal equation. You do not need to reduce the timings taking your personal equation into account, but it should be noted on your observations. A timing accuracy of 1/5th of a second or better is desired, and your personal equation should be established each time you decide to time an occultation, as your state of mind and body can change it slightly.

Grazing occultations are a powerful tool which help astronomers to establish orbital precision, and they also allow accurate mapping of the lunar limb. Grazes require a little more effort since they can only be observed over a narrow track of the Earth, perhaps only a kilometre wide, so you must be prepared to travel to each location. If you were too far north or south of the graze path, you'd see either the Moon pass very close to the star without contact, or a full occultation.

Observing a graze calls for teamwork and co-ordination; a string of observers is spread along the predicted path, so that each person will get slightly different results, enabling a map of the lunar profile to be drawn. Some may only record one or two events, while others may see many. Since a graze can have multiple events, all happening during a short time, it is essential to use a tape recorder with short-wave time signals playing in the background, as you announce each disappearance and reappearance. Providing the start time of the signals on your tape is known, at a later time you can count each 'beep' elapsed to arrive at the exact time of each event.

Amateurs are always willing to relate stories about their experiences at grazes. Since the team could be spread along a country paddock or suburban road, it immediately attracts the quizzical attention of passers-by, with sometimes very funny reactions. Once I attended a graze party where suspicious residents telephoned the police to investigate the strange group of people outside their homes. The police were sympathetic after we explained the importance of our presence, and joined in. It should be a lesson to all: as the general public will naturally be suspicious of things they do not understand, it is a good idea to arrive early, introduce yourself to any householders in the area, and invite them for a look through your telescope before the event. A little courtesy goes a long way.

One of the prettiest celestial displays is an eclipse of the Moon. Lunar eclipses are rarer than their solar counterparts, and each year, on average, about 1.5 lunar eclipses occur compared with 2.4 solar eclipses. Table 5.4 lists lunar eclipses until the turn of the century. The advantage of lunar eclipses over solar eclipses is that they can be seen from anywhere on the night side of the Earth, and, as the event takes several hours, you have time to take a break during your observations.

As the Moon's orbit is tilted to the ecliptic by about 5°, we can only have an eclipse when the Moon is at or near one of its nodes. At this time, the Moon will be at full phase with the Earth between it and the Sun. The conical umbral shadow cast by the Earth extends about 1.35 million kilometres into space (this value varies due to the Earth's elliptical orbit), and in the region of the Moon's orbit the shadow width is about 9000 km. The 3480 km diameter Moon will take some time to pass through the Earth's umbral shadow; if conditions are ideal and the Sun and Moon both exactly at the node, the length of totality will be close to 1.75 hours. The entire event, including the partial phases, can last up to four hours.

If the Moon is situated at the right place we will have a total lunar eclipse; the Moon will travel through the Earth's penumbral shadow before and after totality in the umbra. If situated on one side of the node, the Moon will pass through the penumbra and a part of the umbra, creating a partial lunar eclipse. If the Moon is too far from the node it may only travel though the penumbra. This is known as a penumbral eclipse.

A total lunar eclipse can be classified into five main stages, as shown in Figure 5.24. As the penumbra is entered, you will witness a gradual dimming of the Moon's light, but a casual glance up at the Moon wouldn't show anything unusual. As contact is made with the umbra, an arc of darkness is seen to creep across the lunar landscape. As the arc nearly engulfs the Moon, a curious thing happens: a dull red or coppery hue is visible in the darkened area. After a period the coloured globe again begins to move into the penumbra, the colour fading as more and more of the surface brightens.

Why does the Moon appear reddish and not totally dark when in the Earth's umbral shadow? The reason is that some light still reaches its surface. Sunlight striking the Earth's atmosphere is refracted or bent and scattered into the umbral shadow, feebly illuminating the Moon. Since the short blue wavelengths of the spectrum are absorbed and scattered by particles in the Earth's atmosphere (this is the reason we have a blue sky), they do not reach the Moon. The longer red wavelengths penetrate better,

**Table 5.4 Lunar eclipses**

| Date | Type | Duration | Mid-Longitude |
|---|---|---|---|
| 9 February 1990 | Total | 46 m | 76° E |
| 6 August 1990 | Partial | | 149° E |
| 21 December 1991 | Partial | | 159° W |
| 15 June 1992 | Partial | | 74° W |
| 10 December 1992 | Total | 74 m | 3° E |
| 4 June 1993 | Total | 98 m | 165° E |
| 29 November 1993 | Total | 50 m | 99° W |
| 25 May 1994 | Partial | | 53° W |
| 15 April 1995 | Partial | | 176° E |
| 4 April 1996 | Total | 84 m | 1° W |
| 27 September 1996 | Total | 72 m | 46° W |
| 24 March 1997 | Partial | | 69° W |
| 16 September 1997 | Total | 66 m | 77° E |
| 28 July 1999 | Partial | | 172° W |

The Mid-Longitude given is the longitude at which the Moon is at the zenith at mid-eclipse. Using this data it is possible to calculate whether the eclipse will be visible from a given location. Any location within 60° east or west of the mid-longitude will see all the eclipse; those latitudes within 120° of the mid-latitude will see at least some of the eclipse.

The time of the eclipse can be estimated from the mid-longitude value. The middle of the eclipse will be one hour after midnight for every 15° east of the mid-longitude the observer is located. For every 15° west of the mid-longitude an observer is, the mid-eclipse will be one hour before midnight. The start and end of the eclipse are roughly two hours before and after this time. This is a rough guide only. For full details of the visibility of any eclipse it is best to consult your local astronomical society.

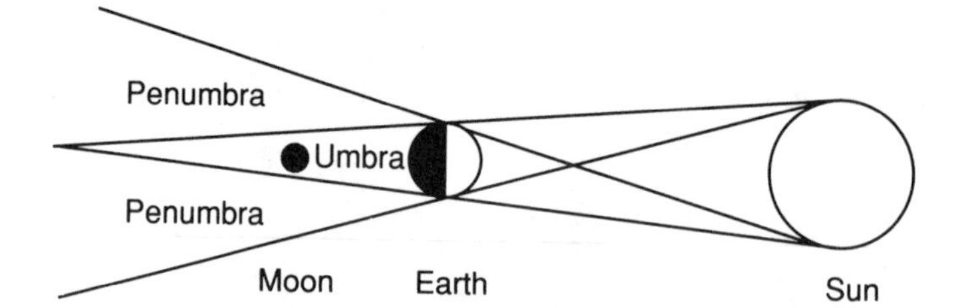


**5.23** The Earth's shadow consists of two parts, the *umbra*, or region of total shadow, and the *penumbra*, where only partial shadow is found. This diagram shows the Moon during a total eclipse fully immersed in the umbra.

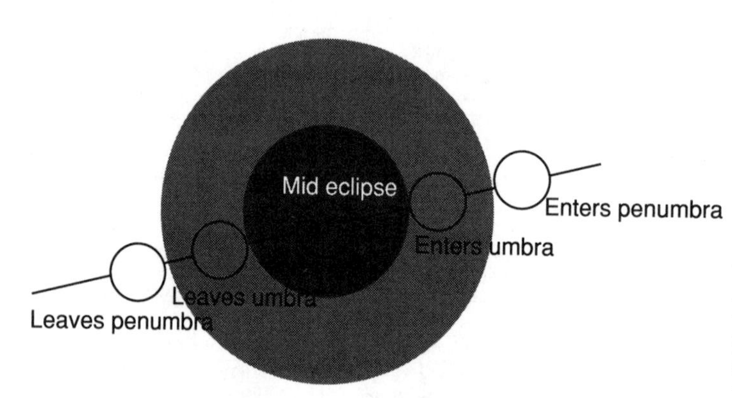


**5.24** The stages of a lunar eclipse. As the eclipse starts, the moon touches the Earth's penumbra; as the eclipse progresses it travels deeper into the penumbra until it reaches the umbra. If the eclipse is total, the Moon will travel completely into the umbra before emerging into the penumbra again, and then into full sunlight as the eclipse ends.

and lead to the colour we see (the reason the Sun appears red at sunset). Some eclipses are darker than others—it all depends on the Earth's cloud cover and atmospheric dust content. The explosion of Krakatoa in 1883 sent a huge volume of dust into the upper atmosphere, and the subsequent 1884 eclipse was unusually dark. The eclipses of 1963 and 1964 were also very dark, after the eruption of the Agunk volcano in Bali.

Apart from appreciating the eclipse from an asthetic point of view, there is much the amateur can contribute. The aforementioned occultations are ideally suited to lunar eclipse conditions, as the dulled disc will permit fainter stars to be timed. Estimates of the brightness of mid-eclipse, called *Danjon estimates*, provide scientists with information about the atmospheric dust and haze content.

Crater-contact timings have long been a part of the amateur's realm, and can provide valuable data on the oblatness and enlargement of the umbra. The method is simple: as the umbral shadow contacts selected craters the exact time is noted, as many timings as possible being recorded prior to totality. During totality, it is time to relax, enjoy the view and prepare for the coming exit of the craters from the umbra, when timings again begin in earnest. During your first attempt at crater-contact timings, do not get too ambitious: select maybe ten or twelve familiar craters, spaced well apart, that you can identify easily. The last thing you need is a panic situation created by choosing too many events that occur close to each other. Old hands at the game time contacts and occultations, and even find time to search for transient lunar phenomena, constantly switching back and forth!

One controversial area of lunar study is that of the transient lunar phenomenon (TLP). As the Moon is considered a dead world, many astronomers are reluctant to accept the possibility of change. But the overwhelming proof suggests they exist. Whether they are caused by meteor impacts, moonquakes, or the venting of internal gases, or some such similar natural phenomenon, the puzzle remains to be solved.

The search for TLPs by the amateur is an arduous task that may never show any result. TLPs manifest themselves to the observer as bright spots, coloured areas or hazy patches. The TLPs cannot be dismissed too lightly, as thousands of sightings have been made by reliable observers. Even the Apollo 11 astronauts

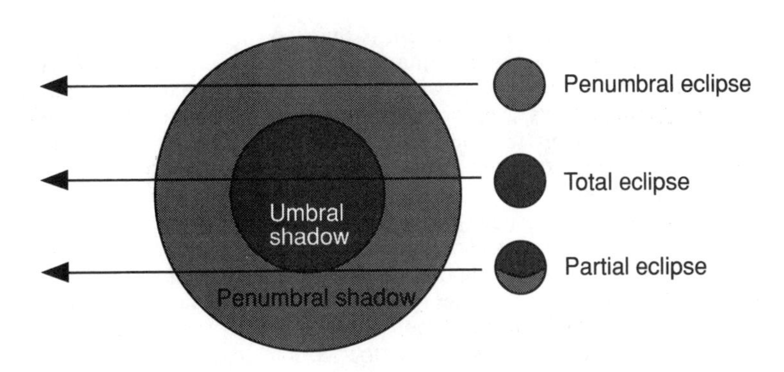


**5.25** A penumbral eclipse happens when the Moon travels only through the penumbra of the Earth's shadow; a partial eclipse happens when at least some of the Moon enters the Earth's umbra. A total eclipse happens only when all of the Moon travels into the umbra.

**5.26** Currently the most popular theory of lunar formation involves the young Earth being struck by a large object which itself had begun to form a planet. The impact joined the two bodies together and ejected a large amount of material into orbit around the Earth. This material reformed itself into the Moon, the core material of the impacting body coalescing with the Earth's core.

saw a brightening in the crater Aristarchus that was also observed by two Earth-based observers. The ideal time to search for TLPs is during a lunar eclipse when the darkened surface will present an ideal background for any such event. Caution should be exercised in your observations as even Herschel thought he saw volcanic activity when observing the darkened lunar hemisphere; in reality the embarrassed Herschel was viewing the brightest lunar feature, the crater Aristarchus.

The origin of the Moon was one of the mysteries lunar exploration was meant to solve, yet now, more than twenty years since it was first visited and the study of hundreds of kilograms of lunar rocks, the question is still unanswered. The samples of the Moon returned to Earth show that they are much the same as equivalent terrestrial samples. Not only are the minerals and their proportions similar to those here, but measurements such as the isotopic abundance of various oxygen atoms and other elements are also the same. Scientists thus suspect that the Earth and Moon both formed from the same part of the solar nebula.

A problem with the Moon is that it lacks the Earth's metallic core. With a density of only 3.3 g.cm$^{-3}$, the Moon appears to be similar to the material of the Earth's mantle. Combine with this some of the strange element scarcities on the Moon—five times less potassium, but the same proportion of uranium—and the theory that they were formed together from the same material becomes strained. There are three theories of lunar formation which have been fighting it out for nearly a hundred years; it is now known that none of them is completely correct, but perhaps a synthesis is.

The fission theory of formation supposes that the Moon formed from the Earth. It was first proposed by mathematician George Darwin (son of Charles Darwin the biologist), who calculated that a rapidly spinning planet could split and form a double planet, such as the Earth-Moon system. If this happened after the Earth's core had formed, that would explain the

Moon's lack of metal. One problem is that this theory, although explaining the similarities, does not explain the differences between the Earth's mantle and the Moon's. Another difficulty is that the material spun off for the Moon would be much more likely to form a ring around the Earth than a planet.

The second theory suggests that the Moon formed from a ring of material around the Earth in the same way as the moons of the outer planets. Unfortunately this theory does not explain the lack of metal in the Moon at all well.

The final theory suggests that the Moon formed somewhere else in the solar system as an independent planet, and that it was later captured by the Earth when it passed close by. This theory also has a number of flaws. It doesn't account for the similarities in the Earth and Moon and it requires the intervention of a third body; there is no obvious candidate.

A synthesis of these theories has been proposed by a number of scientists. It is a combination of the first and second theories. It proposes that the forming Earth was split by one or more extremely large impacts during the period of planetary accretion, but after enough differentiation had taken place for the Earth's core to have settled to the bottom so most of the material ejected would be mantle material. After this catastrophic collision the ejected material would remain in Earth orbit and condense to form the Moon. The Moon would initially have been much closer to the Earth than it is today, but tidal slowing on the satellite would have moved it to its present position. It is hoped that the study of the other planets of the solar system will shed more light on this still-unsolved riddle.

The idea held by some amateurs that there is little left to achieve with lunar studies is completely wrong. Sure, twelve people have walked upon the surface and lunar orbiting spacecraft have mapped detail invisible from Earth, but many mysteries still remain. The Moon is almost exclusively now the amateur's domain, and work of value can be done with the smallest of instruments. So instead of treating the Moon as an annoying bright light that spoils faint objects, try observing it!

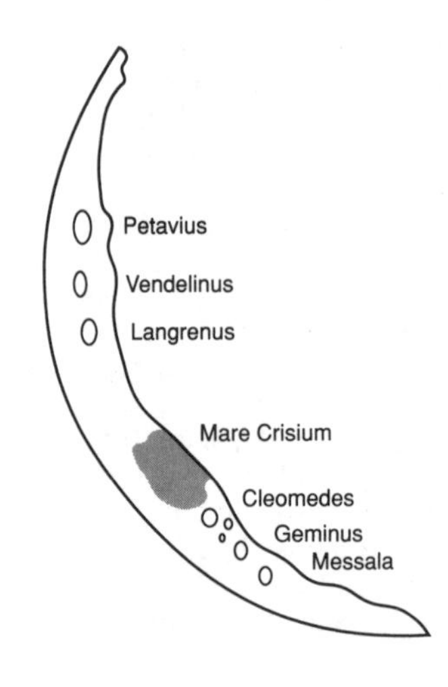


**Moon age 1.5 days** The narrow crescent of the Moon presents only a few objects of interest to the observer. The Mare Crisium is over 500 km from north to south and 450 km east to west. At the south of the sea the Agarum Promontory, over 3 km high, juts out into the sea. North of Mare Crisium is the crater Cleomedes. 130 km in diameter, its central peak is interesting, being divided by three clefts. When looking at the Moon at this time look also at the dark part, as earthshine often allows features to be seen in that hemisphere. Look for the craters Aristarchus and Grimaldi—their contrast with the surrounding terrain makes them easy to pick.

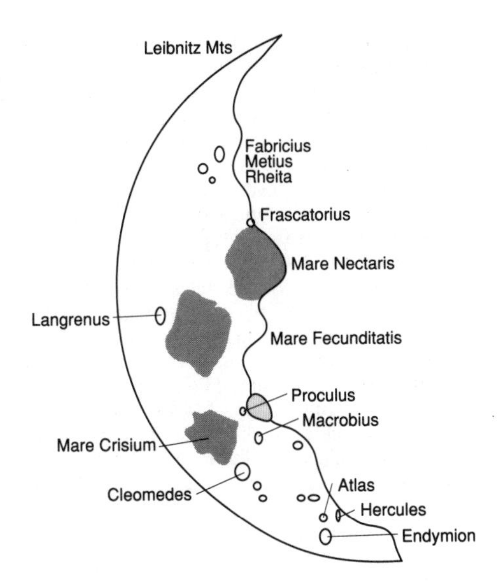


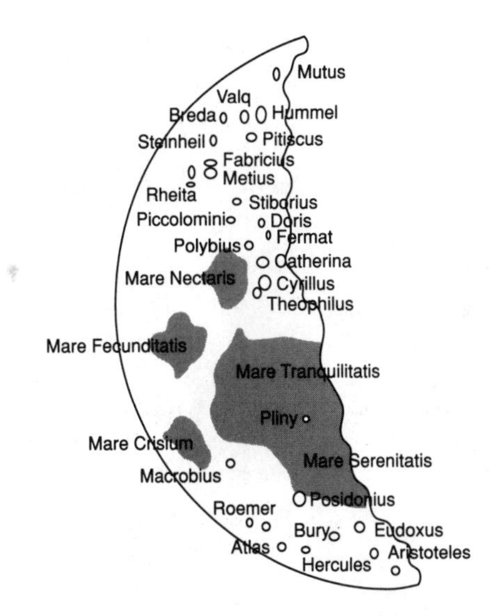


**Moon age 3.5 days** The maria Fecunditatis and Nectaris are both visible now that more of the Moon is illuminated. At the south of Mare Nectaris is the crater Frascatorius with its northern wall broken down by the lava which formed the sea. The crater Endymion near the Moon's northern limb appears elliptical in shape. Often its walls will be seen in sunlight while its floor 4.5 km below remains in shadow. It is worth observing along the southern limb of the Moon as the peaks of the Leibnitz mountains can be seen. Often these peaks will be seen in the dark part of the Moon, their summits catching the dawning sunlight.

---

**Moon age 5.5 days** The Mare Tranquilitatis and Mare Serenitatis are now partially visible separated by the Haemus mountains on the western side of the crater Pliny. Slightly further north the Taurus mountains also curve into the Mare Serenitatis. Much of the highland region in the Moon's southern hemisphere is now visible, with many thousands of craters providing many hours of observing pleasure. To the west of the Mare Nectaris, the three craters Theophilus, Cyrillus and Catherina make a striking sight. In some places the walls of these craters rise 5.5 km above their floors. On the northern shore of Mare Serenitatis is the 100 km wide crater Posidonius. This crater's floor is 800 m below the surrounding area, the crater is best viewed six days after new moon. Careful observation will reveal a ridge running from Posidonius across the floor of Mare Serenitatis to the crater Pliny. This ridge is the remains of a mountain range overwhelmed by lava flooding the Mare Serenitatis' floor.

---

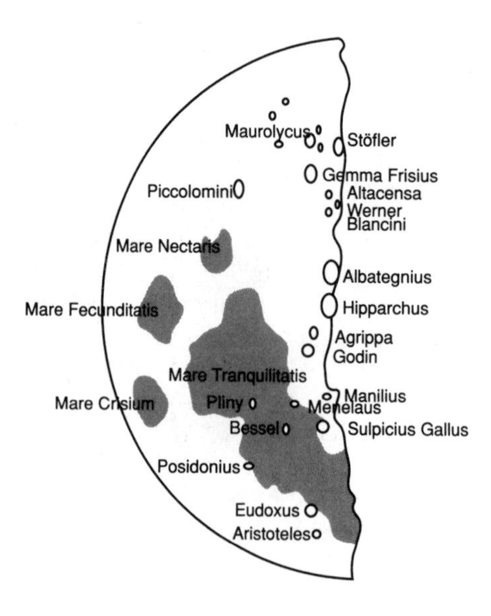


**Moon age 7.5 days** At first quarter moon the best place to begin observation is along the line of the boundary of the shadow. With the entire southern section of the terminator lying in highland regions, there are many more craters to be seen than could listed. In the south Maurolycus and Stöfler are both prominent with peaks in their walls rising to 4.5 km. Near the centre of the Moon, just to the east of the terminator, are the craters Albategnius and Hipparchus. Both craters are about the same size, Hipparchus shows the effects of age with much of its walls demolished, while the younger Albategnius is almost perfect. On the western shore of Mare Serenitatis are the Caucasus Mountains, an interest range to view, dividing Mare Serenitatis from Mare Imbrium. On the floor of Mare Serenitatis you can now see a light streak formed from material thrown out when the crater Tycho, 3000 km away, was formed.

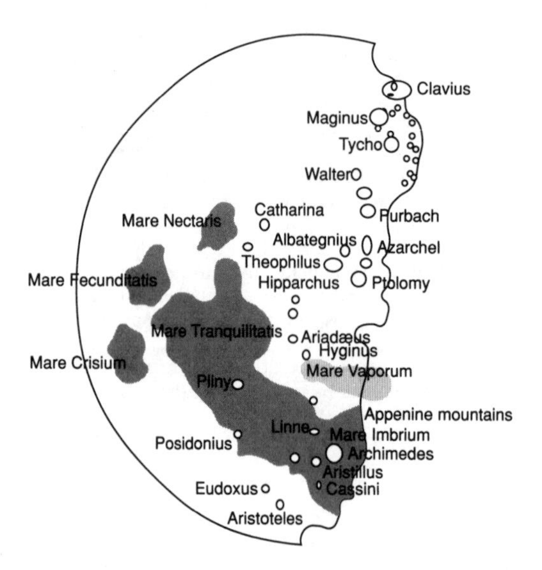


**Moon age 9.5 days** Clavius is a conspicuous feature of the Moon's southern hemisphere once the sunlight reaches it; 225 km across, over 90 smaller craters have been counted within its walls. To the north of Clavius is Tycho, a perfectly formed large crater 87 km in diameter and one of the youngest craters on the Moon. Not very evident until full moon, Tycho is the centre of a system of light-coloured rays stretching across the Moon's surface. North of Tycho are two chains, each of three large craters. Walter, Regiomontanus and Purbach make up one triplet; Arzachel, Alphonsus and Ptolomaeus the other. Around the southern shore of Mare Imbrium are the Apennine mountains, beginning with Mt Hadley, 4.5 km high and stretching 700 km to the crater Eratosthenes. Within the Mare Imbrium are the craters Archemedes and Aristillus. To the north of Aristillus are the lunar alps gouged by the Alpine Valley, a rift 3500 m deep and 5.5 to 10.5 km in width cutting 130 km through the mountains.

**Moon age 11.5 days** To the east of the crater Thebit in the Mare Nubium can be seen the Straight Wall, a fault line 100 km long and 300 m high. Note the crater Gassendi: there are many features in the walls of this crater to make careful viewing worthwhile. Copernicus, placed in the midst of the Ocean of Storms, perhaps as a comment on his ideas, is one of the most interesting sites on the Moon; there are eight central peaks, one being 800 m high. The inner walls of the crater are terraced and the crater is the centre of a ray system which spreads across the Oceanus Procellarum. North of Copernicus are the Carpathian mountains separating the Oceanus Procellarum and Mare Imbrium. Plato, on the northern shores of Mare Imbrium, is one of the darkest spots on the Moon. The lava filling the crater floor is much darker than the surrounding area. On the north-west side of Mare Imbrium is the Sinus Iridum surrounded by the Jura mountains. To the south of Tycho, in amongst the confusion of the southern highlands, is the crater Newton, 23 km across and 7.3 km deep, the deepest crater on the Moon.

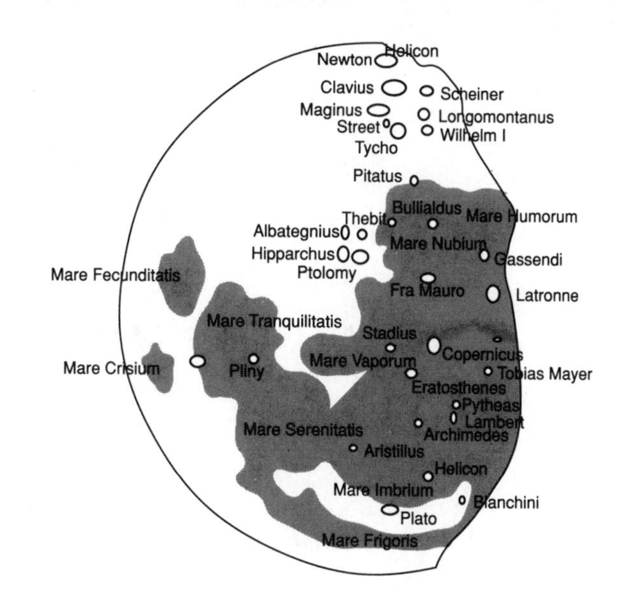


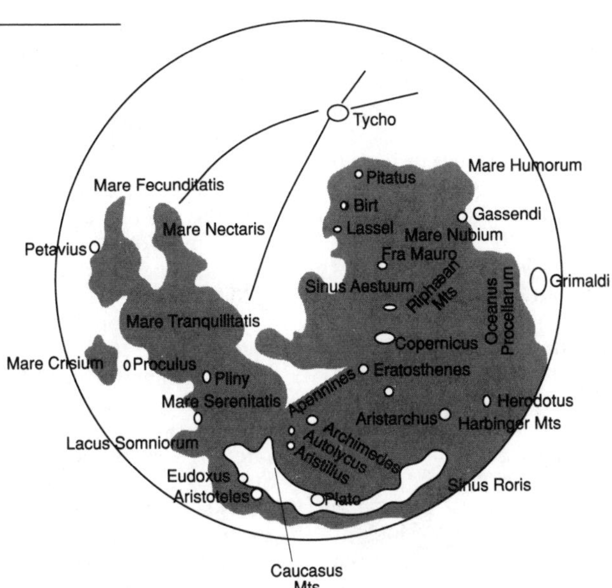


**Moon age 14.5 days** Full moon is the worst time for lunar observation. With the Sun directly behind the observer there are no shadows visible to provide contrast when looking for features. Grimaldi, a crater 240 km long near the Moon's western edge, is the darkest spot on the Moon and so shows up well at this time. Now is also the best time to trace the rays left by the craters Tycho and Copernicus and to map the extent of the Moon's system of maria which, apart from Mare Crisium, are all interconnected.

# 6

# The Sun

All life on Earth depends upon the Sun for heat and light, and also for energy. At the base of all the food chains in the world, except for a few at great depths below the ocean, are plants, and it is from the Sun that plants get their energy. But from where does the Sun get its energy? The Sun, as we have seen, formed from a collapsing cloud of gas some four and a half thousand million years ago. As the cloud collapsed, the centre became more and more condensed and the temperature increased. Once the conditions were right, energy production in the Sun could begin.

The process by which the Sun forms energy is known as *nuclear fusion*. It involves four hydrogen atoms fusing into one atom of helium. In the process a small amount of mass is lost. The mass lost is converted into energy according to Einstein's equation $E = mc^2$. The amount of mass lost by one such reaction is small, just $5.0 \times 10^{-26}$ grams, but when $8 \times 10^{37}$ of them occur each second, the total mass lost by the Sun is $4 \times 10^6$ tonnes per second and the power generated is $3.5 \times 10^{26}$ watts.

The structure of the Sun is fairly simple. At the centre is a core where the nuclear fusion reactions occur; here the temperature of the Sun is $10^7$ K, and the pressures are enormous. Above the core is a layer of gas which is in a constant state of convection, carrying the energy of the core to the upper regions. At the top is the layer of gas which we see when we look at the Sun, the *photosphere*. Here the energy from the nuclear reactions at the centre is radiated into space. Above the photosphere is the *chromosphere*. This layer radiates less energy than the layer below it, so normally it can't be seen from Earth, but when a total solar eclipse takes place the chromosphere is seen as a coloured band of light around the edge of the Moon. The last layer of the Sun is the *corona*; this region of tenuous gas extends many millions of kilometres from the solar surface. Again, the corona is too faint to be seen except during total solar eclipses, but then it shows up as streamers of light extending well away from the Moon's disc.

Solar observation is a very satisfying and rewarding pastime for the amateur; it brings astronomy into the convenient daylight hours. As the Sun is the closest star to the Earth, it presents us with a sizable disc which can be studied in detail with the smallest of telescopes, while all other stars are seen as mere pinpoints of light.

The dangers of observing the Sun cannot be overemphasised. Viewing the Sun with any form of optical aid, be it binoculars or telescope, may lead to instant blindness, or at the very least permanent eye damage. Even looking at the Sun with the unaided eye for one or two seconds will do harm. Your eyes are your greatest asset in astronomy, so look after them!

Most objects viewed through a telescope are faint, and to gain a brighter image the aperture must be increased. The Sun, however, is so overwhelmingly bright that we are faced with exactly the opposite problem: the light must be reduced if we are to view the solar surface in safety. Even the brightest star in the night sky, Sirius (α Canis Majoris), is ten thousand million times fainter than our own Sun as seen from the Earth.

Since the recognised safe observing procedures give excellent results, the observer should not experiment with unsafe practices. Many techniques recommended in old astronomy books do not provide adequate protection; they may dim the Sun to a comfortable level, but still allow the transmission of invisible infrared and ultraviolet radiation. Filters made of glass blackened with a deposit of soot from a candle flame, over-exposed photographic film or any other homemade device should never be used.

Even some purchased Sun-viewing apparatus may be unsafe. You should never use the dark tinted Sun filters or 'suncaps' that fit over the eyepiece. Since these filters are placed in the hottest part of the light path, there's a very real possibility that they may shatter. The instant a filter cracks, a shaft of intensified sunlight will strike the unprotected eye before the observer can react, causing serious injury.

Some textbooks maintain that you can safely use Sun filters on telescopes with a maximum objective aperture of 50 to 80 mm, but is it worth trusting your sight to a piece of coloured glass subjected to extreme heat conditions? There are no second chances when it comes to solar observing. If you own a Sun filter of this type, it is best to destroy it, thereby eliminating any temptation to use it.

Solar observing should be carried out using one of two recognised safe methods: projecting the Sun's image on to a screen behind the eyepiece, or direct viewing with a special pre-filter over the objective. While direct observation with the pre-filter gives the best detail for casual viewing, the projection method is better suited for accurately plotting the positions of sunspots.

Commercial telescope manufacturers supply solar projection screens to suit their telescopes at a moderate cost. Alternatively, you can make a satisfactory screen yourself from materials found around the home. The projection principle is very simple: the telescope objective brings the light to a focus inside the tube. The eyepiece then enlarges the image and projects it on to a screen as shown in Figure 6.2. The further the screen is from the eyepiece the larger the image becomes, but at a sacrifice in the degree of contrast and brightness. Ideally a projected image 150 mm in diameter is best.

Inexpensive eyepieces that don't have cemented elements are best for solar work, like the Ramsden or Huyghen designs. As the eyepiece is located at the hottest region of the light path, using good quality eyepieces is not recommended; the elements may crack or separate under intense heat, and coatings will vaporise, leaving a messy film on the lenses.

One important consideration in the design of a projection screen is that the distance of the screen from the eyepiece be easily adjustable. As the Earth moves around the Sun in an elliptical orbit, its distance varies; it is closest in January and furthest away in July. In January the apparent diameter of the Sun's disc is about 32.5 minutes of arc, whereas in July it is about 31.5 minutes of arc. The difference, although small, calls for an adjustment if the Sun's projected image is to stay constant on your observations.

Since daytime seeing is limited at best to about one arc second, the theoretical limit of a 110 mm telescope, there's generally no gain with larger instruments. The larger the aperture, the more affected it will be by atmospheric disturbances. It is a good idea to limit the aperture of large telescopes to about 100 mm. A circular

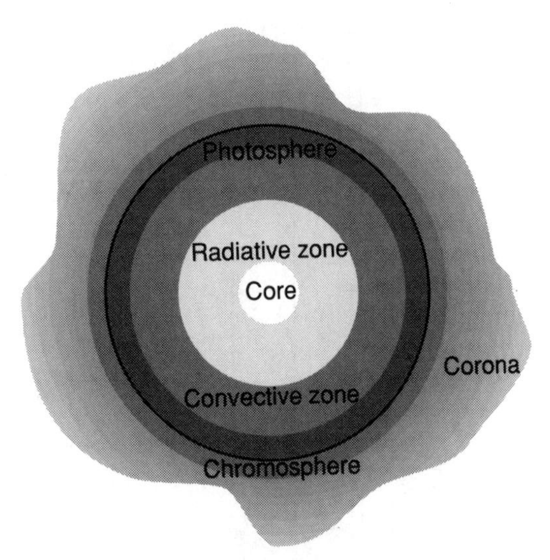


**6.1** It is at the core of the Sun that the nuclear reactions which power that star occur. The energy from these reactions passes through the region above the core in which temperatures are too low for nuclear fusion to occur by radiation, the particles of energy being continually absorbed and re-emitted by the atoms in that region. Above the radiative zone is a region where the energy rises by convection, the gases moving up and down carrying heat away to space. Above this convective zone is the photosphere, the surface of which we see when we look at the Sun. Further out than the photosphere is the chromosphere, a region of tenuous gas visible only during eclipses. Lastly there is the corona, a region of very low-density high-temperature gases which extends millions of kilometres from the Sun.

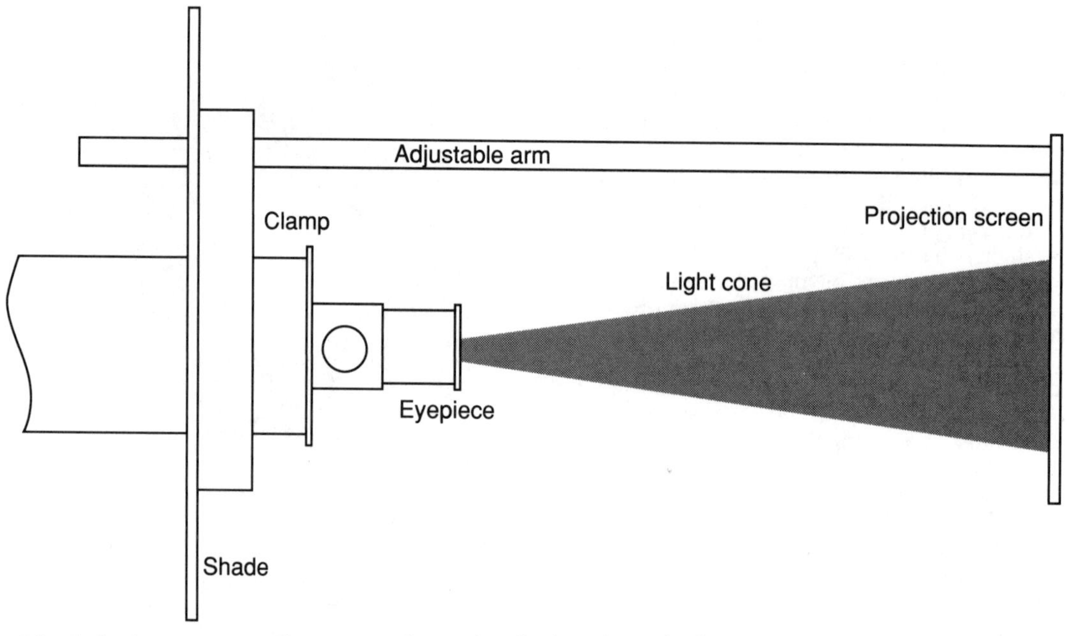


**6.2** A simple arrangement of parts can make a quite effective solar projection screen.

hole cut in a piece of cardboard mounted over the front of the telescope is adequate for this purpose. Several stops with various hole sizes will allow the observer to evaluate the aperture best suited for both telescope and atmosphere.

Where possible, viewing should be carried out by mid-morning, before the air is plagued by thermal currents caused as the Sun warms the ground and surrounding buildings. However, no matter when you observe, it is important to prevent direct sunlight falling on the telescope tube, causing internal currents.

Most amateurs who frequently view the Sun make a permanent record of their observations. Daily diagrams can be made of the entire disc showing the sunspot positions, and a graph made of their number will provide interesting data throughout each solar cycle. If you are not one for the long-term tedium of methodical record-keeping, you may prefer to do a detailed drawing of a noteworthy group's development and decay over a short period.

To accurately plot sunspot positions, it is necessary to draw a grid over a circle of the same diameter of the Sun's projected image. Use a hard, sharp pencil (2H), so as not to obscure any small sunspots. The grid size is a matter of personal choice but 10 mm squares or smaller are preferred, and will allow precise positioning. The horizontal and vertical sides should be numbered to allow easy identification of individual squares.

Drawing the sunspots directly on the projection screen is almost impossible, as the image will shake, so a second identical grid is made, this time with black ink. A blank observation sheet of lightweight paper that allows the backing grid to show through is placed over the grid. The pencilled grid is fastened to the projection screen; any spots or groups observed are then copied on to the observation sheet attached to a clipboard.

The same method is used for enlargements of individual groups. You no longer need a circle, just a square grid of convenient size. The image of the Sun is enlarged by moving the screen further away from the eyepiece, and the magnification increased with a higher power eyepiece until the desired image scale is reached.

A motor-driven equatorial mount is preferred for this work, but an altazimuth mounted telescope can be used if you are prepared to con-

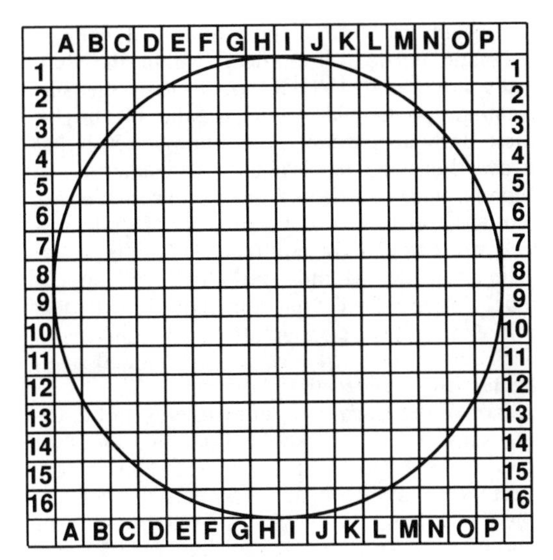


**6.3** Solar projection grid

stantly re-centre the image and occasionally rotate the screen as you do your diagrams. The projection screen must be orientated so the east-west latitude of the Sun is parallel to the horizontal grid lines. Simply turn off the motor drive, choose a spot on the Sun's disc, position the image so the spot is near the edge of the grid, and allow the Earth's rotation to carry it across the grid. Adjustments are then made to the projection apparatus by rotating it around the optical axis until the spot drift is parallel to the grid lines.

The projection method has stood the test of time. For almost 400 years astronomers have viewed the Sun with this method. Galileo himself used projection (discovered by his pupil, Benedetto Castelli) to safely view the Sun. In Galileo's 'Letters on Sunspots' he describes in fine detail the method used, and details the importance of fitting a large paper shade on the telescope tube and darkening the observing room.

If you choose to view the solar surface directly, unlike the projection method, you need to filter the light before it enters the telescope with a pre-filter (or full-aperture filter). There are two basic types of pre-filters available: glass coated with nickel-chrome, or Mylar (plastic) coated with aluminium. The glass filters are the most expensive, and give a pleasant yellow-

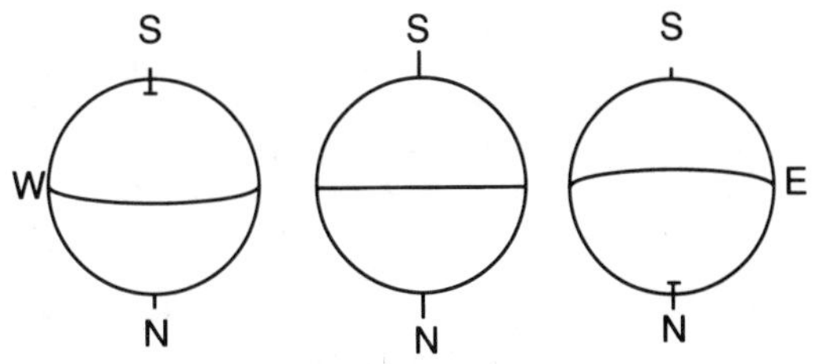


**6.4** Because of the inclination of the Earth's orbit to the Sun's equator, only during June and December is the equator seen as a straight line during the year. At other times the equator is curved and one or other of the poles is hidden from view.

orange image. The Mylar filters give an unnatural blue image, but they are as effective, and cheaper, than their glass counterparts.

Although using a full-aperture filter is considered perfectly safe, it should be checked regularly. The filter should be held up to the Sun and examined for defects like scuff marks and cracks. Small pinholes are common; a few are acceptable, but if they are too numerous, the image contrast will suffer. Pinholes are easily rectified with a small spot of black paint. The filter should be a snug fit on the tube so it cannot be dislodged if the telescope is bumped or a sudden gust of wind develops.

Full-aperture filters are excellent for casual viewing, and the experienced amateur, with practice, can make accurate freehand drawings. Although seldom matching the precision of the projection and grid method, freehand drawing will be assisted with the aid of a low-power eyepiece fitted with cross-hairs; this will divide the disc into four quadrants, helping to determine positions with a fair degree of accuracy.

Since the Sun's equator is inclined to the ecliptic by 7°15′, it follows that the polar axis is not always perpendicular to us, and the tilt varies throughout the year with respect to the Earth. On 5 June and 7 December the equator is seen as a straight line across the disc, with the solar poles at the limb. At other times the equator is curved either upwards or downwards from the disc's apparent centre, and only one pole is visible. Sunspots traversing the disc travel parallel to the equator and, as either of the poles can be depressed behind the limb by up to 7°15′, a noticeable curve in their path will be apparent.

It is useful to know the Sun's heliocentric

(measured from the centre of the Sun) co-ordinates to measure sunspot latitude and longitude. Tables supplying this information are available in publications like the United States Naval Observatory's *Astronomical Almanac* or similar works published by many amateur astronomical groups.

When using these publications, there are three variable solar co-ordinates that you will need to become familiar with: P, $B_0$ and $L_0$. P is the position angle of the polar axis. The north end of the axis moves over an arc of 26.3° east and west of the north point of the disc (a positive number shows that the axis is east of the north point, and west if negative). $B_0$ is the latitude of the Sun's centre. This angle varies by up to 7.15° and, as mentioned earlier, has the effect of displacing the equator and poles. When positive, the equator is found to the south of the disc's centre, and the north pole is tilted to the observer. When negative, the situation is reversed and the south pole is visible. $L_0$ is the longitude of the centre of the disc. Since there's no fixed point on the photosphere (the Sun's visible surface) by which longitude may be measured, an arbitrary zero of longitude was set on 1 January 1854. The value of $L_0$ is then calculated using the sidereal rotation period of 25.38 days from this date.

Long before the invention of the telescope, sunspots were observed with the unaided eye through thick smoke from forest fires, fog or cloud. The records go back to over 2000 years ago and references can be found through the ages in Chinese, Korean and Japanese histories. The earliest recorded sighting is accredited to Theophrastus of Athens in the fourth century BC. However, sunspots were largely ignored by Western science until the invention of the telescope. It is not known with certainty who first observed the spots telescopically, but four names are associated with this discovery: Galileo Galilei in Italy, Christopher Scheiner in Germany, Thomas Harriot in England and Johann Goldsmid (also known by his Latinised name Johann Fabricius) in Holland.

Galileo recognised that the *macchie solari*, or sunspots, were on the surface of the Sun, whereas at first Scheiner believed them to be small planets orbiting around the Sun—a safe line in days when the Sun was considered 'most pure and most lucid'. Galileo wasn't as conservative as his contemporaries and boldly proclaimed that the Sun was 'spotted and not pure'.

Sunspots are the manifestations of colossal magnetic vortices from deep within the Sun. They are about 1000 K cooler than the surrounding photosphere, but are still very bright; their dark appearance is a contrast effect between them and the brilliant photosphere. The Moon's limb silhouetted against the Sun's disc during an eclipse of the Sun is proof that sunspots are bright light emitters; any spots in the vicinity of the Moon's dark limb will appear much brighter in comparison. Developed spots will exhibit two distinct areas: an inner very dark part called the *umbra*, surrounded by a lighter fringe called the *penumbra*.

Part of the fascination of observing and drawing sunspots is the fact that no two spots or groups appear alike; they change daily, sometimes with dramatic developments. In making the Sun a long-term subject of study, one must adopt a method to reduce daily observations into a graph of solar activity.

One simple method is the *mean daily frequency* or MDF. The observer counts 'active areas'; every spot or group, no matter how small, is counted as an active area, so long as it is separated from any neighbour by at least 10 degrees of latitude and longitude. A very large group covering more than ten degrees is still counted as one active area unless it has separate centres of activity at least 10 degrees apart. At the end of each month the daily count is totalled, then divided by the number of days on which observations were made. The resultant figure is the MDF.

Alternatively the *Wolf number* or *Zürich number* can be used as a measure of the Sun's activity. In 1848 Rudolf Wolf proposed this method of determining the relative sunspot numbers, and since 1855 to this day it has been used at the Zürich Observatory. Many observers prefer this method, and direct comparisons can be made with your own and published figures. The relative sunspot number R is derived from the following formula:

$$R = K\,(10\,g + f)$$

where: g is the number of active areas (each single spot or group is counted as an active area);
f is the number of individual spots; and
K is a correction factor that allows for variations such as telescope size, and the observer's proficiency.

If you intend to study the Sun for pleasure and your own satisfaction, the value of K can be taken as 1.

Therefore, if we observed, one group containing two spots plus two individual and isolated spots, the Wolf number would be:

$$g = 30 \text{ (1 group + 2 spots} \times 10)$$
$$f = 4 \text{ (2 spots + 2 spots)}$$
$$R = (1(30 + 4)) = 34$$

In simple terms, each active area scores ten points, and each spot (singular or within a group) scores 1. These two figures are then added to arrive at the value of R.

Since there can be considerable difference between sunspot numbers in the two hemispheres, it makes an interesting comparison to calculate both northern and southern separately. The unevenness of the two hemispheres is mystifying.

All sunspots and groups start from small dark pinpoints known as *pores*. At the limits of telescope visibility they vary in size, but generally 2–5 seconds of arc is average. Pores have no penumbra and the vast majority last for only a few hours; the survivors progress to form sunspots. Collections of several pores will often evolve into a group which can attain areas in excess of many millions of square kilometres.

Spots may occur singularly, but they are more often found in pairs or more complex groups attended by many smaller spots or pores. The development is varied: a few follow classic evolutionary rules and emerge as spectacular large groups; others reach a certain level and regress back. Whatever stage is attained, all sunspots are conceived as pores and ultimately end their life in the same manner.

The Zürich classification places sunspot groups into nine evolutionary classes as shown in Figure 6.5. Large groups will evolve through all classes A to J. Moderate groups may reach classes C or D and return to A or even jump from D to H before returning to A. Many small groups never pass A or B types.

Sunspots vary in size significantly from the smallest, limited only by atmospheric seeing, to giants spanning a considerable portion of the disc. If you use the projection method, it is an easy task to estimate the size of any group. Since the Sun's equatorial diameter is just over 109 times that of the Earth, you divide the projected image size by 109, the resultant figure is equivalent to the Earth's diameter. For example, working with an image size of 165 mm, any spot that is 1.5 mm in diameter will be Earth size (12 760 km). Therefore a 7.5 mm group will span over 95 700 km, and a 0.5 mm spot about 4000 km.

**Table 6.1 Zürich sunspot classifications**

| | |
|---|---|
| Class A: | A single pore or group of pores. All groups begin and end in this class. |
| Class B: | A group of bipolar pores. (A bipolar group is aligned parallel to the equator with each major component having opposite magnetic polarities.) |
| Class C: | Moderate bipolar group consisting of a single spot with penumbra and a collection of pores. |
| Class D: | Moderate bipolar group,. Both main spots are of similar size and have penumbra. The overall length is less than 10° of solar longitude. |
| Class E: | Large bipolar group with complex structure. Both components have penumbra, length is greater than 10°. |
| Class F: | Very large bipolar group, extremely complex. Reaching naked eye visibility, length is greater than 15°. |
| Class G: | Large bipolar group with no small spots between the main components. The preceding spot appears first, and the following spot develops later and declines earlier. Length greater than 10°. |
| Class H: | Single moderate spot with penumbra and some attendant pores, sometimes breaking up into smaller divisions. Diameter greater than 2.5°. |
| Class J: | Single small spot with penumbra which may break up into smaller spots. Diameter less than 2.5°. |

The life of average sunspots is about one week. Smaller ones may only last a day or two while very large groups can persist for 100 days or more. There are no set rules, but generally the larger the area of a group, the longer it will survive. Since the Sun takes about a month to rotate on its axis, a spot observed on the east limb will be carried over the western limb after about two weeks. Should the spot survive it will reappear on the east limb after a further fortnight. The Sun's actual equatorial rotation is 24.65 days, but a spot very near the equator can be followed across the disc for about 13.5 days and reappear 13.5 days later—a total of 27 days. The inconsistency lies in the Earth's motion; as the Sun rotates about its axis the Earth does a little bit of catching up as it moves in its orbit, and effectively lengthens the apparent solar day to 27.28 days.

As the Sun is gaseous it does not rotate as a solid body; the equatorial regions rotate faster than the poles. This differential rotation causes an apparent drift in sunspots of different latitudes as illustrated in Figure 6.6. Near the equator a spot will take about 25 days to com-

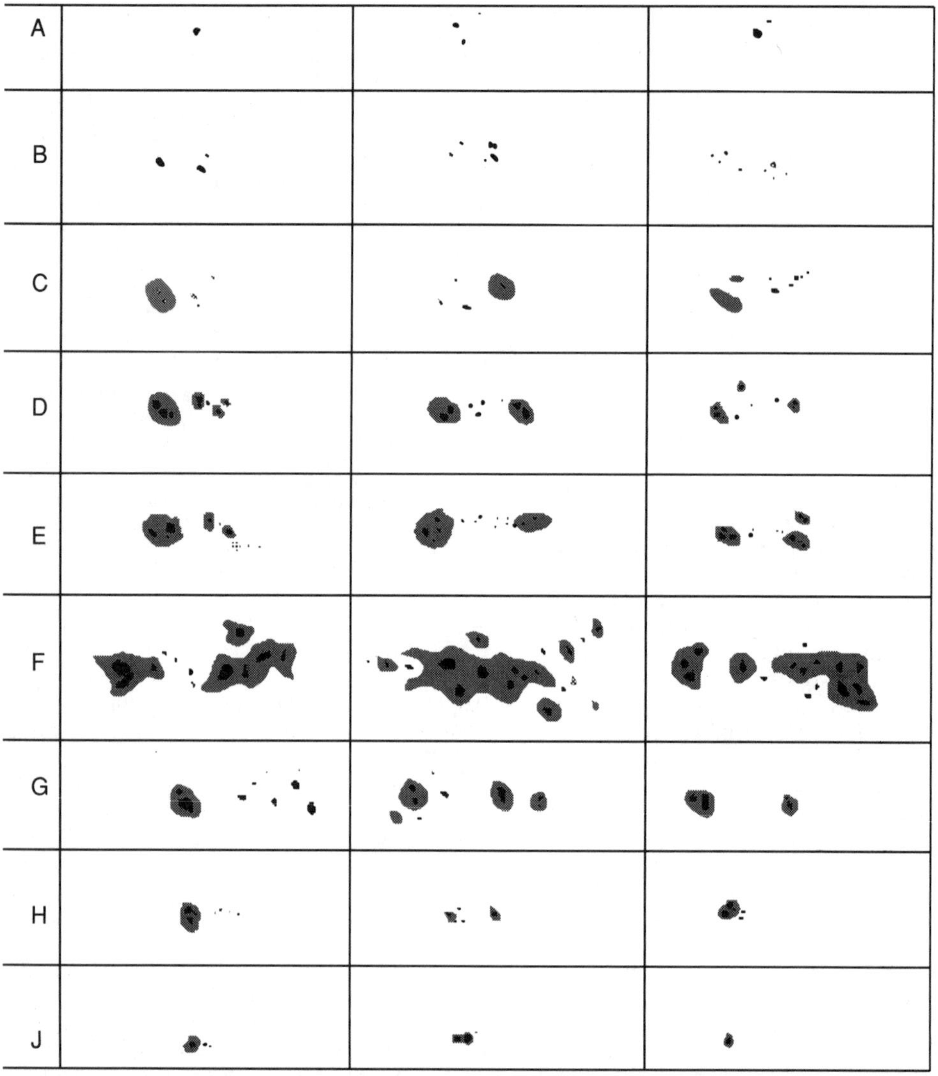

**6.5** Sunspot groups are classified into divisions as shown here. A given group of sunspots may follow the evolutionary path from A to J, or it may skip one or more stages.

plete a rotation, whereas at 40° latitude almost 28 days will pass.

In 1769, Alexander Wilson made the first significant contribution to solar astronomy since the pioneers. While observing a seemingly normal spot with penumbra evenly distributed around the umbra, he noted that as the spot was carried to the limb the umbra appeared displaced. The side nearest the limb displayed a wider cross-section of penumbra than the opposite side. The spot rounded the eastern limb two weeks later, a mirror image of his earlier observations. Wilson concluded that the spots were saucer-shaped depressions in the photosphere, exposing the dark interior of the Sun.

The depression idea was supported by many

astronomers, and evidence that a notch can sometimes be seen when a spot is situated on the limb supplied further proof. In recent times, however, astronomers have come to believe that they are not really true depressions, the notched limb being merely an illusion. As the sunspot material is more transparent than the surrounding photosphere, we are just looking deeper into the Sun.

Over 70 years were to pass after the discovery of the *Wilson effect* before the next exciting development in solar research came. In 1843 a German, Heinrich Schwabe, after many years of constant observing, announced that sunspots appeared to follow a ten-year cycle. Soon afterwards, an Englishman, Richard Carrington, discovered that the spots formed at mid-solar latitudes at the beginning of a cycle, and the average latitude gradually decreased until the end. After researching all available records since Galileo's time, the ten-year cycle was then refined to an average period of 11.1 years by Rudolf Wolf.

With almost 400 years of observations available, it is clear that the eleven-year cycle can be as short as seven years or as long as seventeen. Spots forming at 35° to 40° north and south latitudes mark the beginning of each new cycle, their magnetic polarities opposite to the spots of the preceding cycle. As the cycle progresses, spots form at lesser latitudes. Maxima occur when the spots are within about 15° of the equator, and minima as the spots near 5° in latitude and become less numerous. Spots of the next cycle begin to erupt again at higher latitudes, even before the low latitude ones disappear. The spot migration is not a physical movement downwards by individual spots, but rather the average distribution in latitude of spots over the cycle. The classic butterfly diagram shown in Figure 6.8 illustrates the spot movement over several cycles.

The peak which each cycle reaches varies enormously. The most active on record was the cycle culminating in 1957. The rise to maxima generally takes 3.5 to four years, followed by a gradual decline over seven to 7.5 years. Since the last minimum in 1986, the latest cycle is the most rapidly rising on record, and could surpass the 1957 maximum for activity. Figure 6.9 is the result of Sydney amateur astronomer Colin Martin's patient daily observations using a 60 mm refractor and the projection method. Colin's graph clearly illustrates the decline to minima in 1986 and the subsequent rapid rise in 1988.

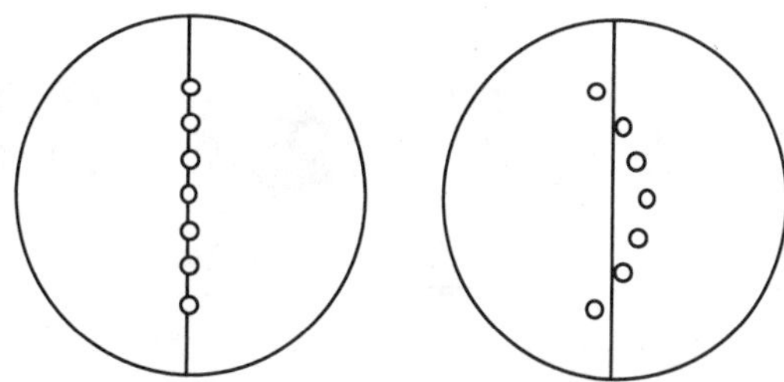

**6.6** Because the Sun does not rotate as a solid body, but more quickly at the equator than the poles, spots seen at high latitudes seem to lag behind those of lower latitudes. This diagram shows a theoretical set of sunspots as they would appear at one time and then one rotation later.

The Sun appears a little darker at the edge than across the rest of the disc. Readily observed using the projection method or direct observation; this phenomenon is termed *limb darkening*. At the centre of the disc, we look into the brighter and denser regions of the photosphere, whereas near the limb we are unable to penetrate to the same depth as our view is blocked by the upper and cooler (less bright) reaches of photospheric material.

Associated with sunspots are extensive, irregularly shaped masses of bright material known as *faculae* or *plagues*: hot hydrogen clouds just above the photosphere. It is difficult to see them over the centre regions of the disc as they are of similar brightness to the photosphere. Nearer the limb, where the contrast is better, they stand out unmistakably. The faculae (Latin for a little torch) were named by Scheiner in the seventeenth century. Like serpentine lattice work, the faculae typically cover areas larger than their attendant sunspot area.

Sunspots at the limb are always attended by faculae, although faculae can occur without sunspots. The birth of sunspots is always heralded by faculae appearing a few hours or days beforehand. They persist long after a spot has disappeared, up to several weeks in very large groups. Faculae also appear near the poles, but these tend to be more compact than others in the spot zones.

Occasionally, a brilliant patch may be visible in the vicinity of an active sunspot group. These

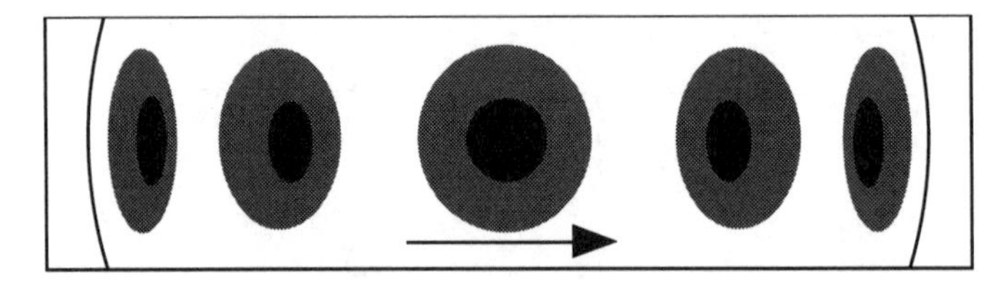

**6.7** Some sunspots, when they appear close to the Sun's limb, appear as saucer-shaped depressions. As the spot approaches the centre of the disc, the effect is lessened.

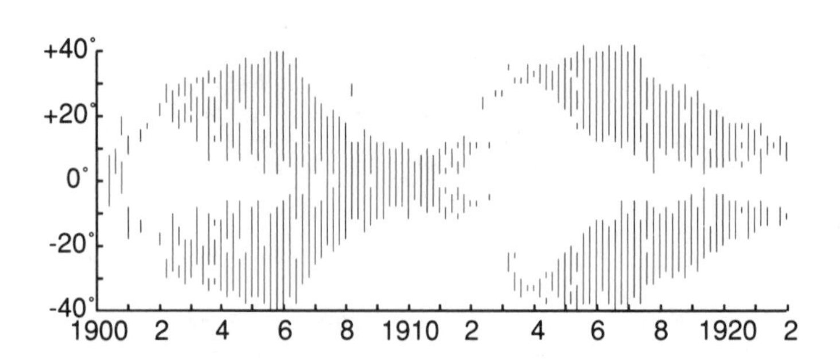


**6.8** Plotting the latitude of sunspots over a 22 year period, as has been done here with data from 1900, gives a distinctively shaped graph, known as a butterfly diagram, showing the eleven-year sunspot cycle.

flares are the result of violent explosive releases of energy in the lower chromosphere (the solar atmosphere immediately above the photosphere) caused when a sunspot's magnetic field is stressed and unstable. In normal white light they are exceptionally rare, but have been observed occasionally by amateurs. The first recorded sighting was made by two English amateurs, Carrington and Hodgson, in 1859. They described the event as a very brilliant star of light which dazzled the eye.

Flares, lasting from a few seconds to an hour or more, can trigger terrestrial auroral displays on Earth, upset the ionosphere and interfere with short-wave radio. It is common for intense flares to repeat themselves in an active area within a few hours or days.

If the seeing is steady, the Sun will display a mottled texture over the entire disc, often described as having the appearance of rice pudding or lemon peel. The granulation appears as small bright spots of irregular shape, often like polygons, bounded by darker material. The average size of each granule is about one or two seconds of arc, equivalent to about 700 to 1500 km across, larger ones may reach five seconds of arc or 3500 km.

The granulation is evidence that the solar surface is in constant motion: we are seeing the tops of rising columns or cells of hot gas from within the photosphere. The granules are short lived, typically from one to ten minutes.

There are many features on the Sun not visible in normal white light; these can be viewed by the amateur with the aid of special narrow-band interference filters. Hydrogen alpha (H$\alpha$) filters open whole new fields of study. Unfortunately H$\alpha$ filters are very expensive, but there are some cheaper wide band width models available which allow good views of detail projected beyond the limb, although at a sacrifice in surface detail.

The narrow-band H$\alpha$ filter allows only a certain wavelength (656.3 nm) of light to pass to the observer, effectively blocking the flood of light at other wavelengths. It is in this region that the spectacular great looped masses and columns of hydrogen gas known as prominences are brightest. When silhouetted against the disc, prominences appear as dark, sinuous lanes called *filaments*. The otherwise rare flares are bright at this wavelength and are commonly seen, as are the faculae. Indeed, the whole disc in monochromatic light shows a wealth of detail hidden to the white light observer.

Eruptive prominences may eject material far into the chromosphere and beyond, but most prominences, although giving the appearance of great, violent eruptions, are just cool clouds of dense material condensing out of the hot chromosphere. This material then descends along magnetic fields to active areas on the photosphere below.

The amateur who is willing to travel can view

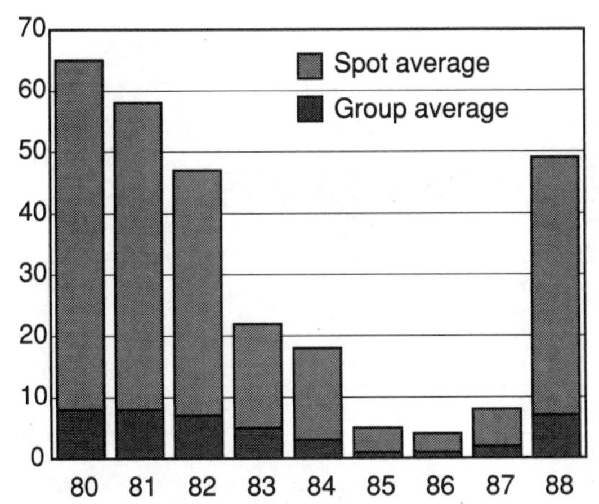


**6.9** Solar activity over a nine-year period from 1980. This graph shows the dramatic decrease in solar activity in 1985 and 1986, and the quick rise in 1988. The data in this graph were prepared from observations made by Sydney amateur astronomer, Colin Martin.

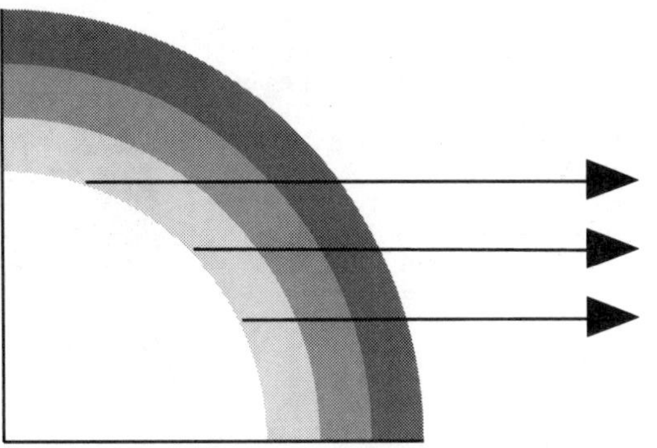

**6.10** When viewing near the limb of the Sun, the light from the photosphere must pass through a greater amount of cooler and darker gases; hence the limb appears darker.

some rare solar features during an eclipse of the Sun. Solar eclipse viewing is addictive, and amateurs chase eclipses throughout the world just to experience the awe and wonder of the event, occasionally in dangerous and almost inaccessible places. What draws these people throughout the world together for a few precious minutes of celestial splendour? Some come just for the adventure and aesthetic pleasure, others to collect useful scientific data, but no matter what the purpose, all are touched by an emotional awe at one of nature's most spectacular displays.

The Earth's Moon is unique: no other planet can lay claim to a satellite that so closely matches the angular size of the Sun in the sky. This relationship is purely coincidental—the Sun's diameter is about 400 times the diameter of the Moon—but, since the Sun is about 400 times further away their diameters match nicely.

The elliptical orbits of both the Moon and Earth determine the type and duration of each eclipse. From the Earth, the Sun's diameter varies from 31′31″ to 32′35″. The Moon's diameter ranges from 29′22″ to 33′31″. It is easy to see that when the Moon is close to the Earth it will completely blanket the solar disc, causing a total eclipse. At its most distant point, the solar disc will not be completely hidden, and a bright ring of sunlight will be visible around the edge of the Moon; this is known as an *annular eclipse*.

As the Moon orbits the Earth in about 28 days, you would expect to see an eclipse of the Sun each month as the Moon passed between the Earth and the Sun. It follows that an eclipse of the Moon should also occur two weeks later when the Earth happened to lie between the Sun and the Moon. Unfortunately this does not happen since the lunar orbit is tilted by just over 5° to the ecliptic and the orbit itself rotates, resulting from gravitational influences of the Sun, taking 18.61 years to complete a circuit. For an eclipse to happen, the Moon must cross the plain of the Earth's orbit (two points known as the *nodes*) at new moon for a solar eclipse, and at full moon for a lunar eclipse.

Since most people have at some time seen a lunar eclipse, it is generally thought that they are more common than solar eclipses. The reverse is the case: more people see lunar eclipses because they are visible from the entire night side of the Earth, whereas total solar eclipses are only visible from a very narrow track across the globe. On average about 2.4 solar eclipses happen each year, and these will be almost equally divided into total, partial and annular. Lunar eclipses, on the other hand, only occur on average about 1.5 times each year.

The umbral shadow cast by the Moon during a solar eclipse just reaches the Earth and at most is only about 280 km wide. The closer the Moon to the Earth, the wider the shadow will

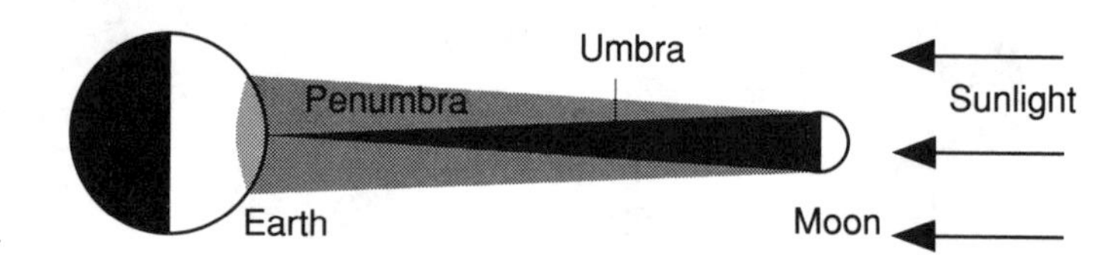


**6.11** The arrangement of the Earth, Sun and Moon shown here only results in a solar eclipse because both the Moon and Sun appear to be the same size when viewed from Earth. If the Moon were a little further from the Earth, there would be no solar eclipses as its disc would be too small to completely cover the Sun.

**Table 6.2 Solar eclipses**

| Date | Type | Duration or magnitude | Location |
|---|---|---|---|
| 22 July 1990 | Total | 153 s | Scandinavia, USSR, Nth Pacific |
| 15 January 1991 | Annular | 475 s | Australia, New Zealand, Sth Pacific |
| 11 July 1991 | Total | 414 s | Nth Pacific, Central America |
| 4 January 1992 | Annular | 702 s | Central Pacific |
| 30 June 1992 | Total | Very short | Sth Atlantic |
| 24 December 1992 | Partial | 84% | Arctic |
| 21 May 1993 | Partial | 74% | Arctic |
| 13 November 1993 | Partial | 93% | Antarctic |
| 10 May 1994 | Annular | 374 s | East Pacific, Nth America, Atlantic |
| 3 November 1994 | Total | 263 s | Nth part Sth America, Atlantic |
| 29 April 1995 | Annular | 398 s | Sth Pacific, Nth part Sth America |
| 24 October 1995 | Total | 125 s | Middle East, Nth Pacific |
| 17 April 1996 | Partial | 88% | Antarctic |
| 12 October 1996 | Partial | 76% | Arctic |
| 9 March 1997 | Total | 170 s | USSR, Arctic |
| 2 September 1997 | Partial | 90% | Antarctic |
| 26 February 1998 | Total | 236 s | Pacific, Central America, Nth Atlantic |
| 22 August 1998 | Annular | 194 s | Indian Ocean, Sth Pacific |
| 16 February 1999 | Annular | 143 s | Indian Ocean, Australia, Sth Pacific |
| 11 August 1999 | Annular | 143 s | Nth Atlantic, Europe, India |

be, ensuring an eclipse of long duration, up to 7.5 minutes.

For a total eclipse to happen, the Moon's umbral shadow must reach the Earth. Observers outside this area will only see the event as partial. As an observer moves further from the totality zone, a lesser percentage of the solar disc is obscured by the Moon. Once outside the penumbral shadow, no eclipse is visible.

During an annular eclipse, the umbral shadow does not quite reach the Earth. On rare occasions, it is possible to have a combined total and annular eclipse simultaneously. As the annular eclipse progresses, with the apex of the shadow cone above the Earth's surface, a brief moment of totality will happen if the cone grazes a mountain range, or encounters higher land due to the Earth's curvature. A return to the annular phase again happens as the umbra again comes to a focus in the Earth's atmosphere as the land drops away. This type of eclipse, known as a *central eclipse* is shown in Figure 6.12.

There is much useful work an amateur can do during a total solar eclipse, but if it is your first, it is recommended that you enjoy the spectacle to the utmost and not commit yourself to the rigours of serious observation. Because so much happens during totality in such a short timespan, early preparation is a must. If you intend to photograph the event, spend time perfecting techniques and become thoroughly familiar with your equipment. Where possible, take a backup for any piece of equipment which may fail.

Make a habit in the months before an eclipse of photographing the Sun. Use a stop-watch to ensure you can take as many photographs as possible within the allocated time. The Sun should be at a similar altitude to that at which it

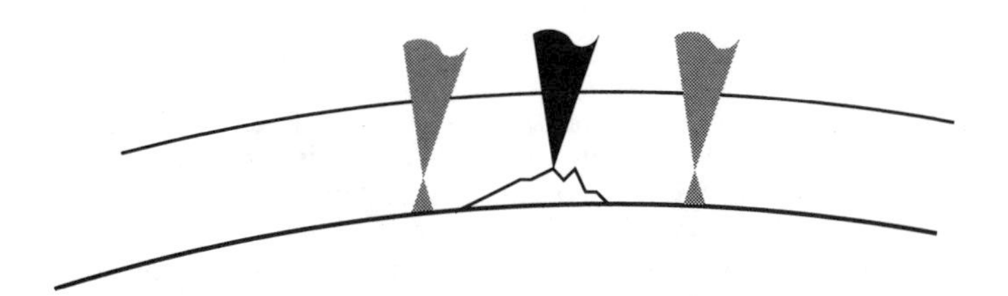

**6.12** In a central eclipse the focus of the Moon's shadow cone briefly touches a high point on the Earth's surface giving a total eclipse at that location. At other places the cone comes to a point in the Earth's atmosphere making an annular eclipse.

will be on the day, so that if anything does go wrong you will be familiar with the positioning of your equipment, enabling you to change filters, film or lenses automatically. While your adrenalin is running high, you must be able to control any panic and swiftly adapt or change plans midstream.

As an eclipse progresses, the partial phases may be viewed only with the same protection needed for normal solar viewing. Near to totality, with the Moon covering most of the Sun's disc, the surrounding landscape will start to turn an eerie grey, with shadows disappearing. The temperature will now begin to drop and a breeze may develop. Thousands of tiny crescents will be visible on the ground beneath the trees, small gaps in the foliage acting in the same manner as a pin-hole camera.

In the last few seconds before totality, the Sun's limb breaks into a series of tiny beads. These are the *Baily's beads*, they are formed as the Sun shines between mountain peaks on the Moon. As the last of the beads vanish, the Sun's faint, tenuous outer atmosphere, the corona, appears, surrounding the Moon's black silhouette. The pearly white corona varies from one eclipse to another; at solar maximum it displays a symmetrical petalled structure, much like a flower. At minimum, long equatorial streamers with small polar bristles are seen. The elusive prominences are now easily seen, projecting out from the limb in all their red and pink splendour.

All too soon the Moon has moved enough to allow a bright bead to be visible in a deep valley, then suddenly, in a burst of glory, a bright flash, known as the diamond ring appears as the photosphere streaks around the western limb. The uneventful, anticlimactic partial phases again take over, and animals and birds that a few minutes earlier were confused by a false night return.

During totality the eclipse may be viewed in complete safety with the unaided eye, telescope or binoculars. Since the corona is visible out to one or two solar diameters, the field of view in telescopes or telephoto lenses should cover at least 5°. There is nothing worse than a photograph which misses the fine outer detail—it's like taking a photograph of a friend and cropping their head.

The most important contribution an amateur can make at solar eclipses is observation of the Baily's beads. By noting the exact time each of the beads forms and disappears, valuable information will be gathered that will help in obtaining very precise positions of the Sun and Moon, and aid in the determination of the solar oblateness, or polar flattening. The equipment needed for these experiments is simple: any small telescope, a short-wave radio tuned into a station transmitting time signals, and a portable tape recorder. As each bead is noted, the observer makes a running commentary, with the time signals simultaneously being recorded.

Astronomy clubs regularly organise eclipse expeditions, as do commercial tour operators. Preparation for these tours is planned years in advance. Weather prospects at stations along the path are known, as are the local accommodation facilities and amenities (if any). It is advisable to contact these groups for advice, and join in on a party of experts or eclipse regulars to maximise those precious few minutes. Table 6.2 gives the forthcoming eclipses until the turn of the century. The 1991 eclipse will be the longest total eclipse until the year 2132.

A final word: the Sun will provide countless hours of pleasure for the amateur, but it can be dangerous. Treat it with care and respect and never let your guard down.

## *Nuclear fusion*

Nuclear fusion is the process whereby a small amount of mass from atomic particles is destroyed and turned into energy. It was Albert Einstein who first alerted the world to the possibility of turning matter into energy with his special theory of relativity. The nuclear fusion used by stars such as the Sun involves four hydrogen nuclei being forced together to form a helium nucleus. It is a three-stage process.

In the first stage, two protons come together, and one of the protons is converted into a neutron, giving rise to two small particles, a positron and a neutrino. The particle formed from one proton and the new neutron is called a deuteron. With an electron a deuteron is deuterium, a form of naturally occurring hydrogen. (A)

In the next stage the deuteron is struck by another proton, the two combine creating the nucleus of a form of helium, $^{3}$He. At the same time some energy is released in the form of a gamma ray. (B)

In the final stage, two $^{3}$He nuclei come together to form a nucleus of normal helium, $^{4}$He, and two protons are released along with more energy in the form of another gamma ray. (C)

The total amount of energy released by the process is just $4.5 \times 10^{-12}$ joules and the mass lost $5.0 \times 10^{-26}$ grams, but in the Sun these reactions occur at the rate of $8 \times 10^{37}$ per second so four million tonnes of mass is destroyed and $3.5 \times 10^{26}$ watts of power is made.

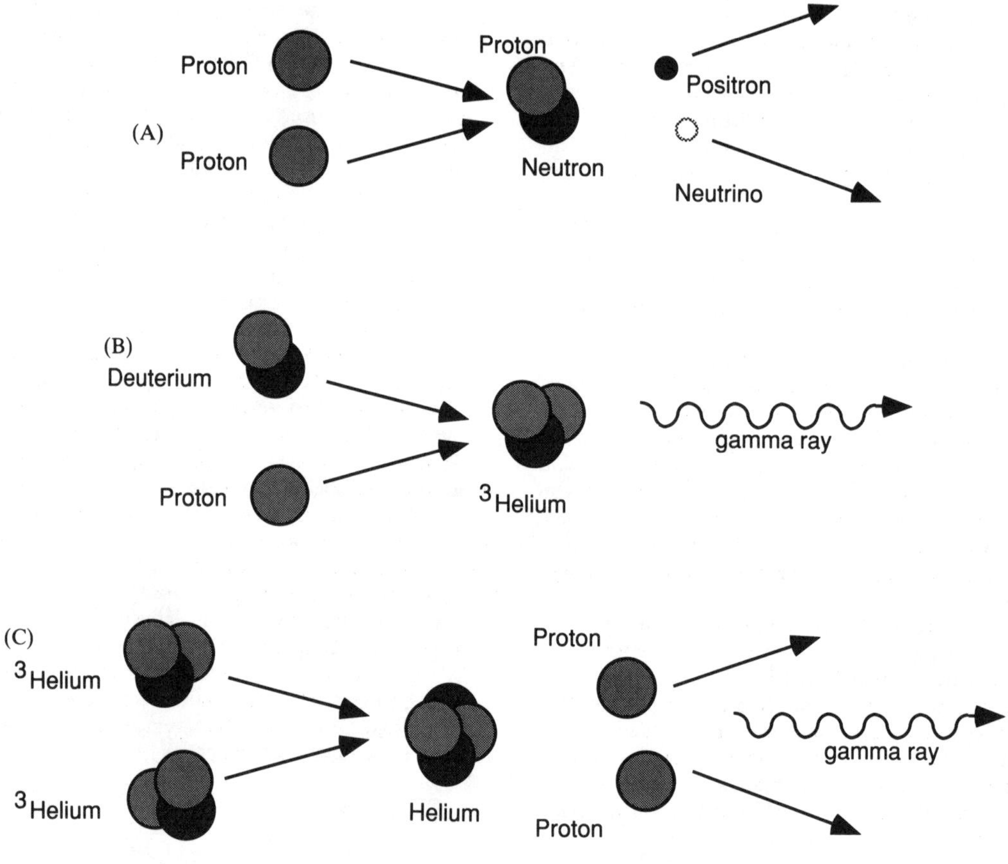

(A)
Proton
Proton
Proton
Neutron
Positron
Neutrino

(B)
Deuterium
Proton
$^{3}$Helium
gamma ray

(C)
$^{3}$Helium
$^{3}$Helium
Helium
Proton
Proton
gamma ray

# 7

# Mercury

Mercury was the fleet-footed messenger of the gods, so it is an apt name for the closest planet to our Sun as it sweeps through its orbit at nearly 50 km per second, faster than any other planet. As the closest planet to the Sun, Mercury is an elusive object to see. Of all the five planets known since antiquity it is perhaps the least studied by amateur astronomers. The reason is simple: because it can be as close as 46 million kilometres to the Sun, it never appears far from it in the sky, so it is never seen in really dark skies. Scientists too have seen very little of Mercury because, although it is much closer to Earth than any of the gas giant planets, it has been viewed by only one spacecraft, the United States' Mariner 10 which passed by the planet three times in 1974 and 1975.

The best time to search for Mercury is during the time of greatest *elongation*. An elongation is the angular distance of an inner planet east or west of the Sun, and it follows that Mercury's greatest eastern or western elongation is the time when it is situated at its maximum separation east or west of the Sun. Aspects of an inner planet's orbit are shown in Figure 7.1.

The chance of a good view of Mercury will depend on a combination of two things: the degree of elongation and where you reside on Earth. In tropical latitudes, at maximum elongation, Mercury can be viewed against a dark sky for a short period. In temperate latitudes where twilight is considerably longer and the ecliptic is tilted to the horizon, you are faced with hunting down a planet low in altitude against a bright sky.

Table 7.1 gives Mercury's greatest elongations from 1990 to 2000. It is unnecessary to observe the planet on the exact day of greatest elongation, but you must keep in mind the rapid motion of the planet in the sky, and observe within five or six days either side of the elongation. At eastern elongation the planet will be to the right of the Sun in the southern hemisphere (left in the northern hemisphere), and be visible in the evening twilight. At western elongation it will be to the left of the Sun in the southern hemisphere (right in the northern hemisphere), and rise before the Sun in morning twilight.

Armed with this information, you are ready to track the planet down. As it may be difficult to pick Mercury out in the light sky with the unaided eye, binoculars may be used, but a word of warning: never use binoculars in your search if the Sun is above the horizon. The normal practice is to carefully 'sweep' the sky, and as binoculars have a large field of view there is a very real possibility that you may chance upon the Sun, causing irreparable eye damage.

Once found, you may be surprised at how bright Mercury really is; you may even wonder why you had any difficulty in finding it in the first place. At greatest elongation the planet can match the brightest stars. Only Sirius (α Canis Majoris) and Canopus (α Carinae) exceed its brilliance. Even at an elongation that happens when the planet is at aphelion, its furthest distance from the Sun, it ranks about tenth against the scale of brightest stars. As Mercury reaches aphelion, southern hemisphere observers obtain the best view, as the planet is south of the ecliptic at this time.

If you locate Mercury in a semi-dark sky, it will naturally be close to the horizon, at a point where you are looking through the maximum atmosphere, and where conditions are likely to be poor, resulting in a distorted, boiling image. Observations made while Mercury is at a greater height (with the Sun above the horizon) will often aid image quality, but the planet will be more difficult, and more dangerous to find.

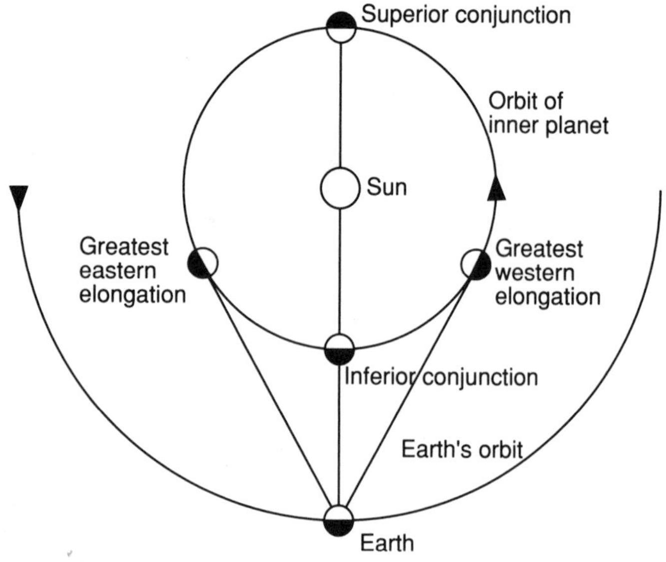

**7.1** As a planet closer to the Sun than the Earth travels, there are a number of significant places in its orbit. At inferior conjunction it is possible for the planet to pass across the face of the Sun. At greatest elongation east and west the planet is furthest from the Sun in the sky and hence easiest to observe. For Mercury the greatest elongation is an angle of 28°, for Venus it is 47°.

To view Mercury during daylight hours it is best to use an equatorial telescope, correctly polar aligned, with setting circles. Choose a date within four or five days of an elongation and consult an almanac to find the right ascension and declination of both Sun and Mercury for that date. Then, using either the projection method or a full-aperture filter, centre the Sun, then set the circles to read the Sun's position. Now swing the telescope so the circles read Mercury's position and remove the projection screen or filter, taking care not to jar the instrument. Mercury will be visible within a low-power eyepiece field or the finder scope; if not, the polar axis alignment may need some adjustment. Never sweep the sky if Mercury is not visible at your first try. Replace the solar projection device or full-aperture filter, go back to the Sun and start again.

Mercury's small disc may seem a disappointing target. On the opposite side of the Sun at superior conjunction, it is only five arc seconds in diameter. At inferior conjunction the disc reaches almost 13 arc seconds and, when on either side of the Sun near greatest elongation, you will be working with an image around seven arc seconds in diameter; a magnification of at least 60x will be needed to see the disc. Because the atmosphere is constantly working against us, surface features are virtually impossible to see.

**Table 7.1 Elongations of Mercury**

| Eastern elongation | | Western elongation | |
|---|---|---|---|
| | | 1 February 1990 | 25° |
| 13 April 1990 | 20° | 31 May 1990 | 25° |
| 11 August 1990 | 27° | 24 September 1990 | 18° |
| 6 December 1990 | 21° | 14 January 1991 | 24° |
| 27 March 1991 | 19° | 12 May 1991 | 26° |
| 25 July 1991 | 27° | 7 September 1991 | 18° |
| 19 November 1991 | 22° | 27 December 1991 | 22° |
| 9 March 1992 | 18° | 23 April 1992 | 27° |
| 6 July 1992 | 26° | 21 August 1992 | 18° |
| 31 October 1992 | 24° | 9 December 1992 | 21° |
| 21 February 1993 | 18° | 5 April 1993 | 28° |
| 17 June 1993 | 25° | 4 August 1993 | 19° |
| 14 October 1993 | 25° | 22 November 1993 | 20° |
| 4 February 1994 | 18° | 19 March 1994 | 28° |
| 30 May 1994 | 23° | 17 July 1994 | 21° |
| 26 September 1994 | 26° | 6 November 1994 | 19° |
| 19 January 1995 | 19° | 1 March 1995 | 27° |
| 4 May 1995 | 22° | 29 June 1995 | 22° |
| 9 September 1995 | 27° | 20 October 1995 | 18° |
| 2 January 1996 | 20° | 11 February 1996 | 26° |
| 23 April 1996 | 20° | 10 June 1996 | 24° |
| 21 August 1996 | 27° | 3 October 1996 | 18° |
| 15 December 1996 | 20° | 24 January 1997 | 24° |
| 6 April 1997 | 19° | 22 May 1997 | 25° |
| 4 August 1997 | 27° | 16 September 1997 | 18° |
| 28 November 1997 | 22° | 6 January 1998 | 23° |
| 20 March 1998 | 18° | 4 May 1998 | 27° |
| 17 July 1998 | 27° | 31 August 1998 | 18° |
| 11 November 1998 | 23° | 20 December 1998 | 22° |
| 3 March 1999 | 18° | 16 April 1999 | 28° |
| 28 June 1999 | 26° | 14 August 1999 | 19° |
| 24 October 1999 | 24° | 2 December 1999 | 20° |
| 15 February 2000 | 18° | 28 March 2000 | 28° |
| 9 June 2000 | 24° | 27 July 2000 | 20° |
| 6 October 2000 | 26° | 15 November 2000 | 19° |

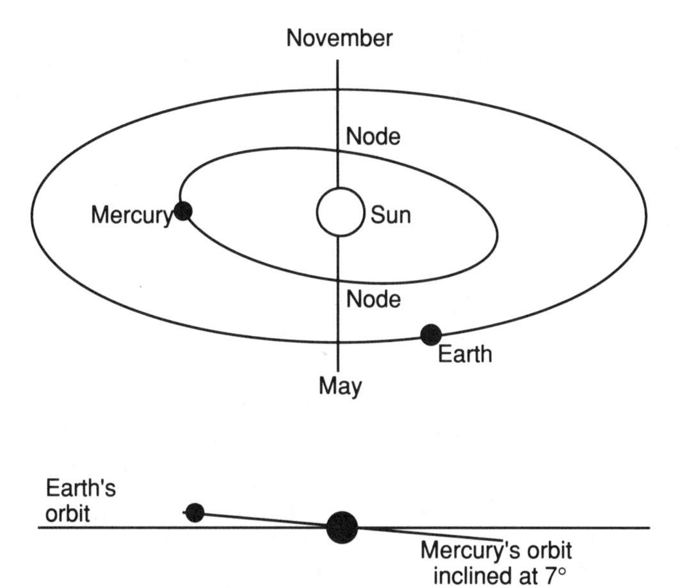


**7.2** Transits of Mercury across the face of the Sun are only possible when inferior conjunction occurs when Mercury is at one of the nodes of its orbit. Mercury crosses the nodes of its orbit only in May and November, so transits are infrequent.

In large telescopes, dark areas may be observed; however, the amateur using a small instrument will have to be satisfied with the achievement of capturing the planet in the eyepiece and viewing the phases.

One thing that will be apparent is that Mercury undergoes phases like the Moon. When at inferior conjunction it will be seen in the full phase. This rapidly changes from gibbous to first quarter as elongation is approached, and lastly to a thin crescent near superior conjunction.

Because of the apparent closeness of the Sun we usually cannot see Mercury near superior conjunction, except perhaps during a total solar eclipse. At inferior conjunction it will also be invisible, as the night side is facing us, although occasionally we will see the tiny disc silhouetted against the Sun.

The passage of Mercury across the Sun at inferior conjunction is known as a *transit*. Because of Mercury's 7° orbital tilt to the ecliptic, transits are not very frequent, as the planet usually passes above or below the Sun. Mercury will be seen to transit the Sun's disc only about thirteen times each century, at intervals of about three, seven, ten or thirteen years between successive transits. Like lunar and solar eclipses, transits can only happen when Mercury crosses the ecliptic, or the nodes of its orbit, at inferior conjunction, as illustrated in Figure 7.2.

The last transit happened on 12 November 1986 and the coming events into the twenty-first century are as follows: 6 November 1993, 15 November 1999, 7 May 2003, 8 November 2006 and 9 May 2016.

The duration of a transit may last from a few minutes if Mercury just skims the solar disc to almost nine hours across the Sun's equatorial regions. The progress of a transit may be monitored using normal solar observing techniques: the small disc will appear much darker than any nearby sunspots. Unlike solar eclipses, Mercurian transits can be viewed from anywhere on the sunlit side of the Earth, unless the planet just grazes the solar limb, in which case it may only be seen from polar regions.

In the nineteenth century, many astronomers studied Mercury in the hope of finding out more about the planet. At that time a number of maps were drawn showing indistinct markings on the surface. Based on these markings, astronomers believed that Mercury was another case of synchronised rotation; that the planet kept one face turned permanently towards the Sun. In the mid-1960s astronomers used a new technique to measure the distance to, and rotation rates of, the planets. By bouncing radar signals off a planet, not only could the distance to the planet be calculated to a high degree of accuracy, but frequency change in the returning signal could be analysed and the rotation rate of the planet

**7.3** A technique developed in the 1960s uses a high-powered radio transmitter to send pulses of radio waves to a planet and then watch for the return of the signal. When the signal returns two times can be measured. From the time taken for the signal to reach the planet and return, a very accurate distance to the planet can be calculated. From the width of the returned pulse, caused by the Doppler effect, the speed of the planet's rotation can be determined.

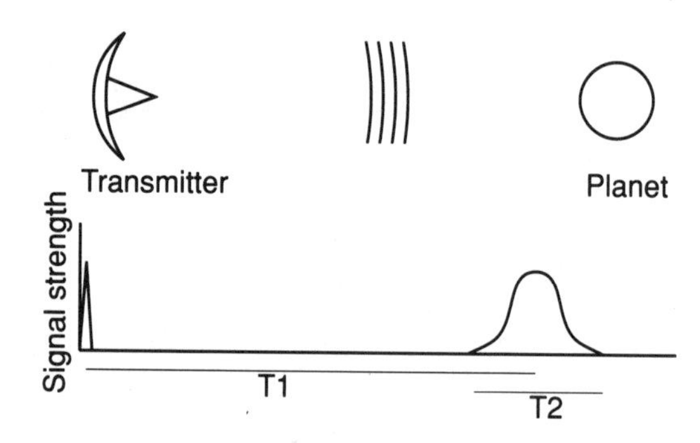


calculated. When this technique was applied to Mercury, the results did not match expectations. Instead of an 88-day period, matching the rotation, radar measurements gave a period of 59 days.

Further investigation found that the period was in fact 58.65 days, exactly two-thirds of the time taken for the planet to orbit the Sun once. It is confusing when trying to describe days and years on other planets. On Earth the day is the time between successive noons, 24 hours, and the year is the time taken to orbit the Sun once, 365.25 days. Another measurement of a day might involve the time taken for the Earth to rotate once about its axis with respect to the stars—the sidereal day. On Earth this is 23h56m04s.

On Mercury the days and years are confusing. The sidereal day is the period measured by the radar, 58 and two-thirds days, the year is 88 days, but the solar day is 176 Earth days, or exactly two Mercurian years. Why? Imagine you are an astronaut standing on the planet at sunrise, with your watch running to Earth time. Over the next 88 Earth days the planet will spin on its axis one and a half times, but because you are orbiting the Sun in the same direction as the rotation, this motion cancels and the planet makes only one half a turn with respect to the Sun, so after eighty-eight days standing around, the Sun is just setting. A further one and a half revolutions with respect to the stars (88 days) and the Sun is once again rising—176 days since the last time.

Unlike the Earth, where only the changing latitudes affect the climate, the longitude on Mercury also plays a part because of the strange rotation and orbit. Next to Pluto, Mercury has the most eccentric orbit in the solar system, an orbit which takes it from 45.9 million km from the Sun to 69.7 million km—an eccentricity of 0.2! If you live at one of the longitudes which has the Sun directly overhead at perihelion (there are two), then at sunrise and sunset, half an orbit away, you will be at aphelion. This will mean that the Sun will appear to move faster when rising and setting and remain almost stationary while overhead. On the other hand, if you lived 90° away, then Mercury would be at perihelion at sunrise and sunset and aphelion at noon. In this case the sun would rise slowly, warming the surface, then at noon when its heat was most direct it would be moving most swiftly. At these places the temperature range would be smaller than at the longitudes 90° away.

We have seen with our Moon that a synchronous orbit is caused by the gravitational pull of the larger body acting on mass concentrations in the smaller, eventually slowing the rotation until it is synchronised, so what about Mercury? Is the rotational period caused by some similar effect, or is it fortuitous? Of course not. One of the tenets of astronomy is that you are never lucky. In particular this applies to planetary astronomy. Consider this, we have observed Neptune in detail for a few weeks, the length of time that Voyager 2 was near the planet, yet data gained during that period is being used to determine the history of the planet and explanations of what is going on. Anything Voyager 2 observed at Neptune must be explained in terms which explain the phenomena over a long period of time. Resorting to explanations such as 'Well it happens very rarely, we were there at just the right time . . .' are not acceptable; it may be true, but it is much more likely that the spacecraft arrived at the planet on an unremarkable Neptunian day.

The same goes for Mercury. The chances that we are seeing it at a special time when its orbital

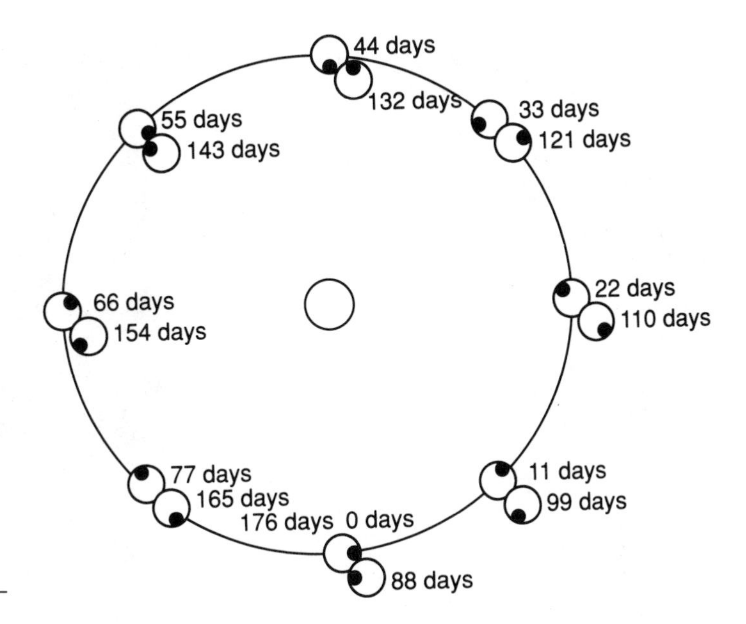


**7.4** Mercury's orbit around the Sun is a complicated affair as a combination of synchronous rotation and the high eccentricity of the planet's orbit makes its day twice as long as its year. Here we see two orbits of Mercury around the Sun, the black spot marking a fixed point on the planet's surface. In the time it takes for Mercury to orbit the Sun once, the planet will turn on its axis one and a half times, so that after 88 days the planet has completed only one-half of a revolution with respect to the Sun.

and rotational periods are coincidentally related as they are is small. A theory must be found to explain what we see. One exists, a variation on the normal synchronised period theory. Like the other planets of the solar system, Mercury began its life with a fairly rapid rotational period. Tides caused by the Sun would act on the planet, tending to slow its rotation significantly; at the same time, the eccentricity of its orbit carried it closer to and further from the Sun.

As Kepler's second law tells us, the speed of the planet when it is near perihelion is much faster than when at aphelion. Because of their nature, tidal forces are very dependent upon distance, much more so than most other forces. It was at perihelion, when the tidal forces were greatest, that the orbits became synchronised. This arrangement, with the planet returning to the same orientation every second orbit, and remaining synchronised at perihelion, gives a minimum energy configuration and so it is stable; there is no need for the planet to slow further to a fully synchronised orbit. Indeed something would need to give the planet extra energy in order to slow it further.

The photographs which Mariner 10 sent back from Mercury show a planet which could quite easily be mistaken for the Moon. It is covered with craters similar to those on the Moon, but there are none of the large lava-flooded plains like the maria on the Moon, except for one small area. This is, of course, not an exhaustive study of the surface of Mercury. Mariner 10 was able to photograph only 45 per cent of the planet's surface; 55 per cent is yet to be seen. Because of the orientation of the lunar maria, it is also possible to photograph 45 per cent of our own Moon's surface and not see any maria either, so what follows must be taken in the light of incomplete knowledge.

Many of the theories about Mercury are based upon knowledge of our Moon. The albedo of Mercury and the colour of its surface match closely with the colour and reflectivity of the lunar surface. The surface must consist of igneous silicate rock, but without samples it is impossible to tell whether the rocks are chemically similar to the silicates of the Moon. Also, without samples it is impossible to measure the age of the Mercurian surface, except by noting that the huge number of craters of all sizes means that it must be fairly old.

When it was thought to have synchronous rotation, Mercury was said to be both the hottest and coldest place in the solar system; it is in fact neither, though the temperatures there are quite extreme. In the latitudes near the equator which have noon at perihelion the temperature can reach 402°C during the daytime, while during the 88-day night the temperature will drop to −183°C.

The densities of craters on Mercury are

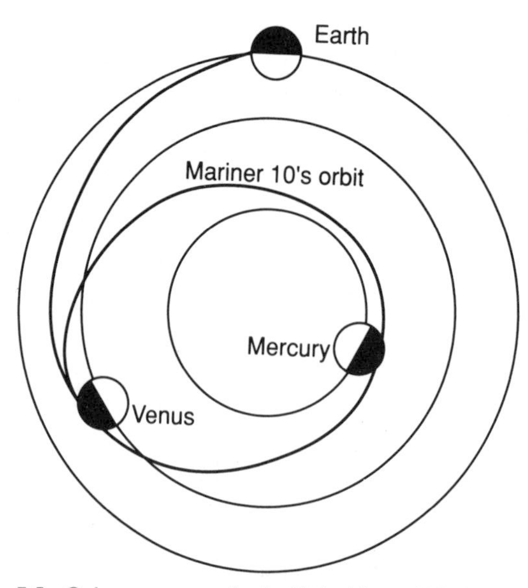


**7.5** Only one spacecraft, the United States' Mariner 10, has visited Mercury. The craft was launched on 3 November 1973 and flew past Venus on 5 February 1974 before reaching Mercury on 29 March of that year. Rather than going into orbit around Mercury, Mariner 10 continued orbiting the Sun, passing by Mercury twice more on 21 September 1974 and 16 March 1975. Because these passes were all 176 days apart and Mercury has a rotation period exactly one-third this long, Mariner saw exactly the same part of Mercury illuminated each time, so we have photographs of only one-half of the Mercurian surface.

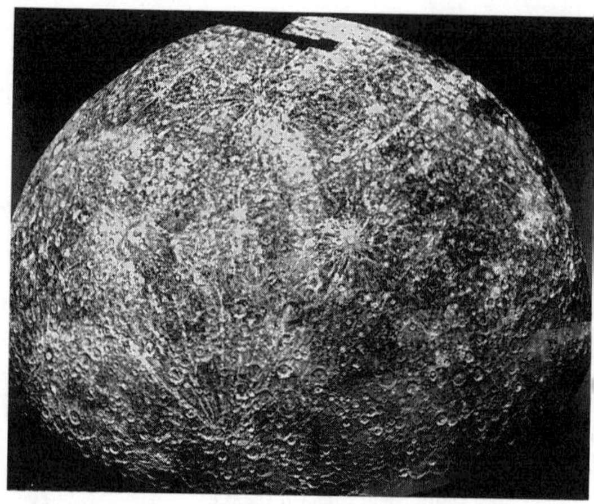

**7.6** This photograph showing one quarter of Mercury's surface is a compilation of photographs taken by Mariner 10 during its passes by the planet in 1974 and 1975. The surface of the planet is heavily marked by craters indicating a very old surface and a planet devoid of any atmosphere to shield it from impacts. PHOTO CREDIT: JPL/NASA

approximately the same as those on the Moon, from a thousand 10 km craters per million square kilometres down to around 100 craters in the same area. The sparsely cratered regions of Mercury are unlike the maria of the Moon; they are not lava plains, but simply areas with fewer craters. They do, however, tell us that Mercury must have had some sort of volcanic activity in the past to erase the older craters. Other evidence of past volcanic activity comes from the ridges, or *scarps*, which are found on the planet. These are the result of the interior of the planet shrinking slightly and the crust collapsing on top. This shrinking and collapsing may be evidence of the tidal slowing of Mercury to its present almost synchronous state.

The density of Mercury is surprisingly high and may account for the differences seen between it and the Moon. The density of the planet is 5.42 $g.cm^{-3}$, second only to the Earth, but when the fact that Mercury is smaller than Earth is taken into account and that the interior rocks are therefore less compressed, the density of Mercury is higher than that of the Earth, or any other planet. It is from a planet's density that we can tell something about what goes on inside the planet. In the case of the Moon, its corrected density is only 3.3 $g.cm^{-3}$ little more than the rocks found on the surface. This tells us that the interior of the Moon is made of similar materials to its surface.

The much higher density of Mercury tells us that what is below the surface is much denser than the lunar-type rocks we see. For the density required, Mercury must be made mostly of metal. Indeed 60 per cent of the planet must be metal. This means that Mercury can be thought of as a large ball-bearing 3500 km in diameter covered with a 700 km layer of rock and dirt. The discovery of a magnetic field around the planet supports such a theory. Mercury's magnetic field is only 1 per cent as strong as the Earth's but its existence shows that the planet must have a small liquid core.

Mercury, because of its low mass and its proximity to the Sun, has no atmosphere to speak

**7.7** This photograph shows the remaining portion of Mercury's surface which has so far been seen. Despite the proximity of the planet to the Earth, it has been visited by only one spacecraft, albeit on three occasions, so we have photographs of only 45 per cent of the planet's surface. The remaining 55 per cent is still a mystery. PHOTO CREDIT: JPL/NASA

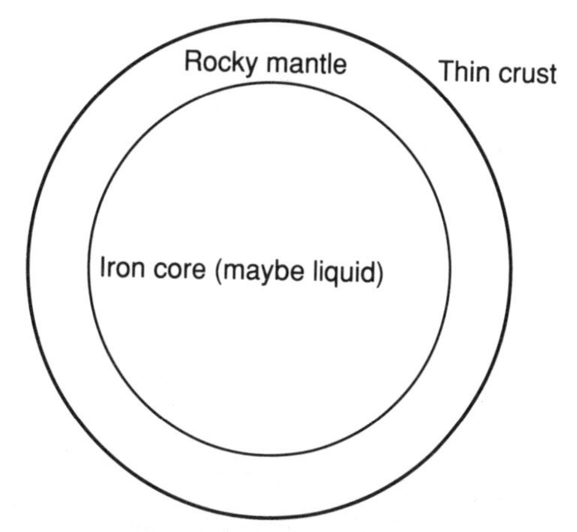


**7.8** Because of its high density (5.5 g.cm$^{-3}$) Mercury must have a large nickel-iron core extending to within 700 km of the surface. Above the core, a layer of silicate materials makes up the mantle and crust of the planet.

of. But there are some gases there, not from the planet, but from the Sun. As the Sun blows out the solar wind, a small amount is slowed by the gravity of Mercury, making a transient hydrogen and helium atmosphere around the planet. Spectroscopic measurements have also found slight traces of sodium and other elements in the tenuous envelope of gases, elements which must have escaped from the rocks of the surface.

There's little the amateur can do to further our knowledge of Mercury, observations are mainly carried out for the sheer joy of locating and viewing this small world.

# 8

# Venus

The Earth's closest neighbour in its journey around the Sun is Venus. At its closest, that planet is just 40 million km away, yet despite its proximity many factors conspired to keep the planet a mystery until very recently. Apart from being the closest planet to Earth, Venus is also the brightest object in the sky, after the Sun and Moon. Venus is at its brightest when high in the morning or evening sky and this has led to its other name, the morning or evening star.

The reason for both Venus' brightness in the sky and for the lack of knowledge about the planet is the same: Venus is covered with a thick blanket of clouds which reflects 75 per cent of the light reaching it. Unlike the Earth, which is generally only 50 per cent covered by clouds, Venus is always totally shrouded; the clouds have never parted enough to allow a view of the surface in all the years it has been observed. Through the telescope Venus presents an unchanging view, a white planet with no visible markings at all.

When comparing Venus and the Earth, it is hard not to consider them as twin planets. They are alike in size and composition, both have a thick atmosphere and clouds, and at the top of the clouds the temperatures are close to those of Earth. These similarities led scientists in the 1930s to deduce that the clouds of Venus were made of water like that on Earth, and that the carbon dioxide which had been detected spectroscopically was a minor component of an atmosphere similar to Earth's. It was even conjectured that, below the clouds, oceans of water could exist and that conditions not too harsh for life might be found.

This idyllic picture changed in the 1950s when radio telescopes were first turned towards the planet. Unlike light, radio waves are not stopped by clouds, so using radio frequencies it is possible to determine much about a planet's surface, as we will see later. By measuring the energy given out by Venus at a number of different frequencies and finding the frequency which has the maximum amount of energy, an exact measurement of temperature is possible. For an object like the Sun, the temperature is so high that the energy maximum is found not in the radio range, but in visible light. In the case of the Earth, its peak emission is at 10 μm giving a temperature of 17°C, but on Venus the measurements showed a temperature of 327°C.

Scientists had expected Venus to be warmer than Earth—it is after all closer to the Sun—but had counted on the higher reflectivity of the clouds to limit this extra warmth to around 15°C. This 1958 result was so unexpected that it was at first dismissed and alternative explanations were sought. Could a layer in the Venusian atmosphere be causing an anomaly? The solution to this problem was one of the primary objectives for the first Venus space probe, the United States' Mariner 2. When the results from that spacecraft were received they showed that the radio radiation really was from the surface and that the planet was indeed extremely hot.

The discovery of the temperature of Venus' surface banished for ever the ideas of a lush tropical rainforest teaming with life. Instead it is an inferno with daily temperatures hot enough to melt tin and other metals. Despite Mercury being closer to the Sun, Venus is the hottest planet in the solar system. Rather than being a paradise, it is more like hell.

The question of course is: how did it get that hot? The answer has been in the news much over recent years: it is the greenhouse effect. You can investigate the effect yourself on a sunny day with a car. If the car is parked in direct sunlight, the light falling on the seats and in-

This photograph of the Earth was taken by Apollo XI astronauts on the way back from the Moon. Through the clouds which cover much of the globe can be seen the land masses of Africa and Arabia. The Red Sea and the Mediterranean stand out in dark blue against the red-brown of the land. It is interesting to compare this photograph, taken 300 000 km from Earth, with similar shots of Venus, Mercury and Mars to see the difference between the four terrestrial planets. PHOTO CREDIT: NASA

This curtain form of the Aurora Australis was photographed from the Australian scientific station at Mawson base in Antarctica. The green colour of the aurora comes from oxygen atoms being ionised by particles from the solar wind. PHOTO CREDIT: AUSTRALIAN ANTARCTIC DIVISION. PHOTOGRAPH BY R. BUTLER

This photograph of a rare Corona Aurora was taken from Macquarie Island, south of Australia. The colours in the photograph come from different gases in the atmosphere being ionised by the charged particles in the solar wind. Ionisation knocks electrons off the atoms of gas in the same way that electricity makes a fluorescent light glow; light is given out when the electrons rejoin the atoms. PHOTO CREDIT: AUSTRALIAN ANTARCTIC DIVISION. PHOTOGRAPH BY A. NUTLEY

This photograph, taken from Apollo XI while in orbit about the Moon, shows a typical scene in the Moon's southern highlands. Craters ranging in size from a few hundreds of metres diameter to the 80 km diameter crater in this photograph overlap to give a confused look to the landscape. PHOTO CREDIT: NASA

This photograph was taken from the western rim of the Barringer Meteorite Crater in Arizona. The far side of the crater is 1.2 km away, and the floor is over 100 m below. The fractured and twisted state of the rocks in the crater's wall can be seen, testifying to the force of the impact which made this hole some 20 000 years ago. At the bottom of the crater is equipment used in the past to search for the remains of the meteorite which are now believed to be buried somewhere under the far wall. PHOTO CREDIT: DAVID REIDY

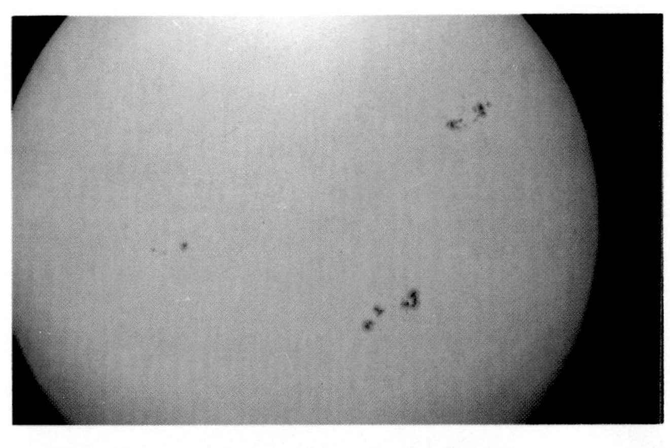

This photograph of the Sun in visible light was taken on 23 February 1989. Three groups of sunspots can be seen on the surface. Notice the limb darkening caused by the Sun's light having to travel through more of the solar atmosphere. On the surface of the Sun a number of lighter patches, floculae, can be seen, especially towards the western limb. PHOTO CREDIT: PAUL WYATT

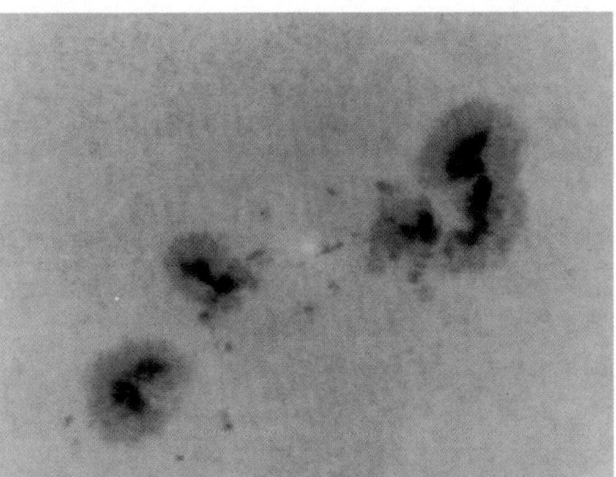

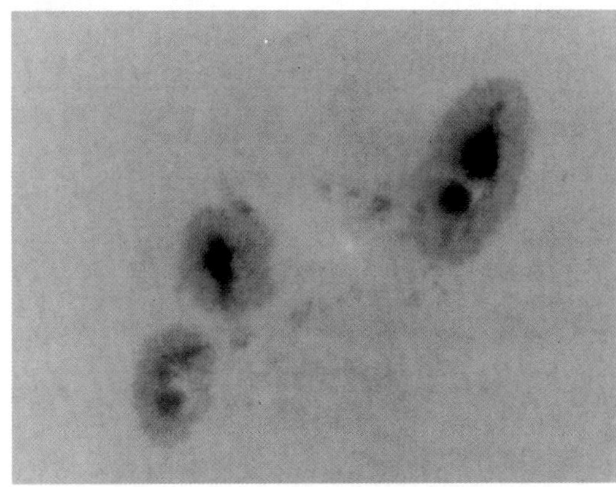

Two photos of a group of sunspots. The photograph on the left was taken on 23 February 1989 and the photograph on the right one day later. In the time between the two photographs many changes can be seen in the group. Notably the number of smaller spots has decreased, indicating that the group is on the decline. This group can also be seen in the photograph of the Sun (top of page) in the bottom right quadrant. PHOTO CREDIT: PAUL WYATT

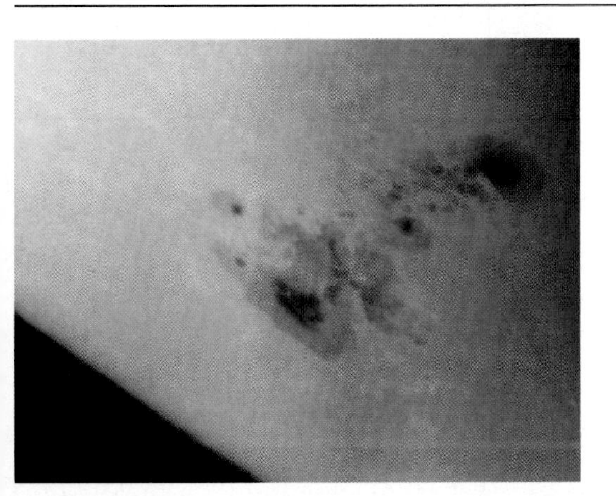

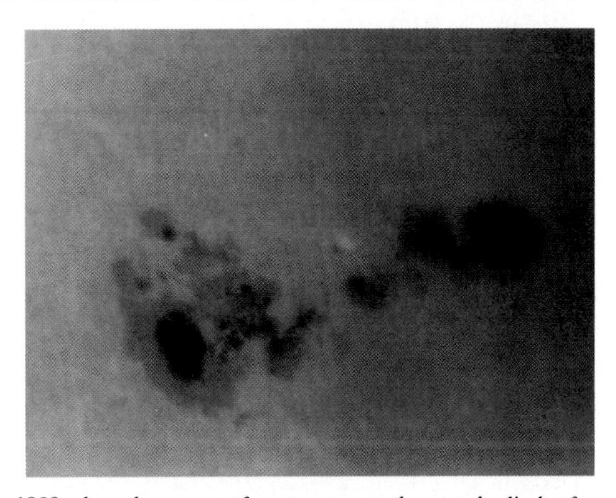

The first photograph (left) of this pair was taken on 7 February 1989 when the group of sunspots was close to the limb of the Sun. Notice how the group appears foreshortened compared with the second picture (right) taken one day later. Both photos show lighter areas on the Sun's surface, floculae, which are often associated with sunspot groups. PHOTO CREDIT: PAUL WYATT

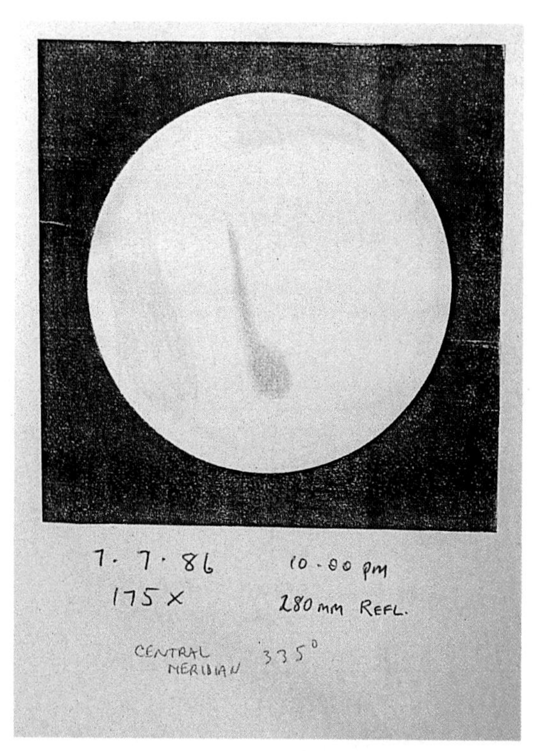


This series of observations was made over the period 7 July 1986 to 13 July 1986 recording the changes seen due to the rotation of the planet. Because each observation was made at around the same time each night, 10 p.m. local time, the central meridian of the planet has changed by around 8° from night to night. The colours in the drawings are those seen at the telescope by an experienced observer. All the observations were made with a 280 mm reflecting telescope using powers between 100 and 300x depending upon the seeing conditions. PHOTO CREDIT: OBSERVATIONS BY CHRIS TOOHEY

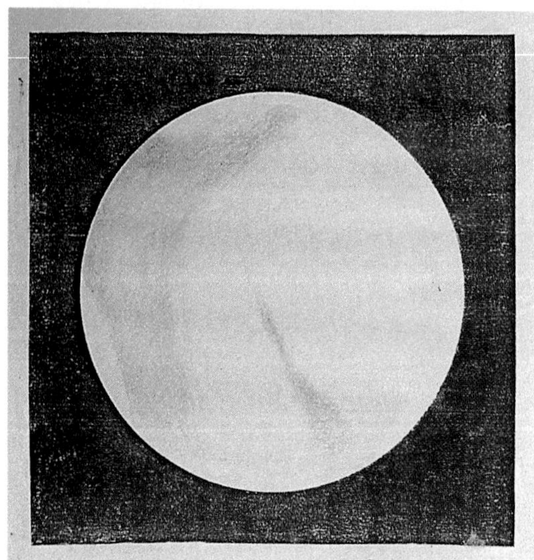

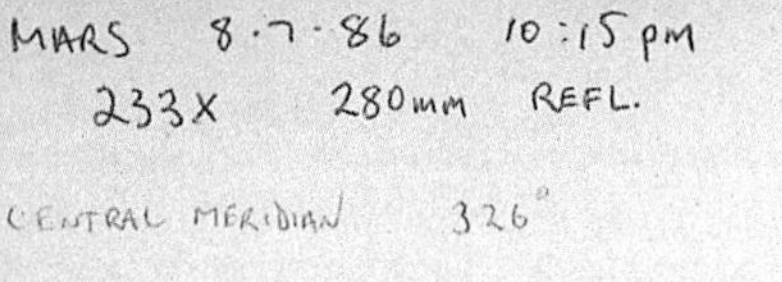

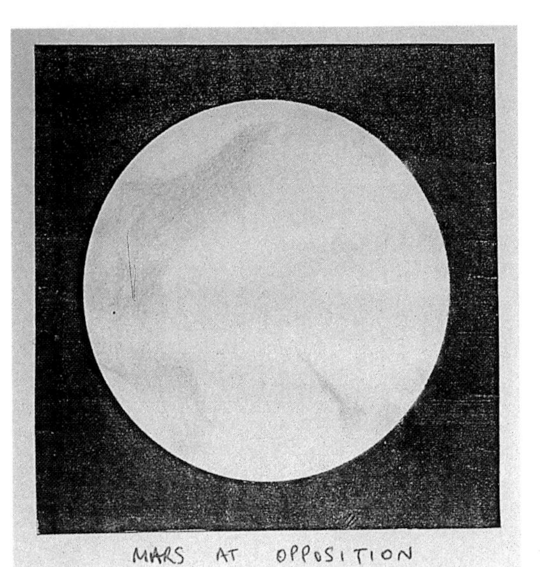

MARS AT OPPOSITION
10.7.86 10:15 pm
280X 280mm REFL.

SEE P 116 OF NOVEMBER ISSUE "ASTRONOMY" FOR A PHOTO OF THIS DRAWING

CENTRAL MERIDIAN 309°

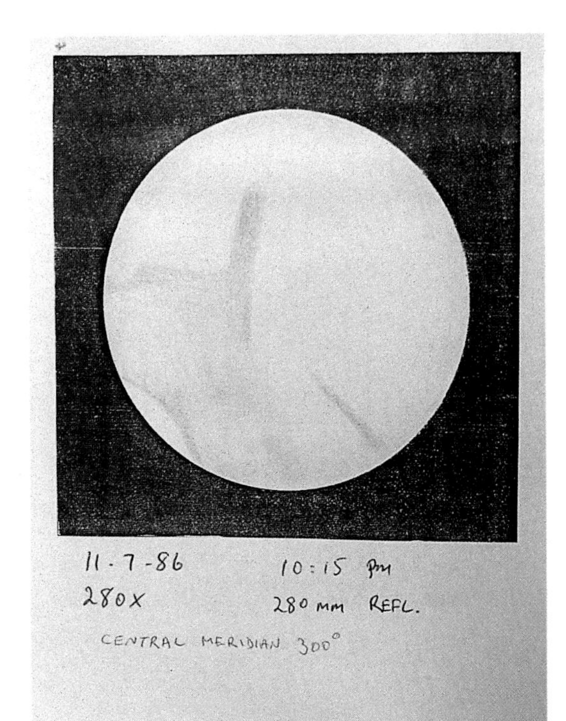

11.7.86 10:15 pm
280X 280 mm REFL.

CENTRAL MERIDIAN 300°

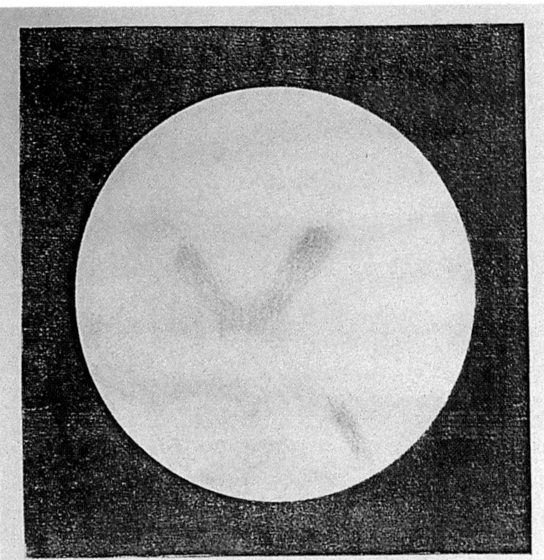

12.7.86 10:15 pm
233X 280 mm REFL.

CENTRAL MERIDIAN 291°

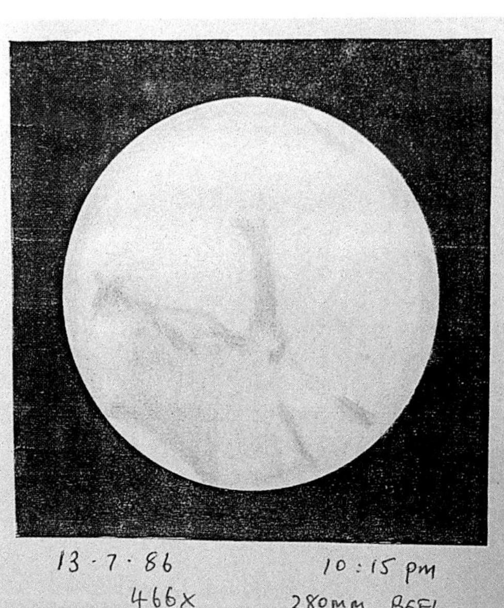

13.7.86 10:15 pm
466X 280mm REFL.

CENTRAL MERIDIAN 282°

To the naked eye Venus looks nothing like this picture. It appears as a boring white ball. This photograph was taken in ultraviolet light by the United States' spacecraft Mariner 10 on its way past Venus to Mercury. In UV light the patterns in the clouds caused by winds are clearly seen. This movement of the atmosphere, circling the planet once every six days or so, combined with the greenhouse effect caused by the gases in the atmosphere, evens out the temperature so that most of the planet is within a few degrees of the same temperature all the time. PHOTO CREDIT: JPL/NASA

This photograph of Mars was made during Viking 1's approach to Mars in June 1976. At the top of the picture (south) is an area of heavily cratered ancient highlands, while towards the bottom are the more recent plains. The Vallis Marinaris can be seen cutting into Mars just north of the equator. PHOTO CREDIT: JPL/NASA

The spacecraft Magellan being checked prior to being loaded into the space shuttle for launch. Magellan will travel to Venus and use a technique known as imaging radar to map the surface of the planet to high resolution. PHOTO CREDIT: JPL/NASA

The Voyager spacecraft, Voyagers 1 and 2, have done more to extend our knowledge of the outer solar system than anything else since the invention of the telescope. Here on a model of the spacecraft can be seen the main 4 m diameter antenna for beaming data back to Earth. Beneath this is the section holding the computers and thruster units. On the right is the scan platform holding the television cameras and other equipment, whilst to the left is the arm holding the thermoelectric nuclear-powered generators to provide power for the craft. Missing is the 13 m long boom arm holding the magnetometer experiment. PHOTO CREDIT: JPL/NASA

This is the first colour picture transmitted back from the Viking lander on the Chryse Planitia of Mars. The similarity between this scene and scenes of deserts on Earth is remarkable. The sandy plain is strewn with small rocks, from pebble size up to boulders a few metres across. The rocks are probably volcanic in origin, which would explain their porous appearance, caused by gases escaping from the rocks as they cooled. The colour in this photograph is, however, quite wrong. Scientists on Earth adjusted the picture to give the sky the same light blue colour as it has on Earth. It was a short time before they noticed that the colours of some wires on the spacecraft were wrong and that the scene should be redder. PHOTO CREDIT: JPL/NASA

This photograph shows the arm of Viking 1 after it has made a trench in the Martian soil to collect a sample for internal analysis. The scene here at Chryse Planitia is similar to the landscape at the Viking 2 site, the same sandy soil strewn with rocks. Both sites chosen for the Viking landers are unremarkable, and were chosen for that characteristic because it is easier to land a spacecraft. Notice that the sky in this photograph is pink in colour. This is the correct hue of the Martian sky. PHOTO CREDIT: JPL/NASA

Jupiter as seen from Voyager 2 on 22 May 1979. The complexity of Jupiter's weather systems can be seen in this photograph. Wisps of clouds from the belts (dark bands of cloud) and the zones (light bands) intermingle and spots are common amongst the clouds. On the right-hand limb of the planet half of the Great Red Spot can be seen. PHOTO CREDIT: JPL/NASA

This photograph of Jupiter was taken with a 250 mm amateur telescope and shows the quality of photograph which can be obtained with a little trial and error. In this photo the Great Red Spot can be seen as a hollow in the south equatorial belt. Near Jupiter's south pole is the shadow of the moon Io as it transits across the planet. PHOTO CREDIT: STEVEN QUIRK

A photograph of Saturn taken with a 250 mm reflecting telescope. The rings can be clearly seen, as can one band on the disk of the planet. Behind the planet some of the planet's shadow can be seen on rings. PHOTO CREDIT: STEVEN QUIRK

Jupiter's Great Red Spot is a huge anticyclonic storm over 25000 km in length in the south equatorial belt of Jupiter's atmosphere. The spot has been visible from Earth for over 300 years, though at times it fades almost to invisibility. The turbulence it creates in the belt is clear from the pattern of clouds to its left. A smaller white spot is below the Great Red Spot, it moves relative to the red spot and is a shorter lived feature. PHOTO CREDIT: JPL/NASA

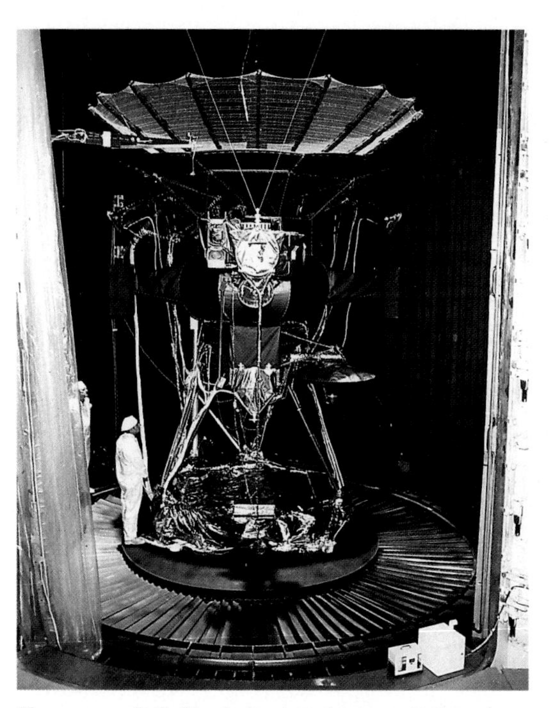

The spacecraft Galileo being tested prior to its launch in October 1989. Galileo will reach Jupiter in 1995 and carry out an extended mission there, investigating the planet, its weather and its family of moons. Galileo will also pass close to two minor planets on its way, giving us our first close-up pictures of those objects. PHOTO CREDIT: JPL/NASA

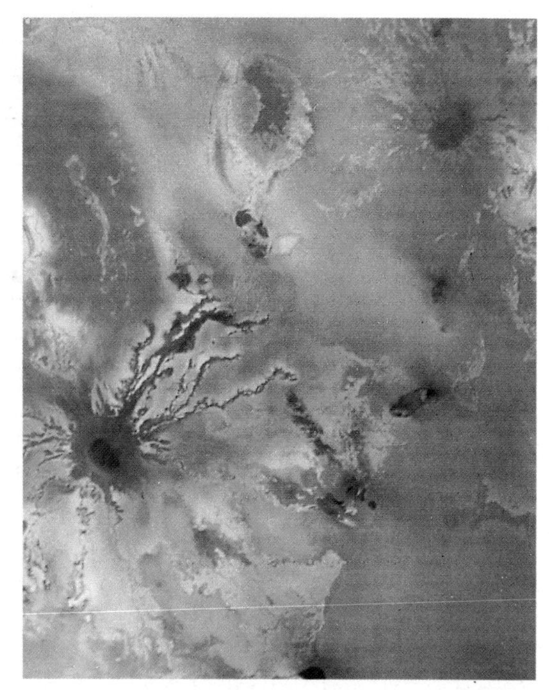

This picture of Jupiter's moon Io was taken by Voyager 1 at a distance of 128 000 km; an area of about one million square kilometres is shown. The features are probably flows of very fluid sulfur lava, spreading from the calderas of Io's volcanoes, here seen as black spots. PHOTO CREDIT: JPL/NASA

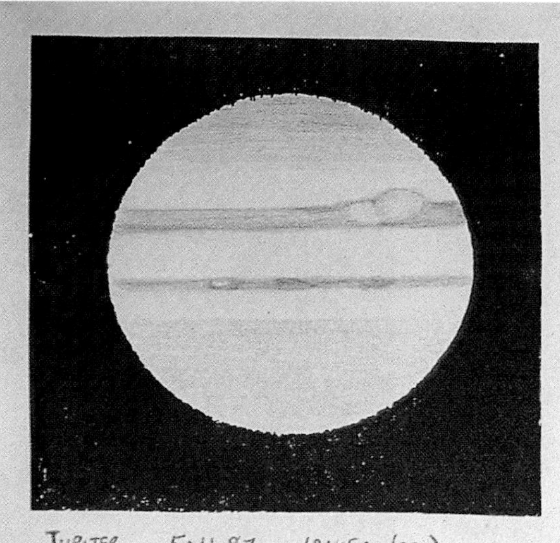


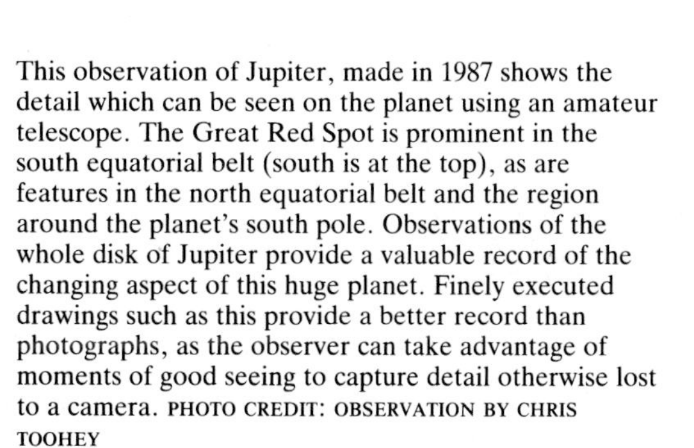

This observation of Jupiter, made in 1987 shows the detail which can be seen on the planet using an amateur telescope. The Great Red Spot is prominent in the south equatorial belt (south is at the top), as are features in the north equatorial belt and the region around the planet's south pole. Observations of the whole disk of Jupiter provide a valuable record of the changing aspect of this huge planet. Finely executed drawings such as this provide a better record than photographs, as the observer can take advantage of moments of good seeing to capture detail otherwise lost to a camera. PHOTO CREDIT: OBSERVATION BY CHRIS TOOHEY

These two observations of Saturn were made two years apart, using the same equipment. The top photograph was made in June 1987 and shows the Cassini division in the rings and banding on the surface of the planet. The C ring can be seen crossing the face of the planet which can be seen through it. The observation on the bottom was made in 1989 when Saturn was at nearly maximum tilt. Banding on the planet is obvious, as is the Cassini division and C ring.
PHOTO CREDIT: OBSERVATIONS BY CHRIS TOOHEY

ıpiter's family of moons reveals a history of the solar system
 miniature. Closest to Jupiter is the tiny moon Amalthea,
·d in colour and irregularly shaped. Io is the closest in of the
alilean moons to Jupiter, permanently locked in a tidal tug
 war with Europa. Io's surface is the youngest in the solar
stem, constantly being reworked by the actions of its sulfur
ılcanoes. Europa is perhaps the smoothest place in the solar
stem, its light coloured surface being crossed by long lines
 darker material which has filled cracks in the ice.
anymede, the largest of Jupiter's moons, has a surface
hich is a combination of old and new. Old regions of heavy
atering are crossed by brighter, newer regions of crust
ımposed of parallel channels running for hundreds of
lometres. Callisto, the furthest out of the Galilean moons,
as a very old surface, heavily cratered all over. Brighter
ots of newer ice are seen where the last of the impacts
ccurred. PHOTO CREDIT: JPL/NASA

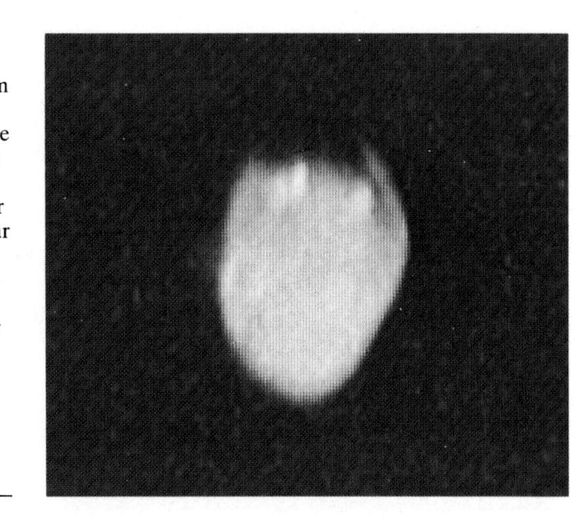

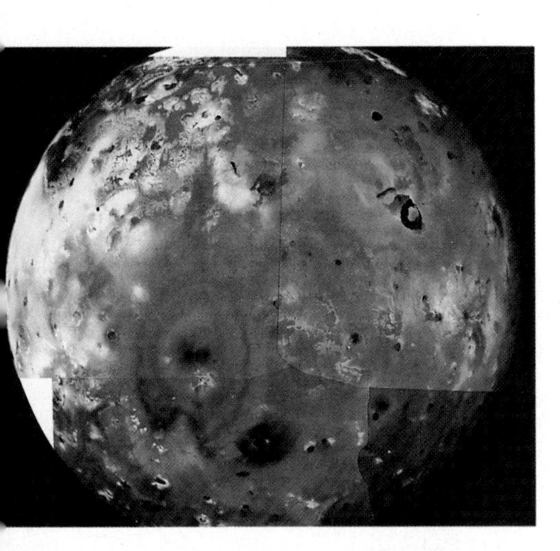

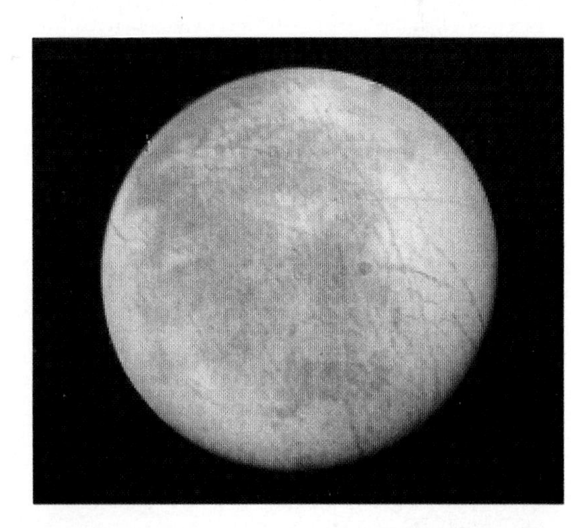

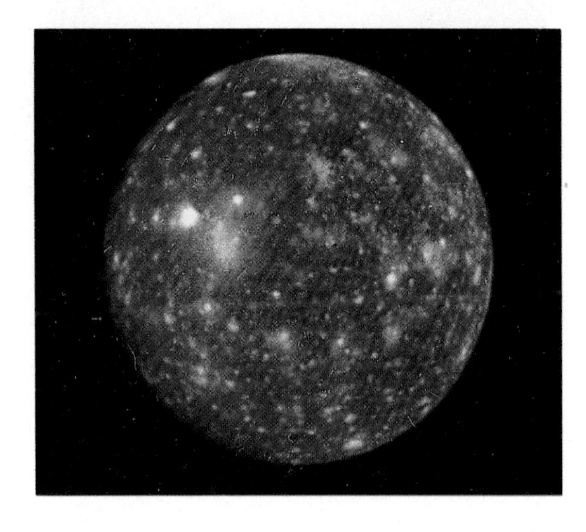

Saturn's rings are seen here from their night side, the sunlight passing through the rings rather than being reflected from them. The C ring is seen closest to the planet. This is normally difficult to see from the Earth, but the small ice covered particles scatter light well, making the ring appear bright. The denser B ring with its larger particles does not scatter light well, and so is seen dark here. Outside the B ring the A ring is brighter, scattering light well. Beyond the A ring, the thin F ring can be seen, looking a little lumpy. PHOTO CREDIT: JPL/NASA

This image of Saturn was taken by Voyager 2 when it was only fourteen million kilometres from the planet. There is a large amount of structure in the rings surrounding Saturn: spokes are clearly visible on the left side of the planet, and the disk of the planet can be seen through the Cassini division. Notice that the rings appear to be cut off behind the planet by the shadow of Saturn falling across them. Three moons are in this image: from left to right, Tethys, Dione and Rhea; the shadow of Tethys is on the planet. The banding of the clouds in Saturn's atmosphere is much less dramatic than the banding about Jupiter, indicating a less energetic weather system. PHOTO CREDIT: JPL/NASA

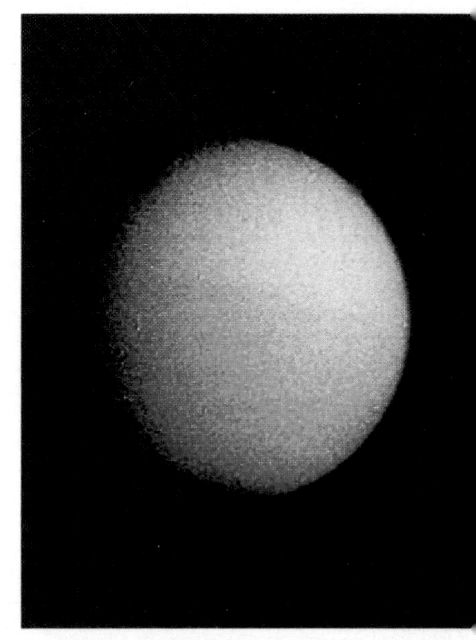

This image of Titan made by Voyager 1 shows very little detail. The surface of Titan is shrouded forever beneath a thick layer of haze high in the moon's nitrogen and argon atmosphere. Here the only feature clearly seen is that the northern hemisphere is lighter than the southern hemisphere. PHOTO CREDIT: JPL/NASA

This observation of Uranus made with a 280 m reflecting telescope on 14 March 1986 shows only subtle shading on the planet's disk. We know from Voyager 2's encounter with Uranus in January 1986 that there are no features to b seen. PHOTO CREDIT: OBSERVATION BY CHRIS TOOHEY

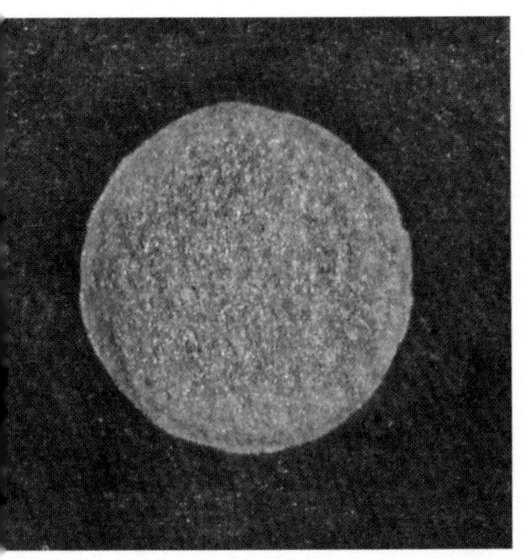

his observation of Neptune made with a 280mm eflecting telescope at 400x magnification shows little etail of the planet, only subtle shadings towards the mb. It is the best that can be expected from Earth ven under exceptionally good conditions. PHOTO REDIT: OBSERVATION BY CHRIS TOOHEY

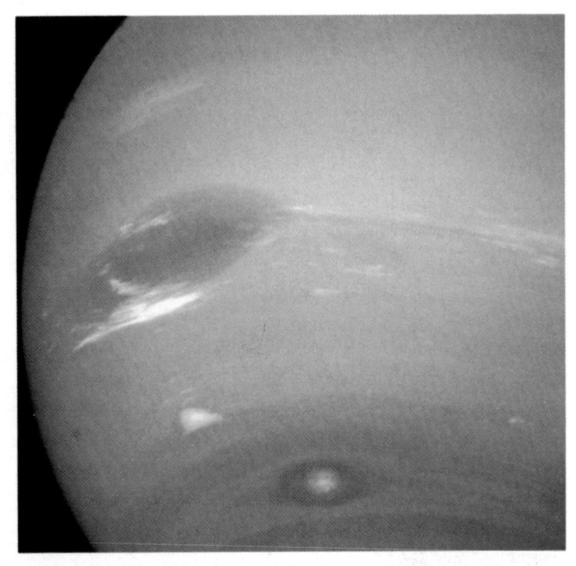

The atmosphere of Neptune proved to be a much more interesting place than the atmosphere of Uranus. Here we see Neptune's Great Dark Spot, a storm almost as large as the Earth, similar to Jupiter's Great Red Spot. Beneath the spot are bright clouds of white methane ice. Further south is a smaller dark spot with a bright centre. Between the two spots is a white spot called the Scooter because it moves through the atmosphere faster than the other features. PHOTO CREDIT: JPL/NASA

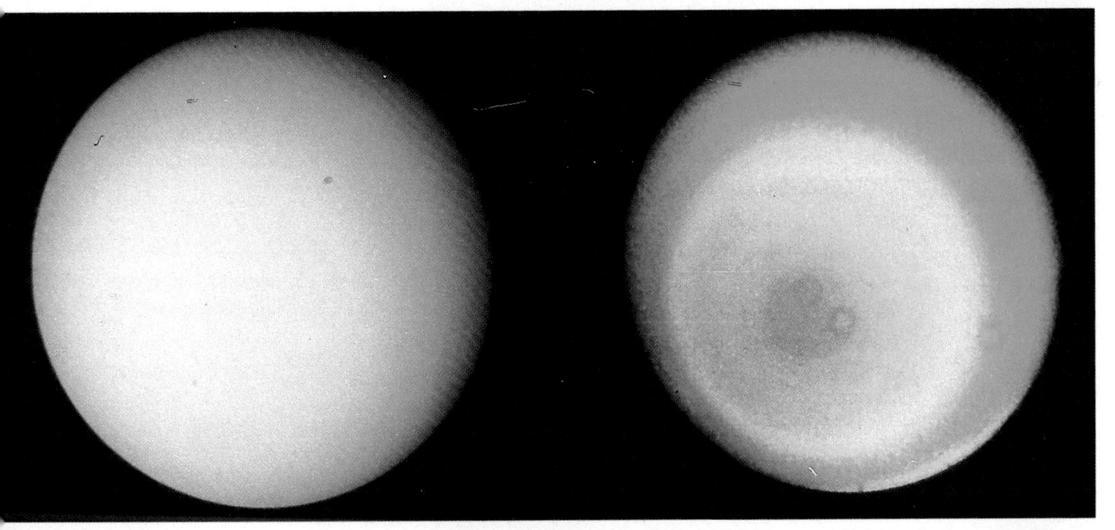

n the left is the image of Uranus Voyager 2 saw as it approached the planet on 17 January 1986. To the lower left of he disk is Uranus' south pole, Voyager having arrived at the planet just after the southern summer. Little detail can be een in Uranus' atmosphere due to the lack of energy from the Sun to drive powerful weather systems and to the presence f a haze layer high in the Uranian atmosphere hiding most of the detail beneath. On the right is a colour-enhanced ersion of the picture which is able to show a little detail in the atmosphere. Even with the help of a computer though, here is still not much to see. The small circles seen on the planet's disk are from small specks of dust on the lens of oyager's camera; they are not on the planet. PHOTO CREDIT: JPL/NASA

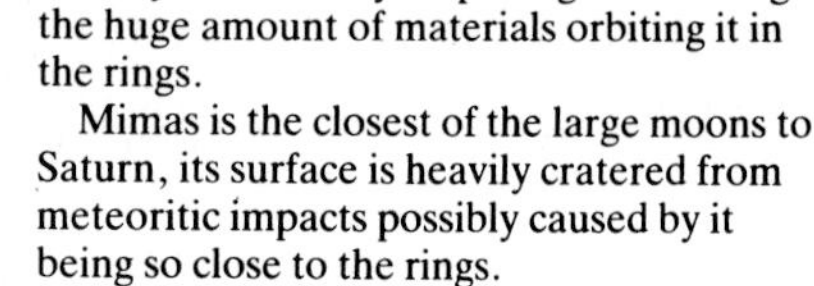

Saturn has the largest family of moons in the solar system, hardly surprising considering the huge amount of materials orbiting it in the rings.

Mimas is the closest of the large moons to Saturn, its surface is heavily cratered from meteoritic impacts possibly caused by it being so close to the rings.

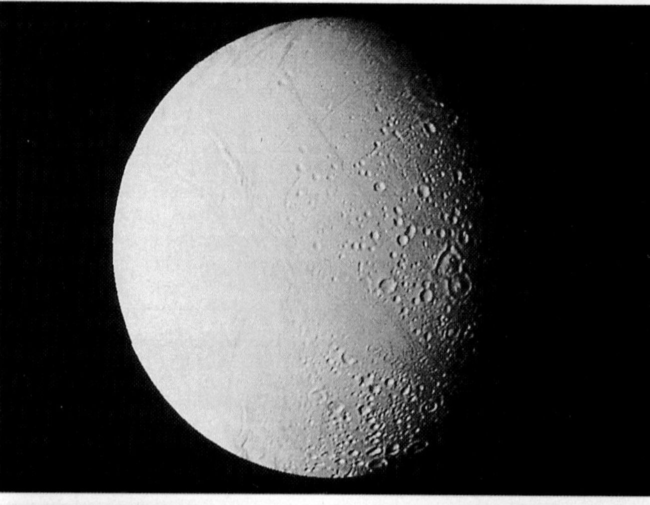

Enceladus is much less cratered than Mimas and part of its surface is completely devoid of craters indicating relatively recent activity on the surface. A huge fault at the lower left and the other linear markings on the surface are signs that the moon has a fairly thin crust over a liquid interior.

The surface of Tethys is old and places on it are as heavily cratered as Mimas. Near to the terminator is the Ithaca Chasma, a huge valley over 2000 km in length travelling three-quarters of the way around the moon. In places the Chasma is 100 km wide and 6 km deep. To the lower left of this image is the Odysseus impact crater, 400 km in diameter. It is possible that Ithaca Chasma is a result of fracturing of the surface by the impact which formed Odysseus, or it may be an unrelated tectonic feature.

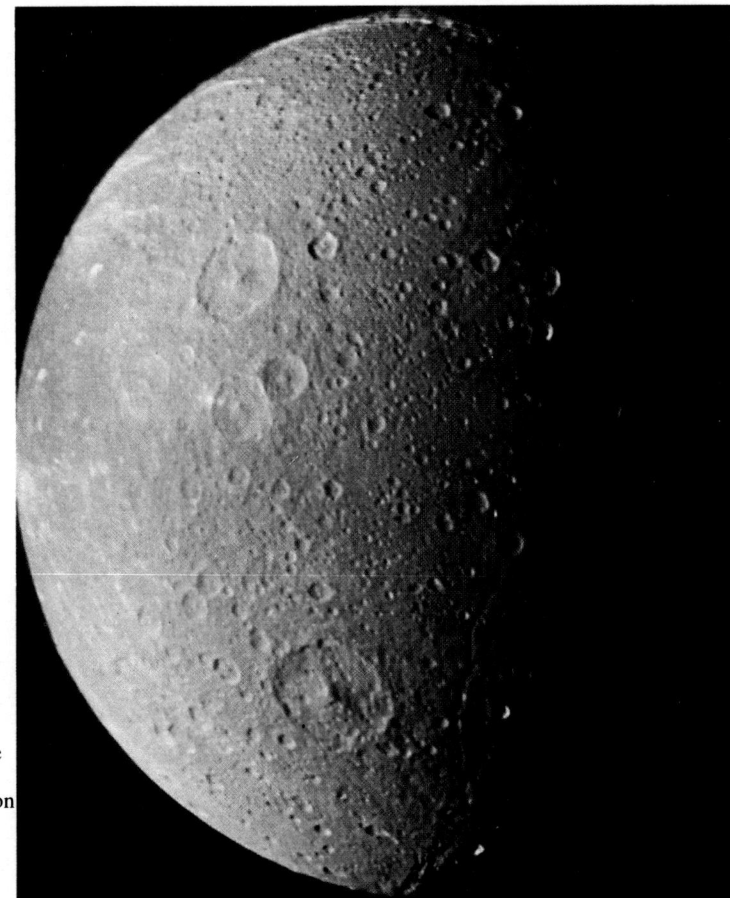

Dione is similar in size to Tethys and like the other Saturnian moons it is heavily cratered. The distribution of craters on Dione is not even, indicating that at some time in the past parts of the surface have been reworked. Only partially apparent on this photo is the trailing hemisphere of Dione, near the terminator, which is considerably darker in colour than the leading hemisphere.

Rhea is Saturn's second largest satellite. Here seen as an enhanced image from Voyager 1, the heavily cratered surface of Rhea makes it one of the oldest surfaces in the Saturnian system.

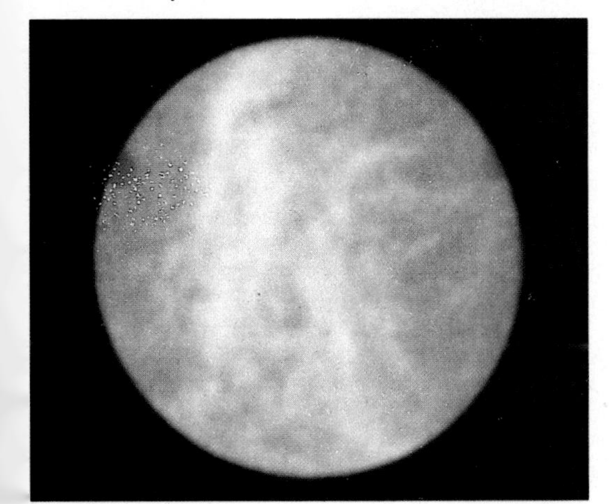

Hyperion is a small, irregularly shaped body which orbits outside the main region of moons. This small rocky world is similar to other small satellites found throughout the solar system. PHOTO CREDIT: JPL/NASA

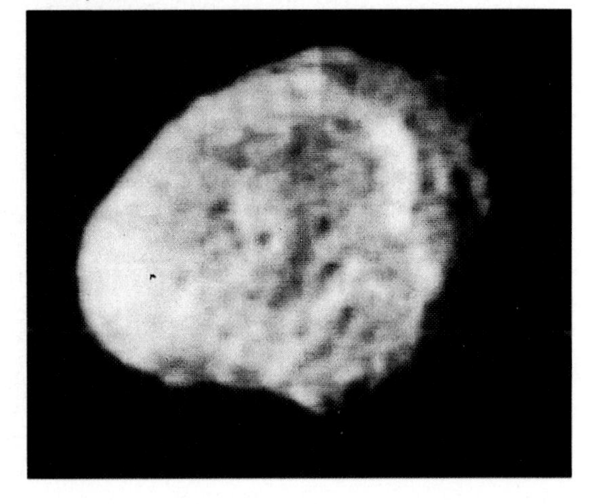

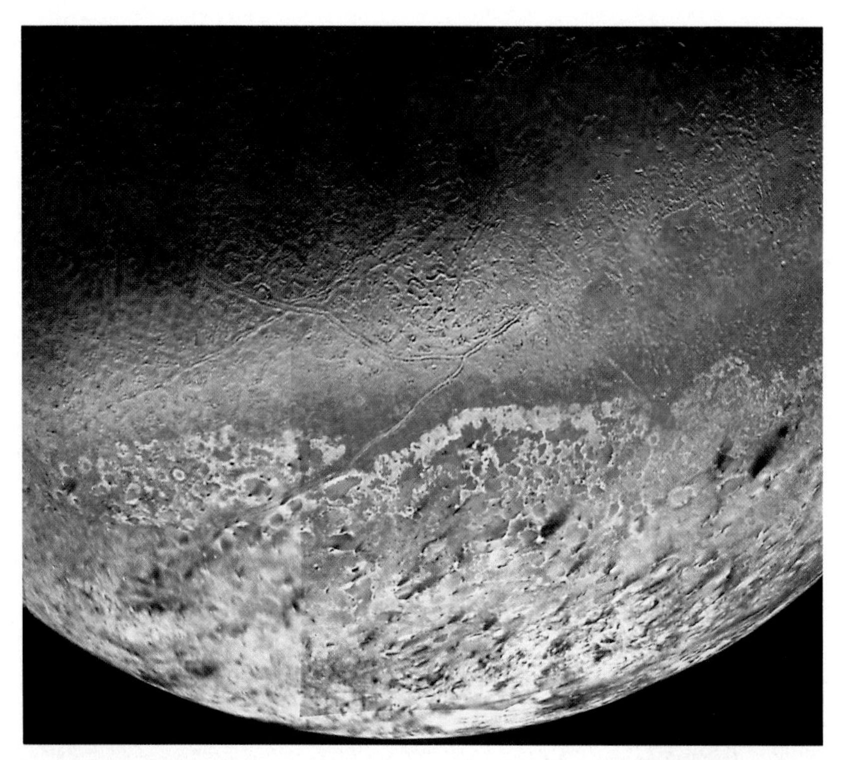

This is a high-resolution mosaic made from images acquired by Voyager 2 as it passed Triton. Here we see a cap of pink methane snow covering the southern hemisphere of the Moon. The dark streaks across the ice are material from nitrogen powered volcanoes on the Moon. The bare, bluish coloured surface of the Moon shows long ridges and troughs and a few impact craters. PHOTO CREDIT: JPL/NASA

These white wispy clouds are around 50 km higher in Neptune's atmosphere than the thick layer of blue beneath them. The height of the clouds can be gauged by the shadows they cast on the cloud layers below. When closer to the terminator, these clouds could be seen extending into the dark half of the globe, catching the last of the sunlight. PHOTO CREDIT: JPL/NASA

**Table 8.1 Venus' atmosphere**

| Gas | | Fraction |
|---|---|---|
| Carbon dioxide | $CO_2$ | 96.5% |
| Nitrogen | $N_2$ | 3.5% |
| Water vapour | $H_2O$ | 0.00015% |
| Sulfur dioxide | $SO_2$ | 0.00015% |
| Argon | Ar | 0.00006% |
| Oxygen | $O_2$ | 0.00003% |
| Carbon monoxide | CO | 0.00002% |
| Neon | Ne | 0.000009% |
| Hydrochloric acid | HCl | 0.0000006% |
| Hydrofluoric acid | HF | 0.000000005% |

terior fittings will heat them up. These objects will then radiate energy, in the same way as all other warm objects do, but not as visible light: they will emit infrared radiation. The glass through which the sunlight passed without problem is opaque to infrared radiation, so it is unable to escape the car. Trapped inside, this radiation is reabsorbed by other parts of the vehicle, raising the temperature further. This cycle continues until in a very short time the temperature of the car's interior is much higher than that of the air outside.

Venus, of course, doesn't have glass to trap the infrared radiation, but gases in the atmosphere can do that. Gases such as oxygen and nitrogen are transparent to both visible and infrared light, but both water vapour and carbon dioxide are opaque to infrared. In the Earth's atmosphere the small amounts of water vapour and carbon dioxide contribute to a small greenhouse effect, raising the average temperature by about 25°C. Clearly the greenhouse effect is significant on Venus, not only from the high temperature but also because the temperature over the planet is constant to within a few degrees, as if it were wrapped in a blanket.

It was known from spectroscopy that there were significant amounts of carbon dioxide in the Venusian atmosphere. In addition to the greenhouse heating, this gas must also make the atmosphere of the planet very heavy. Indeed, measurements made by spacecraft have shown that the pressure on the surface is 90 times greater than the pressure on the Earth. Not only would an astronaut on the planet need good airconditioning, but also heavy armour, such as that on submarines, to withstand the crushing effects of the atmosphere. This high pressure makes the carbon dioxide even more efficient at absorbing infrared radiation, enhancing the greenhouse effect further.

The time of dichotomy (when half of the disc of an inner planet or the Moon is visible) has puzzled observers since the late eighteenth century. In 1797 a German astronomer, Johann Schroter, noted that there was a difference between his observations and the computed time of half phase. The cause seems to be terminator shading, making the phase appear less than it should; when Venus is in the evening sky and its phase is waning, dichotomy would come several days early, and while waxing in the morning sky, it cames later than the predicted time.

The reason is simple. Consider an airless world, like the Moon. The terminator, the line between light and shadow, is sharp; it is simply determined by the lie of the land. On Venus, however, some of the light is refracted through the atmosphere, bending its way around the planet into the region which would otherwise be dark. Hence the boundary between light and dark is further around the planet than would be otherwise expected. As the planet nears greatest elongation west this refraction makes the bright limb of the planet reach the halfway point earlier than expected, making dichotomy appear a few days early. At greatest elongation east, dichotomy occurs as the planet goes from gibbous phase to crescent phase; again because of refraction through the atmosphere, the light appears to extend further into the atmosphere than otherwise. In this case, though, dichotomy appears to be delayed, occurring a few days later than predicted.

Some scientists wondered if, with such a heavy, thick atmosphere, any light at all reached the surface of the planet. Measurements made by the Soviet spacecraft which have landed there show that indeed there is illumination on the surface, about the same amount as there is on a heavily overcast day on Earth. The composition of the atmosphere was also a surprise to scientists who thought of the Earth as the prototypical planet. Instead of a nitrogen-oxygen atmosphere with traces of carbon dioxide and other gases, Venus' atmosphere is almost entirely carbon dioxide with a little nitrogen; oxygen, the gas essential for terrestrial life, is present in only minute quantities. Water, too, which had been expected, is in the atmosphere only as a trace.

Early guesses were that the clouds were water vapour or ice crystals, similar to those of the Earth, but without sufficient water other explanations had to be sought. Hydrocarbons were

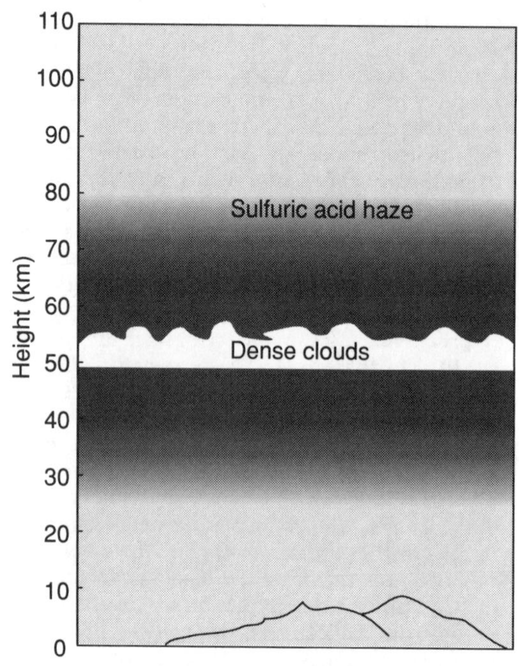


**8.1** The structure of Venus' atmosphere. The clouds which shield the surface from our view are 50 km above the surface of Venus. The clouds consist of several layers of differing compositions. Below the clouds the atmosphere is clear.

**8.2** Venus' greenhouse effect makes it the hottest of the planets. Radiation arriving from the Sun penetrates the upper layers of thick clouds to reach the surface below. At the surface some of the energy is absorbed and re-emitted as infrared radiation (heat); this heat is trapped by the atmosphere which is unable to radiate it back into space, leading to an increase in temperature. Gases such as carbon dioxide are very effective at trapping heat in the atmosphere, causing the temperature to rise to over 320°C.

suggested but no evidence could be found. In the 1970s observations from aircraft of the spectrum of the clouds gave the answer: concentrated sulfuric acid.

The water clouds of Earth are caused by condensation of water gas in the cooler reaches of the atmosphere. On Venus, however, this doesn't account for the clouds. The acid clouds of Venus are formed by photochemistry: chemical reactions which rely upon ultraviolet light from the Sun to occur. On Earth it is photochemistry which produces ozone. In addition to photochemical formation, the clouds of Venus must be supplied with the raw materials, in this case sulfur dioxide. It is thought that this is supplied by volcanism on the planet's surface. Sulfur dioxide often accompanies volcanic activity; it is the chemical which gives the rotten egg odour to thermal springs and other volcanic features on Earth.

Although the clouds of Venus appear featureless in visible light, if photographs are taken using an ultraviolet filter, then dark markings do show up in the cloud layer. These dark markings are quite stable and they have allowed astronomers to follow their passage around the planet. The atmosphere of Venus rotates every four days, travelling from east to west at speeds of around 360 km per hour. The weather on Earth results from two main effects on our atmosphere: the Earth's rotation and the localised heating of the Sun. On Venus only one of these accounts for the weather: the heating of the Sun as Venus rotates on its axis very slowly.

When scientists first measured the rotation rate of Venus using radar in the 1960s, they expected a fairly fast rotation rate, similar to the Earth's and Mars'. What they found, though, was a rotation taking 243 days, longer than the planet's year and in the wrong direction. Normal rotation in the solar system is from west to east, but Venus turns from east to west: retrograde rotation. Explanations of this retrograde motion have been sought, and some have in-

**8.3** Because Venus' atmosphere is so thick, it is able to refract light quite a noticeable distance around the limb of the planet. This means that at a given time more of the planet appears illuminated that would be expected if the light was simply cut off by the limb of the planet (dashed line). It is this bending of the light which accounts for the anomalies in the predictions of dichotomy for the planet.

**Table 8.2 Elongations of Venus**

| Eastern elongation | Inferior conjunction | Western Elongation |
|---|---|---|
| 8 November 1989 | 18 January 1990 | 30 March 1990 |
| 13 June 1991 | 22 August 1991 | 2 November 1991 |
| 19 January 1993 | 1 April 1993 | 10 June 1993 |
| 25 August 1994 | 2 November 1994 | 13 January 1995 |
| 1 April 1996 | 10 June 1996 | 19 August 1996 |
| 6 November 1997 | 16 January 1998 | 27 March 1998 |
| 11 June 1999 | 20 August 1999 | 31 October 1999 |

volved quite unlikely events. The most plausible involves a collision with a large object near the end of the planet's formation. If the object struck the planet in the direction opposite to the way it was spinning it is possible that it could stop the planet's rotation and make it reverse direction. Such an impact would have come close to splitting the planet in two.

As on Mercury, the day on Venus is strange indeed. The time taken for the planet to rotate once with respect to the stars is 243.08 days. The time between successive noons, if it were possible to see when the Sun was overhead, is only 116.67 days. A day on Venus would start with the Sun rising in the west. It would make its way slowly across the sky setting 58 days later in the east.

The orbits of Venus and Earth have an interesting connection. The time between inferior conjunctions of Venus—that is, the times when it is closest to Earth—is 584 days. This is equal to exactly two revolutions of the planet, give or take two hours. This means that whenever the planets are closest, the same part of Venus is facing the Earth. Unlike our Moon's rotation or Mercury's, this effect cannot be due to tidal forces, or any other known effect; it appears to be just a coincidence.

If Venus is forever cloaked in cloud, how is it that we know about its surface? Radio waves, as we have seen, can pass through the cloud cover and, as radar uses radio waves, it can be used to map the surface. The first use of radar mapping at Venus was with the United States' Pioneer Venus spacecraft. It used radar to determine its altitude above the planet's surface, so over some hundreds of orbits a map of the surface could be built up. That map gave the first global look at the Venusian landscape.

Once a map of the surface of Venus was available, scientists looked for points of similarity and difference between it and Earth. Heights on Earth are measured with respect to mean sea level, but with no seas on Venus, scientists simply adopted a mean land height, corresponding to a planetary radius of 6051.4 km. The first thing that is noticeable about the topography of Venus is the lack of large continental areas. The earth has six large land masses, Venus has but two: Aphrodite Terra and Ishtar Terra. The names for features on Venus come from the women of history and mythology, Aphrodite and Ishtar being the Greek and Babylonian equivalents of the Roman goddess Venus. Only 8 per cent of Venus is continental crust, the rest is made of plains (65 per cent) and lowlands (27 per cent).

The next features which are noticeable on Venus are craters, or rather the lack of them. This is not wholly unexpected. Venus is a large planet, so unlike the Moon and Mercury it will not have cooled quickly; internal heating and activity will have had time to rework the surface, removing evidence of early impacts. The Soviet spacecraft Venera 15 and 16 have shown us much more of the surface of Venus. Instead of having simple altitude-measuring radar, both these craft were equipped with imaging radar which is able to produce pictures of the surface similar to those taken with an ordinary camera. Imaging radar is the type used in military satellites to keep track of the movement of arms and ships under all weather conditions. The imaging radar on the Venera spacecraft is able to make out features as small as 2 km. These studies show that the craters which do exist on Venus are small and at around 15 per cent of the density of craters on the Moon's maria.This suggests that the surface is less than a thousand million years old.

Imagining the conditions on the surface of Venus is difficult; even the temperatures are hard to imagine. A typical home oven will give temperatures of 250°C, 80° cooler than a warm day on Venus, and the pressure is almost as much as 1 km under our oceans. Despite these most inhospitable conditions seven Soviet spacecraft have successfully landed on the planet's surface. Some of them even continued working for over two hours.

The main work of the Soviet surface probes has been to analyse the rocks around them and collect samples for internal chemical analysis. The results depended upon where the spacecraft landed. Probes landing on the plains found that the rocks there were similar to the Earth's basalts and granites and showed evidence of extensive chemical change. Other landers found similar rocks, but with more potassium—something which occurs in only very old terrestrial rocks. The two probes which landed on Aphrodite Terra both returned similar results, the surface being mainly basalts, but this time richer in sulfur than those of Earth.

In addition to chemical analysis, the Soviet spacecraft were equipped with cameras to photograph the landscape. Some colour photos and many black and white ones have been taken. The first pictures came from Veneras 9 and 10, which landed in October 1975. They both landed near Beta Regio and returned pictures showing a rocky landscape. Veneras 13 and 14 had much more success, lasting two hours and one hour on the surface respectively. Both spacecraft returned panoramic views of the surface. Venera 14 landed on a plain showing a flat landscape covered with plates of rock which showed layering and ripple marks. Venera 13 landed 500 m higher and shows the same plate-like rocks, but also many more small stones and fine dirt; it seems likely that these would have been formed by chemical weathering breaking down the rocks.

The internal structure of Venus is still uncertain. Information is scarce as no seismometers have yet made it to the surface. Evidence of the different sorts of rocks on the Venusian surface indicates that the planet must have evolved chemically, making a dense core and a less dense crust. The similarity in size and density to the Earth also suggests this. The problem is that Venus does not have a magnetic field, something which would be expected if the core had liquid within it, but as a magnetic field from a liquid core depends upon the planet rotating, it could be that the slow rotation of Venus does not provide enough energy to make the field. The solution will have to wait for further exploration of the planet.

Next to the Sun and Moon, Venus is the brightest object in the heavens, so bright that it outshines the brightest star, Sirius (α Canis Majoris), by a factor of twelve. Few people can have failed to notice Venus blazing gloriously against the morning or evening twilight, and it is responsible for many unidentified flying object sightings from those unfamiliar with the sky.

During the evening when at its brightest (magnitude −4.4), Venus can cast a perceptible shadow. This will be best observed in the country, away from other sources of illumination. It may also surprise you to know that Venus can also be seen in broad daylight, provided you know precisely where and when to look.

Since Venus lies within the Earth's orbit, it presents phases and elongations like Mercury. Unlike Mercury, you will not have any difficulty in finding Venus: it is brighter, nearer and larger. Of course, the same rules apply when viewing both Mercury or Venus near the Sun: use extreme caution.

While Mercury has both favourable and unfavourable elongations resulting from its highly elliptical orbit, all elongations of Venus can be said to be favourable. The orbit of Venus is nearer to a perfect circle than any other planet, and therefore its greatest elongations vary from only 45° to 47°. This large separation from the Sun means that Venus can be observed long after sunset or before sunrise in a darkened sky. A dark background is not always advantageous for viewing, however, as the glare can be overwhelming. Most amateurs prefer to view the planet with the Sun above the horizon, where it will appear less bright and at a greater altitude.

To find Venus in daylight, it is necessary to look up the positions of both the planet and Sun in an almanac. If using an equatorial telescope with setting circles, offset from the Sun as described in Chapter 7. The positions in the almanac will also aid in finding the planet through an altazimuth telescope or binoculars. First, compare the right ascension of Venus to the Sun; if greater, the planet will be to the east of and set after the Sun. If the figure is smaller, Venus will be to the west of and rise before the Sun.

By subtracting the right ascensions, the angular distance of Venus from the Sun will be found. For example, if the difference in right ascension is two hours, Venus will be 30° from the Sun, as each hour of right ascension is equal to 15°. At arm's length, the outstretched hand will cover about 20° of sky, so by taking one and a half hands you will come to the point at which your search should begin. Using binoculars focused for distance, you should pan up and down from this location, as Venus could be above or below the ecliptic by several degrees.

Once found, move the binoculars away from your eyes, and try to see Venus unaided; it is a good idea to use a local object—tree or rooftop—to sight along. Near greatest brilliancy it is obvious to the eye, if you are looking at the right location. Once this challenge has been mastered you may try to find Jupiter, or even some bright stars.

The phases of Venus can be quite striking, and they are visible though the smallest of telescopes or binoculars. At inferior conjunction the planet is at its largest, 64.5″, with its night side turned to us. As it moves towards western elongation to become the morning star, the eastern limb will begin to show as a slender crescent. The crescent increases in width as the diameter decreases, through to half disc and the gibbous phases until finally at superior conjunction the planet is fully illuminated, and at its smallest size of 9.9″. Venus then approaches the Earth as the evening star undergoing a reversal of the process back to inferior conjunction. Its greatest brilliancy occurs about 39° east or west of inferior conjunction. Since Venus completes its orbit in 225 days, and the Earth lags behind, it takes 584 days between successive conjunctions. This is known as the synodic period. Table 8.2 details the elongations and conjunctions until 2000.

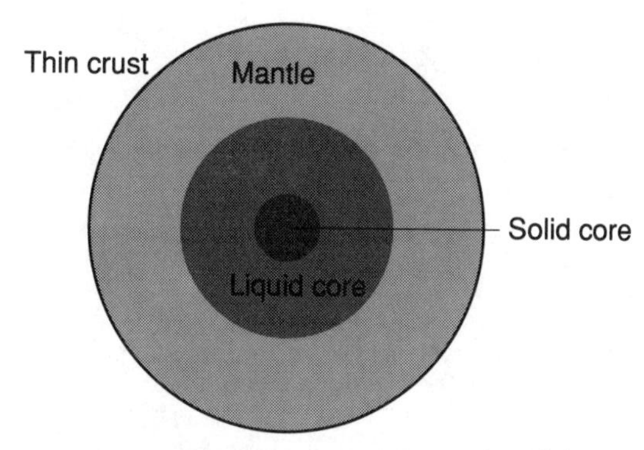


**8.4** The structure of Venus is similar to that of the Earth, a core made of a liquid outer layer and a solid inner core, both made of iron and nickel. The core is surrounded by a silicate-rich mantle with a thin rocky crust making the surface. This model is based upon the best guesses of scientists, as so far no seismometer has been successfully landed on the planet.

The inclination of Venus' orbit to the ecliptic is 3.24° and like Mercury, it transits the Sun when it intersects the ecliptic at inferior conjunction. Unfortunately, transits of Venus are extremely rare events—the last two were on 9 December 1874, and 6 December 1882. We can however, look forward to two more early in the next century: on 8 June 2004 and 6 June 2012. We must then wait until the twenty-second century for the next pair, due in the years 2117 and 2125. Figure 8.5 shows the predicted paths of Venus across the Sun's disc in 2004 and 2012.

Before the invention of radar and other methods of measuring distances, the orbit of Venus was the primary reference tool in finding the size of the solar system. By making measurements such as the greatest elongation a planet has from the Sun, it is possible to come up with a relative scale for the solar system. For example, knowing that Venus has a maximum elongation of 46° from the Sun, you can draw a diagram such as Figure 8.6 and calculate the relative distances of the Earth and Venus. What was needed, though, was an absolute distance scale for the solar system. Just one measurement was required from which all the rest could be calculated. That is where Venus came in. By measuring the position of Venus during a transit from different parts of the world, an absolute distance can be calculated.

As we have seen, transits of Venus are rare. Although they occur in pairs just eight years

**8.5** The apparent paths of Venus across the face of the Sun during the next two transits.

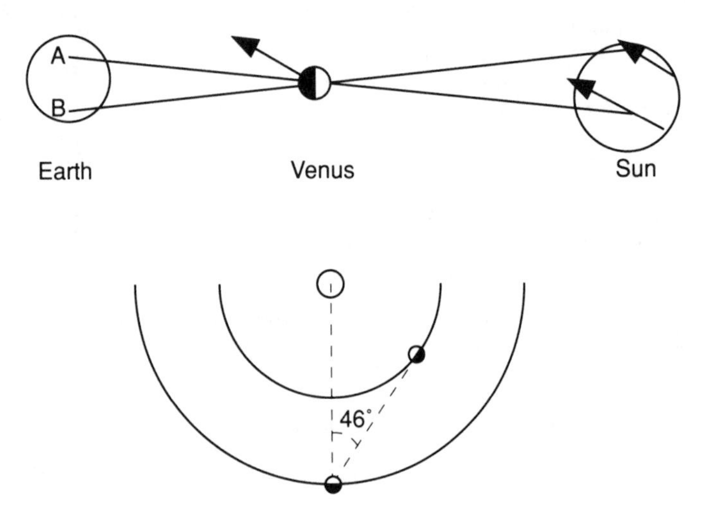


**8.6** By knowing the angle of greatest elongation from the Sun, it is a simple matter to calculate the relative sizes of the orbits of Venus and the Earth, but it is impossible without further information to calculate a size in kilometres. By observing the transit of Venus across the face of the Sun from two separated points on the Earth, it is possible to calculate the exact distance from Earth to Venus, and hence the size of the Earth's orbit and from there the sizes of the orbits of all the other planets.

apart, the pairs are separated by 105 years. One such pair occurred in the mid-1700s when experimental science was at its peak as a gentleman's pastime, so it was natural that observations would be carried out. The first transit was in 1761, but poor weather conditions meant that the results from it were imprecise. Luckily there was another transit in 1769. Unfortunately Europe was in darkness during the transit so it was necessary to send astronomers to other parts of the world to make the necessary observations.

The English government decided to send an expedition to the island of Otahiti to observe the transit. It equipped a coal bark and renamed it *Endeavour* and placed a lieutenant, James Cook, in charge of the expedition. Cook sailed to Tahiti, as it is now known, with astronomer Charles Green to observe the transit. After various encounters with the Tahitian people, including one in which the telescope was stolen, Cook and Green successfully observed the transit on the morning of 3 June from Point Venus. After the observations were complete, the expedition packed up and began the second part of its voyage. Towards the end of 1769 Cook sailed to New Zealand, from there observing the transit of Mercury on 9 November. Having mapped New Zealand, Cook sailed further westwards and discovered the east coast of Australia early the following year.

Because of the *black drop effect*, shown in Figure 8.9, the observations taken at the transit were not of much use. Venus' atmosphere intervened to make the measurements inaccurate, but still they led to a better idea of the solar system's size. The next pair of transits of Venus will be of little interest to professional astronom-

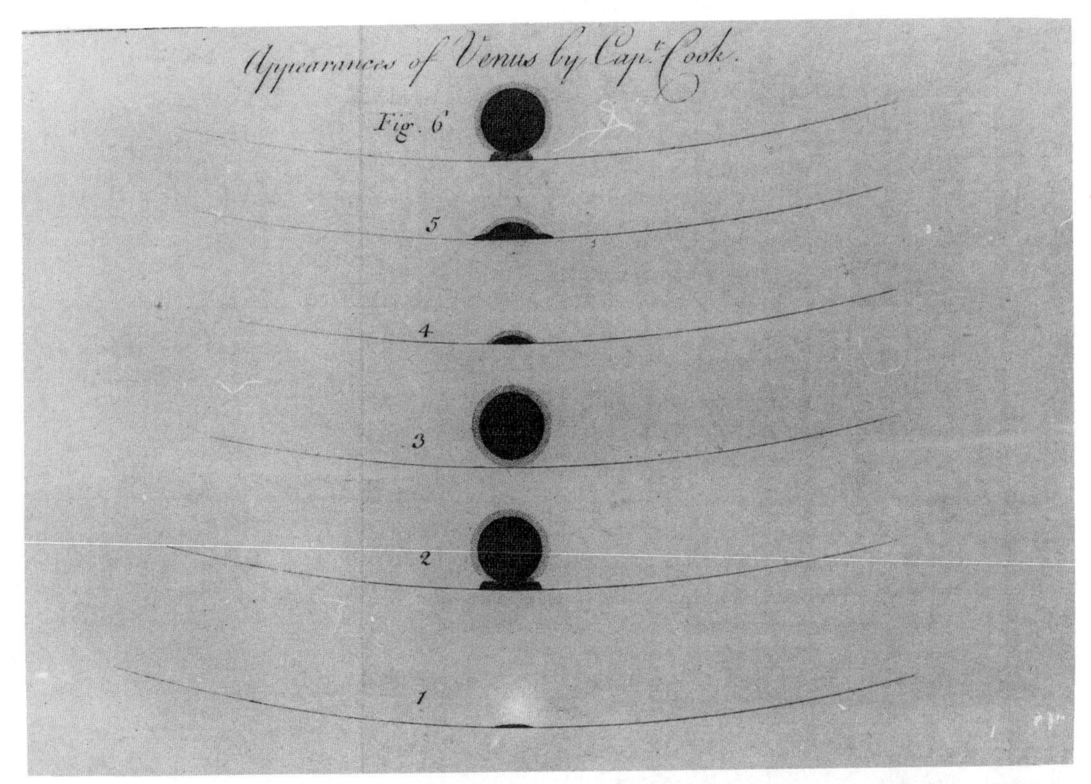


**8.7** Captain James Cook made these sketches of the 1769 transit of Venus across the Sun from his observing site in Otahiti. The observations show the ingress of Venus on to the disc and are the figures referred to in Cook's report shown in Figure 8.8.

ers, as radar measurements now give much more accurate distances to the planets than any observation of a transit can, but they will be watched with interest by many amateur astronomers.

The first telescopic observations of the planet were made by Galileo in 1610. The discovery that Venus (and Mercury) exhibited phases convinced Galileo that the planets must revolve around the Sun and not the Earth. In 1643 Franciscus Fontana noted dusky markings, and Giovanni Cassini observed a bright spot in 1667. Early attempts were made to correlate observed markings into a rotation period, but it was not until the nineteenth century that astronomers concluded that the entire planet was enveloped in an impenetrable dense cloud cover.

Although surface features on the planet are subtle and frequently non-existent, the amateur can make significant contributions by sketching any detail observed. A negative observation showing no visible markings is as important as a positive one. A circle, 50 mm in diameter, in an observation book is ideal for sketching Venus. Though the apparent size of the planet will vary according to the phase, the same 50 mm diameter circle should be used despite the planet's size. A soft lead pencil (2B or 4B) is ideally suited for recording any markings or detail.

One of the most puzzling aspects of Venus is the so-called ashen light. Its existence and cause has been a subject of controversy for close to 350 years. The ashen light is seen as a glow, or slight illumination of the darkened hemisphere of the planet; it has been likened to the earth-shine effect on the three- or four-day-old Moon. Nowadays it is accepted that this phenomenon is

Tranſit of Venus by Mr. Green, with a reflecting teleſcope of 2 feet focus, magnifying power 140 times.

| Time per clock h ′ ″ | | App. time June 2 |
|---|---|---|
| 9 21 45 | Light thus on the ☉'s limb, TAB. XIV. fig. 1. | 21 25 40 |
| 22 00 | Certain, fig. 2. | 21 25 55 |
| 39 20 | Firſt internal contact of ♀'s limb and the ☉ ſee fig. 4. | 21 43 15 |
| 40 00 | Penumbra and ☉'s limb in contact, ſee fig. 5. | 21 43 55 |
| | | June 3 |
| 3 10 05 | Firſt contact of penumbra, undulating, but the thread of light viſible and inviſible alternately | 3 14 3 |
| 10 53 | Second internal contact of the bodies | 3 14 51 |
| 27 30 | Second external contact | 3 31 28 |
| 28 16 | Total egreſs of penumbra, ☉'s limb perfect | 3 32 14 |

Tranſit of Venus by Capt. Cook, with a reflecting teleſcope of 2 feet focus, and the magnifying power 140.

| Time per clock h ′ ″ | | App. time June 2 |
|---|---|---|
| 9 21 50 | The firſt viſible appearance of ♀ on the ☉'s limb, ſee fig. 1. | 21 25 45 |
| 39 20 | Firſt internal contact, or the limb of ♀ ſeemed to coincide with the ☉'s, fig. 2. | 21 43 15 |
| 40 20 | A ſmall thread of light ſeen below the penumbra, fig. 3. | 21 44 15 |
| | | June 3 |
| 3 10 15 | Second internal contact of the penumbra, or the thread of light wholly broke | 3 14 13 |
| 10 47 | Second internal contact of the bodies, and appeared as in the firſt | 3 14 45 |
| 27 24 | Second external contact of the bodies | 3 31 22 |
| 28 04 | Total egreſs of penumbra, dubious | 3 32 2 |

The firſt appearance of Venus on the Sun, was certainly only the penumbra, and the contact of the limbs did not happen till ſeveral ſeconds after, and then it appeared as in fig. the 4th; this appearance was obſerved both by Mr. Green and me; but the time it happened was not noted by either of us; it ap-

7 peared

**8.8** This table was presented by Captain Cook to the Royal Society summarising his observations of the 1769 transit of Venus. The observations were made with a 'reflecting telescope of 2 feet focus and the magnifying power 140'.

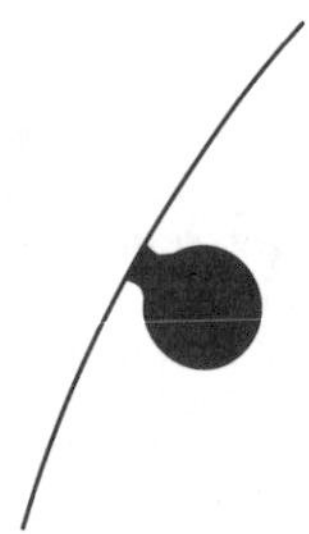

**8.9** Because of refraction through Venus' thick atmosphere and contrast effects, when a transit of Venus across the Sun's disc begins and ends the planet does not appear to cleanly enter the disc. A trail, looking like a drop, joins the planet to the Sun's limb for a short time after the planet has crossed fully on to the Sun's disc and before it has started to exit. It is this black drop effect which spoiled the attempts of astronomers to make accurate measurements during transits.

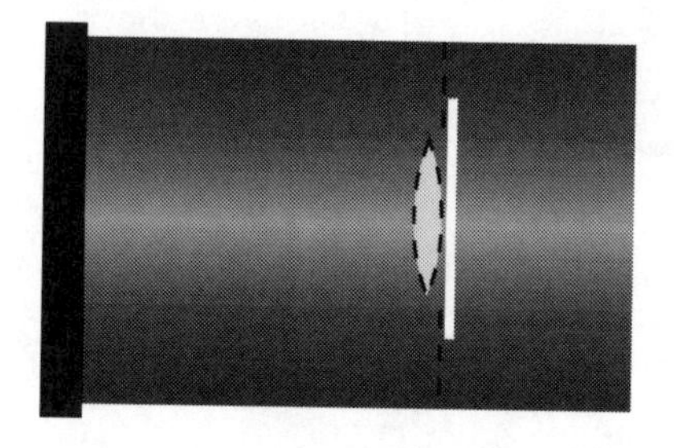

**8.10** A simple occulting bar can be made with a thin piece of paper or metal sheet. Making sure that the edge of the cardboard or metal is straight, tape the piece to the field stop of the eyepiece so that half the lens is obscured.

real, although optical defects and the scattering of light by secondary mirror supports in Newtonian telescopes can confuse the inexperienced observer.

The ashen light is difficult to see, as observations must be made against a dark sky, when observers are disadvantaged by Venus' brilliance. By blocking the light from the bright crescent it is possible to ascertain whether suspected ashen light is real or illusory. It is a simple matter to fabricate an occulting bar for this purpose: a strip of paper should be glued in front of the front lens of your eyepiece at a point where it will be sharply in focus. Eyepieces such as the Kellner, Orthoscopic and Plössls are suitable for this purpose, the strip of paper being mounted at the field stop as shown in Figure 8.10.

Occurring at irregular and infrequent intervals, there is some suspicion that the ashen light may be related to solar activity. It could be similar to terrestrial aurorae which intensify during solar outbursts. As Venus does not posses a magnetic field, some astronomers feel that lightning discharges in the atmosphere are responsible.

There is little hope of observing most other features in detail on the planet against a darkened sky. Daylight viewing is almost essential, or at the very least observations should be made during twilight. Dusky markings on Venus are regularly recorded by patient observers, and are best observed near the greatest elongations when the disc is large. You should not expect to see these markings each time you observe; they can be elusive and only appear as vague greyish, poorly defined areas. They are not permanent markings, merely transitory features of the permanent cloud cover. The bright spots as observed by Cassini are rare and difficult to see, and may in most cases only be a contrast effect between dusky regions.

Another regular feature of the planet that may be observed is the so-called cusp caps, bright regions that appear near the cusps (the pointed part of the crescent). Remaining visible for several days or even weeks, they are frequently seen to have a dark collar separating them from the rest of the disc; this again may only be the result of a contrast effect. The cusp caps are best seen during the crescent phase when the planet presents a good-sized disc to the observer.

Other features to note during observations are irregularities in the terminator. Sometimes the normally smooth curve will appear jagged or notched. As sunlight is scattered by the atmosphere of Venus, the cusps will appear to be projected more and more around the darkened portion of the disc as the planet nears inferior conjunction, until finally it is encircled by a ring of light. As Venus moves past inferior conjunction, the ring will diminish as the crescent broadens. The extensions of the cusps can only be seen during daytime, as Venus is close to the Sun and special care must be taken when observing under these conditions.

Observations with filters will often improve contrast and enhance detail on the planet. A red filter, because it absorbs blue light, will make Venus stand out against a blue sky, aiding daylight observation, and a blue filter will enhance any dark markings on the disc. There are some observers who question the value of filters, but it is worth experimenting with them to establish if they improve your view. When using filters it is a good idea to sketch any detail observed, noting the type of filter, and then make a second drawing without the filter for comparison.

Radar studies, ultra-violet photography and space probes have increased our understanding of the veiled planet, but there is still much work the dedicated amateur can do. The casual observer will also find much enjoyment in the planet which is Earth's 'next-door neighbour' in the solar system.

# 9

# Mars

Of all the planets in the solar system, Mars has long been the most fascinating. Others are bigger, better, closer and brighter, but Mars holds a place in people's imaginations. For hundreds of years, indeed even today, Mars has been seen as the possible abode of life. In H.G. Wells' *The War of the Worlds*, Mars was populated with a dying race of technologically advanced beings which enviously eyed the Earth as a place to live. To us it may seem far fetched, but remember the hysteria which occured in the United States when Orson Wells broadcast his dramatisation of the novel and many believed it to be real. It is the resemblances between Mars and the Earth which make it a fascinating planet, one which has caused considerable controversy throughout the years, as we will see. Today both the United States and the Soviet Union are making plans to visit this planet once more, this time with people. It could become another race like that to explore the Moon, but hopefully it will lead instead to co-operation in space.

It was Percival Lowell, director of the Flagstaff Observatory in Arizona, who did more to bring Mars to public attention than any other person. His vehement support for the existence of canals on the planet kept him in the newspapers and close to controversy. Though it caused great debate at the time, the issue of canals on Mars is now dead: photographs from spacecraft have shown that there are no long straight canals and no large patches of irrigated vegetation. The experience of Lowell and his canals does, however, point to some of the dangers inherent in observational astronomy.

Percival Lowell did not discover the canals of Mars; other observers had seen what they thought were long straight features connecting the dark areas of the Martian surface. It was 1877 when the Italian astronomer Giovanni Schiaparelli first recorded faint linear markings which he called *canali*. Schiaparelli recorded the markings near the limit of visibility. Percival Lowell had read the writings of Schiaparelli and when he learnt of Schiaparelli's retirement due to failing eyesight, he was motivated to undertake his own study of these faint features. Unfortunately, due to a mistranslation, *canali* became canals, suggesting that the features had been manufactured, rather than the more correct translation, channels, which implies natural features.

For many years the major work undertaken by astronomers at Lowell's observatory was to observe and map the canals of Mars at every opportunity. In addition, Lowell attempted to find the solar system's ninth planet, as Chapter 15 relates. Over the years Lowell and his assistants were able to build up incredibly detailed maps of the network of canals stretching across the Martian surface. There were sections of double canals, two running parallel to carry more water, junction lakes where systems of canals intersected and vast areas of vegetation which were irrigated by the system. Lowell was also able to deduce the system of irrigation apparently being used by the Martians by watching which parts of the surface became more verdant during the year.

There was one flaw with Lowell's system of canals. Very few other observers were able to see the canals at all. The dark markings everyone agreed upon, but as for the interconnecting lines, few could agree. Some people thought they could glimpse the lines occasionally under very good seeing conditions, but only Lowell could consistently map them. Debate raged as to whether the features were real or not, many dismissing Lowell as a rich crackpot.

Now that we know the truth, the question

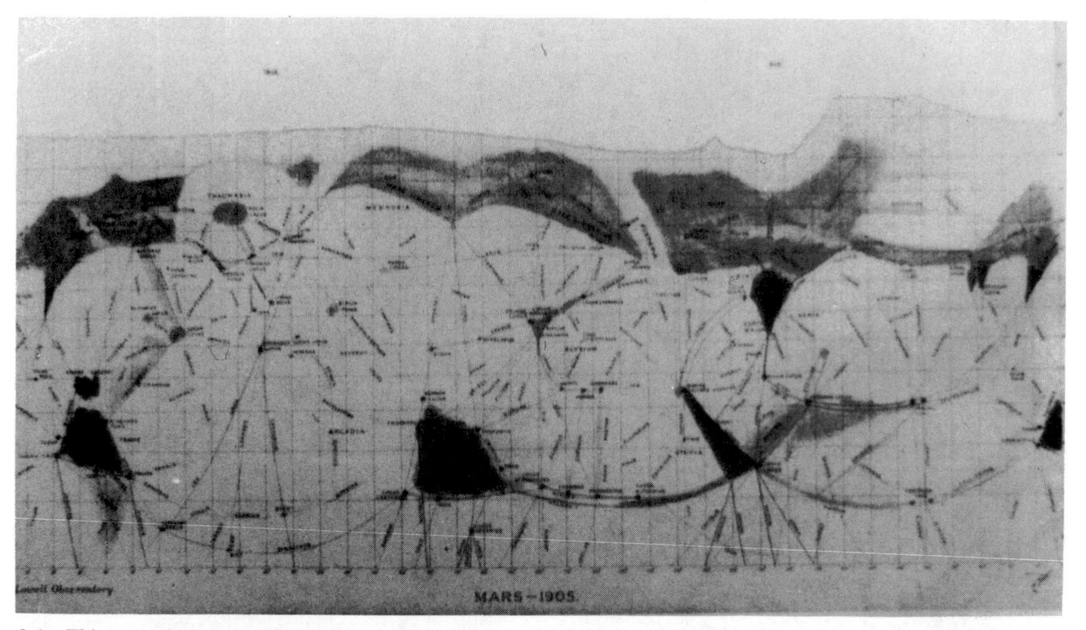


**9.1** This map of Mars was drawn by Percival Lowell from the observations he made during the 1905 opposition of the planet. The canals linking the major features of the Martian surface can be seen as can the extensive system of names Lowell attributed to them.

arises: was Percival Lowell simply a liar or did he really believe that the canals were real? From his writings it seems clear that Lowell was genuinely convinced that he was correct and that the canals and the Martians who built and maintained them were real. The problem lies in the fact that even Lowell admitted the canals were very near to the limits of resolution and seeing, and that they could only be glimpsed for a few moments at a time, also explaining why they were never successfully photographed.

In his early work Lowell shows few canals, only large straight features connecting darker areas, but as the years progressed his drawings of the system became more and more detailed, with new features, such as the double canals, appearing. It is usual for observers to increase in skill over the years, but this doesn't explain the extra details seen, as at some oppositions the planet was much further away than at earlier times. What may explain the new features is that Lowell knew what he wanted to see. At the limits of resolution it is difficult to see anything, and sometimes you simply can't be sure if a feature has been seen or not. If you know what you are looking for then it is sometimes easier to see. Lowell knew where to expect the canals, and so he saw them.

This points out a common pitfall many astronomers fall into: thinking that a feature is likely to be seen, they will see it, especially if the conditions make visibility marginal. For example, if you are going to observe Jupiter, never work out whether the red spot will be visible before observing. If you see it, or think you see it, record the observation, then work out if it was visible to confirm your sighting.

Even though there was controversy over the Martian canals, the idea that an intelligent race had constructed them took hold of the public's imagination. Science fiction writers have long used Mars as the source of their aliens; some even suggest that Martians have visited the Earth in the past (some suggest that they still do). Despite the fact that there is no evidence that life exists or ever existed on Mars, there are some who perpetuate the myth. A recent book resurrected the idea once more, claiming that a mountain was really a large human-type face carved by a now extinct Martian civilisation.

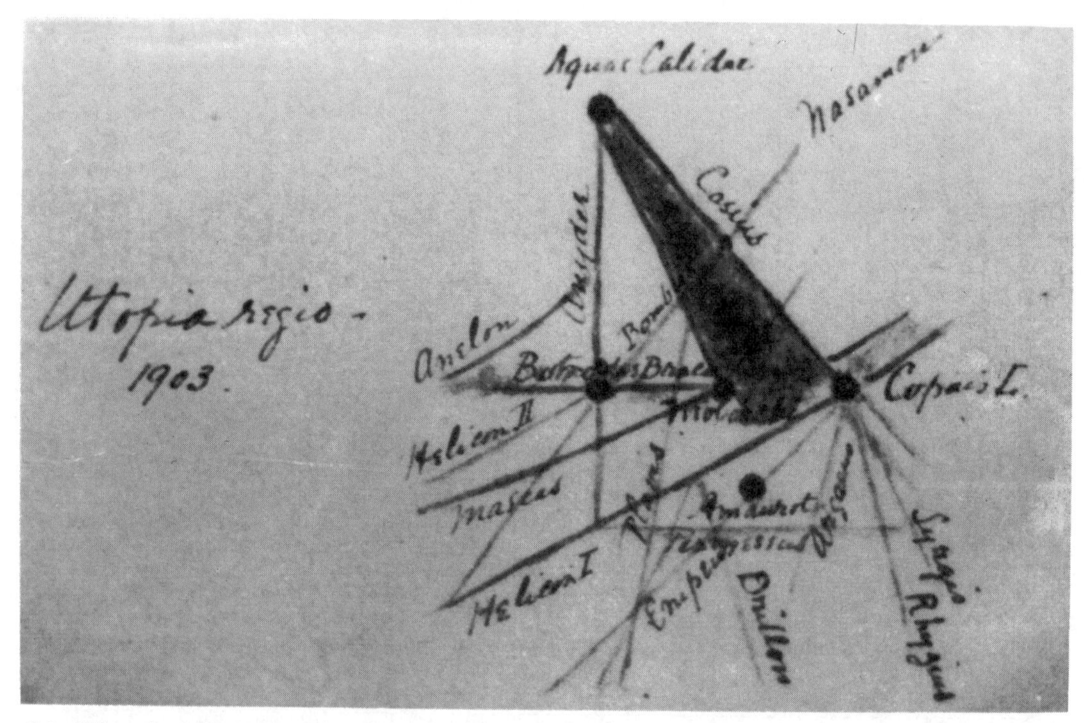


**9.2** This region, called Utopia Regio, is in the southern hemisphere of Mars. Lowell made a detailed map of this area during the 1903 opposition. Here we can see Lowell's system of main canals and tributary canals and their interconnections. Having seen such, alas non-existent, features on Mars, is it any wonder that Lowell considered the planet to be populated by intelligent beings?

Provided that the mountain is illuminated by the Sun in just the right way, it does look vaguely like a human face, but it doesn't look at all like one if lit differently. It is once more a question of seeing exactly what you expect to see, instead of being objective.

Why is it that Mars has attracted the attention of people as a possible abode of life more than any other planet? It is because Mars is the planet most like the Earth—but not by much. There are a number of similarities between Earth and Mars, most of them superficial. Both planets have days of approximately the same length—24h37m for Mars—and the tilts of their axes are close—24° for Mars—so that Mars experiences seasons similar to those of the Earth, but twice as long. The planet also has polar caps and a thin atmosphere and the temperatures are not too far removed from those found on Earth. If life were to exist on another planet, it would probably be Mars.

Mars has also had more Earthly visitors than any other planet. In July 1965 the United States' Mariner 4, arrived at Mars transmitting back 22 views of the surface. These pictures shocked scientists. The only other planet visited until this time was Venus and its thick clouds had prevented the surface being seen. Mars showed craters similar to those of the Moon. Planetologists had been expecting a landscape similar to the Earth's valleys, plains and mountains with some volcanoes, not craters.

In 1969 two more United States probes, Mariners 6 and 7, reached the planet. This time the pictures were of much higher quality showing details as small as half a kilometre; again the surface was mainly cratered. In November 1971 Mariner 9 reached Mars and went into orbit about the planet, the first spacecraft ever to orbit another world. The aim of the mission was to photograph 70 per cent of the planet's surface at a resolution of about 1 km. Unfortunately

**9.3** This photograph returned from Mariner 6 on 4 August 1969, shows a heavily cratered region in the southern hemisphere of the planet at Meridiani Sinus. It was photographs such as this one that initially led scientists to believe that the planet was cratered all over like our Moon. It wasn't until later missions, when spacecraft were in orbit around the planet, that scientists got the whole picture. PHOTO CREDIT: JPL/ NASA

**9.4** This view from Viking 1 shows the surface of Mars in the Chryse Planitia Basin where the spacecraft landed. The fine sand of the area is littered with small and large rocks. PHOTO CREDIT: JPL/NASA

there was nothing to see: Mariner 9 discovered that Mars was sometimes blanketed by planet-wide dust storms which obscured the surface for months. Luckily the spacecraft was in orbit about the planet so the scientists could afford to wait for the storm to abate. When it did, a new world was seen for the first time.

The earlier missions which had seen only cratered land had been unlucky. Mariner 9 showed that indeed there were craters, but each new orbit revealed much more. Mariner 9 found Olympus Mons, the largest volcano in the solar system and Vallis Marinaris, the biggest canyon system. It even found drainage channels which would dwarf many a good-sized river on Earth.

The experience of the first three probes sounds a warning still relevant today. With only a few images of the moons of the outer planets, we should not be too hasty in drawing conclusions. In the case of both Neptune and Uranus, only one half of the moons have been seen; the secrets hiding in the dark will no doubt surprise future scientists as much as the Mariner 9 pictures did in 1971.

The Soviet Union has also been active in the area of Mars exploration, but has been unlucky. Their first two probes, Mars 2 and 3, were sent at the same time as Mariner 9; they too arrived in the midst of the dust storm but had to land anyway, so neither probe returned much data. Four probes were launched in 1973; Mars 4 and 5 were orbiters, Mars 6 and 7 were landers. Mars 4 and 7 missed the planet because of engine failures; the Mars 6 probe failed during descent; only Mars 5 achieved an orbit and returned data and pictures.

The most detailed study of Mars has been done with the two United States' Viking spacecraft. Both Viking 1 and 2 consist of an orbiter unit which was to photograph the planet's surface and relay data back to Earth and a landing section designed to analyse the soil and look for life. Viking 1 was to land on Mars on 4 July 1976 as part of the United States Bicentennial celebrations, but a study of the proposed landing site showed that it was too rocky to be suitable, so a further sixteen days passed looking for a new area. On 20 July 1976 the Viking 1 lander made a successful landing in the Chryse Planitia and two months later the Viking 2 lander set down in the Utopia Planitia.

Although the landers returned detailed panoramic views of the surface of Mars, their primary mission was the search for life. The search took three forms, the first being the cameras. These were such that they would see any creature big enough to be seen by a human eye. Also, by looking at the same area over many months, changes caused by the growth of plant life would be seen. The second instrument was a gas chromatograph which could analyse the soil and air for any traces of the molecules we associate with life. The last item was the most sophisticated, the Viking biology instrument.

The biology experiment consisted of a box about 300 mm in size filled with three separate experiments: the gas exchange, labelled release and pyrolytic release experiments. The gas exchange experiment relied on feeding any Martian organisms. It was hoped that if the organisms used any of the dozen nutrients provided they would change the composition of gases in the chamber sufficiently to be detected by the gas chromatograph. The labelled release experiment assumed that Martian life, like that of Earth, would be based on carbon atoms, so nutrients labelled with radioactive carbon atoms were fed to any potential organisms. These would absorb some and then excrete them again into the air where the radioactive atoms could then be detected.

A problem with the first two experiments was that the organisms mightn't like the food and therefore not absorb it; also the presence of water, foreign on the Martian surface, might have a detrimental effect on the organisms. The third experiment worked by keeping the conditions inside the experiment the same as those outside, but with radioactive gases added which would be absorbed by the Martian organisms. Later the soil was heated to 750°C; by testing the gases given off it would be possible to tell if any of the radioactive gases had been absorbed by the Martian creatures.

The results of these experiments were at first promising. The first data from the gas exchange and labelled release experiments showed positive results. Further testing over the ensuing days showed, however, that the results were not biological in nature but could be explained by chemical reactions within the soil. The pyrolytic release experiment which took five days to work also showed positive results initially, but further work showed that this too was the result of chemical, not biological activity. As the results stand, there is no life on the Martian surface, but there is some interesting soil.

If the search for life was disappointing, the examination of the Martian atmosphere and weather made up for it. The various instruments aboard the landers were able to take readings during their descent through the atmosphere, and once on the surface could continue their work. Also, because the landers remained working for such a long time, over six years, extensive studies of the Martian weather patterns could be made.

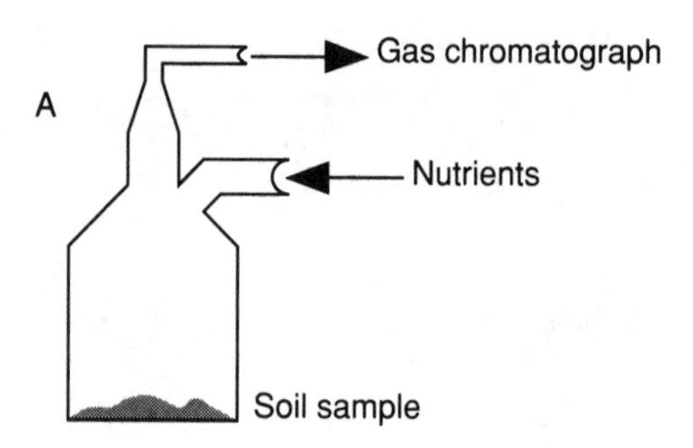


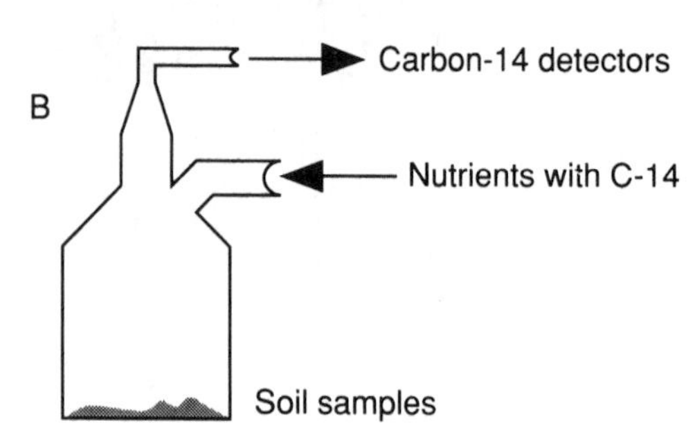


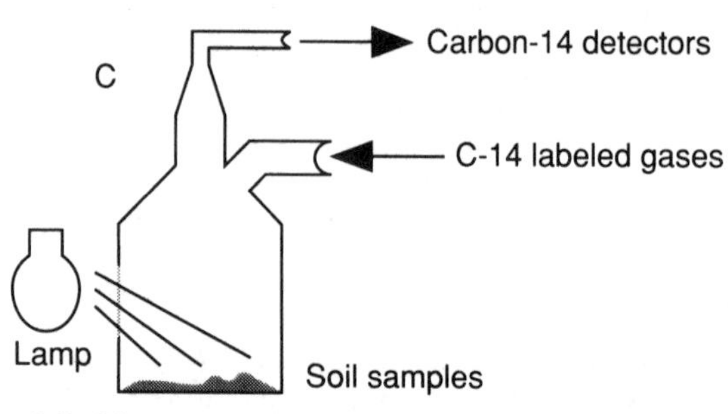


**9.5** Three experiments were contained on the Viking spacecraft in an attempt to detect life on Mars. The experiments were:
A The gas exchange experiment;
B The labelled release experiment; and
C The pyrolytic release experiment.

The Martian atmosphere is similar to the atmosphere of Venus, except that it is much

**Table 9.1 Mars' atmosphere**

| Gas | | Fraction |
|---|---|---|
| Carbon Dioxide | $CO_2$ | 93.5% |
| Nitrogen | $N_2$ | 2.7% |
| Argon | Ar | 1.6% |
| Oxygen | $O_2$ | 0.07% |
| Water | $H_2O$ | 0.03% (variable) |
| Neon | Ne | 0.00025% |
| Krypton | Kr | 0.00003% |
| Xenon | Xe | 0.000008% |
| Ozone | $O_3$ | 0.000004% |

thinner—only 0.77 kPa, or 1/130th that of the Earth. It has been calculated that if the Earth had no life, the composition of its atmosphere would be similar to Mars and Venus too. Because of the tenuous nature of the Martian atmosphere, it provides little blanketing for the planet and no greenhouse effect to increase temperatures, so Mars is a cold place. The Viking landers measured the average daytime air temperature at −30°C and the land temperature at a few degrees above zero in summer. At night the temperature of both the land and air dropped to around −100°C.

The temperatures on the planet and its fairly brisk rotation are the determining factors in the Martian weather. These are similar to the forces which drive terrestrial weather, but there the similarity ends. Because Mars has no seas, its weather is much less complex than Earth's, the major changes being the amount of carbon dioxide in the atmosphere. Mars' store of carbon dioxide is its polar caps where it is cold enough for carbon dioxide to exist as a solid. As the temperature at the Martian poles changes, carbon dioxide is either frozen out of or released into the atmosphere. The amount of gas given off can cause a variation in air pressure of as much as 20 per cent. Winds blow from the polar caps towards the equator, taking clouds of carbon dioxide crystals and water vapour with them.

It is only during summer in the southern hemisphere that conditions change markedly. Because it is highly elliptical, the orbit of Mars takes the planet significantly closer to the Sun at perihelion than at aphelion; perihelion occurs at the beginning of the southern hemisphere summer. This extra heating makes the south polar cap shrink dramatically and also causes continuing high-speed winds to blow towards the equator. These winds whip up dust into the

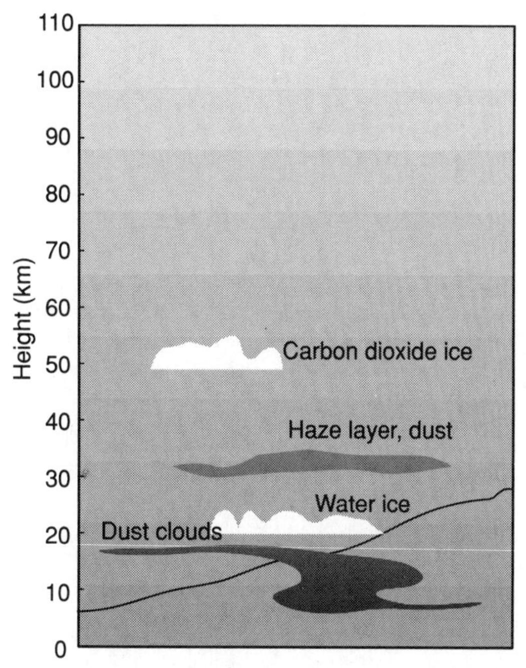


**9.6** The structure of the Martian atmosphere shows clouds of dust, water ice and carbon dioxide ice at various heights. Dust clouds enable the atmosphere to absorb more energy from the Sun, warming it considerably and leading to the Martian weather systems.

atmosphere, creating a global dust storm which can continue through summer (six Earth months). The winds which do this must blow at 150 km/h or faster to be able to pick up the particles of the Martian surface. Even when it is not summer in the south, large numbers of tiny dust particles are in the Martian atmosphere, making the sky a pink colour. In the same way, small particles of dust in Earth's atmosphere give our sky its familiar blue colour.

The 1988 opposition was almost free of dust storms, at least until late November some two months after opposition. The southern hemisphere was obscured by dust for two weeks until the storm abated, but fortunately it happened many weeks past the best observing period. Although you may feel cheated if a dust storm interrupts your observations, you should utilise the opportunity to study the storm. Details of where storms originate, their duration and coverage are important facts that will help in the

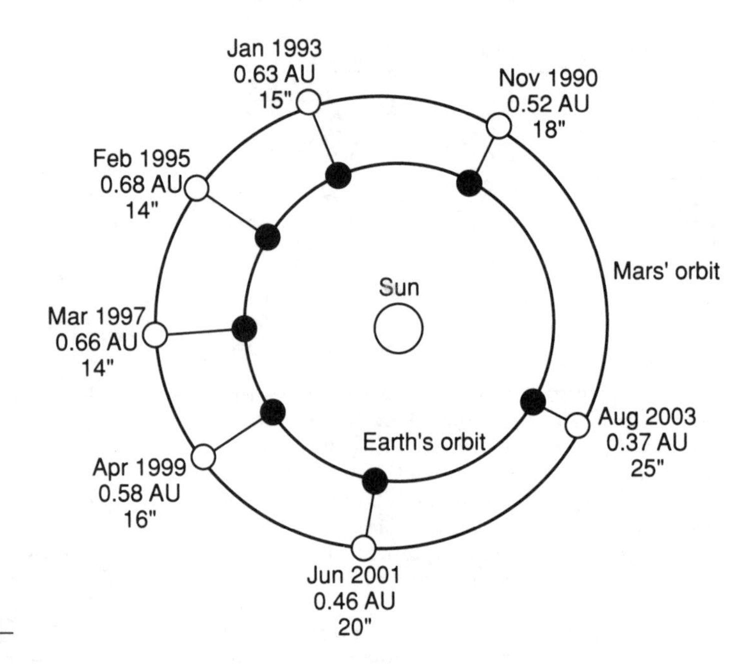


**9.7** Each opposition from 1990 to 2003 is shown with its date and distance from the Earth in astronomical units. The apparent diameter of Mars in arc seconds is also given.

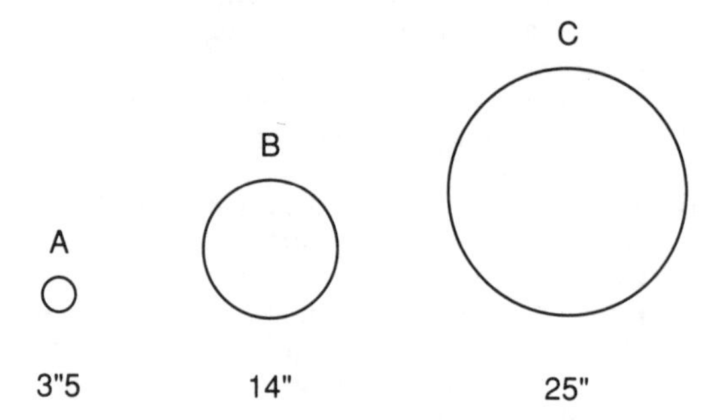


**9.8** The comparative sizes of Mars at: A—conjunction; B—poor opposition; C—favourable opposition.

placement of future probes on the planet. The use of yellow, red or magenta filters will aid in defining storm-cloud boundaries.

Earth-based observers trying to unravel the Martian mysteries are faced with the problem of its elliptical orbit. The average distance from the Sun of 228 million km varies by a considerable 42 million km. The relative motions of Earth and Mars enable us to get a close view of the Red Planet every 22 months, but only once every fifteen or seventeen years do we pass Mars when it is closest to the Sun at perihelion, affording us a good close-up view.

At these close approaches, or oppositions, the Martian disc enlarges appreciably. At a favourable opposition, the disc will have a diameter of about 25 seconds of arc. Even at a poor opposition the disc is about fourteen seconds of arc, still large enough for a good view. When near conjunction, the planet will only be about 3.5 seconds of arc in diameter; the tiny disc, smaller than far-flung Uranus, is hardly worth telescope time. Figure 9.8 shows the comparative disc sizes from conjunction to poor and favourable oppositions.

It is easy to tell when Mars and any of the other outer planets are at opposition: they rise in the east as the Sun sets in the western sky. By midnight, at their highest altitude, they cross the celestial meridian, with the Sun directly below your feet. The last good perihelic oppositions occurred in 1971 and 1988, and the next happens in 2003. The oppositions up until the year 2003 are detailed in Figure 9.7. Fortunately for

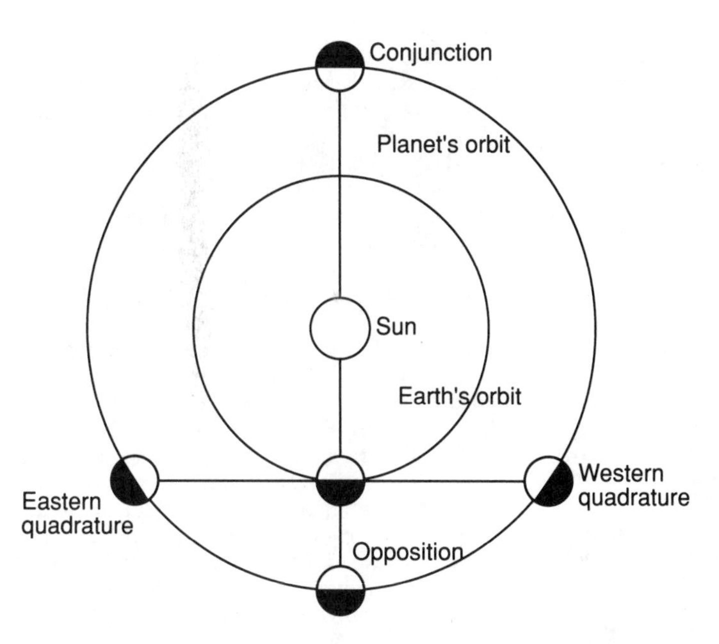


**9.9** In the orbit of an outer planet four positions are important. At eastern and western quadrature the phase of the planet is seen at its greatest, though even for Mars the size of the phase is small. At conjunction the planet is hidden from Earth as it passes into the glare surrounding the Sun. At opposition the planet is at its best for viewing; not only is it in the sky throughout the whole night, but it is also at its closest to the Earth.

southern hemisphere observers, the planet is always well placed for observation at perihelic opposition, its declination placing it high in the southern sky.

The *synodic period*, the interval between successive oppositions, is 780 days (the sidereal period, the Martian year, is 687 days). This is the longest synodic period of the outer planets. All others, because of their slower movement, come to opposition once each year.

The outer planets can appear in any position along the ecliptic, travelling in a west to east direction, and are not restricted to the twilight sky. The inner planets, however, appear to oscillate from one side of the Sun to the other. Sometimes the passage of an outer planet against the background stars seems to reverse direction for a while, make a loop, and then resume its easterly movement. The reason for this seemingly complicated motion is shown in Figure 4.4. As the Earth catches up and overtakes a slow moving outer planet that planet will appear to slow in its eastward motion, stop, move backwards in a westerly direction, stop again, and finally return to its eastward motion. If the Earth overtakes a planet near one of its nodes, the loop becomes an 'S' bend.

The outer planets display phases, although these never grow less than the gibbous stage. The further the planet, the lesser the effect. Mars is the most obvious, with 89 per cent of the planet illuminated at minimum phase; the more distant Jupiter's phase is hardy detectable. At conjunction Mars is in full phase. During its westward motion the western limb will show a narrow dark crescent, the crescent growing until reaching maximum at quadrature, when Mars forms a right angle with the Earth and the Sun, about three months before and after opposition. It then begins to diminish, finally disappearing as the planet comes into opposition; the process is reversed after opposition. The Martian phases are easily seen—Galileo even noted them with his tiny telescope.

During favourable oppositions, Mars reaches magnitude −2.5, outshining all stars and even Jupiter. When opposition happens near aphelion, Mars is still a respectable magnitude −1.0, dropping down to magnitude 2 at conjunction, but still easily identifiable by its colour. You will find that Mars is bright one year and fainter the following year.

Mars is the only planet whose surface details can be seen from Earth. Because of this it has probably been observed more than any other planet. But, why observe Mars after the attentive scrutiny of the space probes? Can we learn any more? The answer is yes. The probes are no longer in operation, and we need to know as much as possible about any surface feature changes, dust storms and clouds. As there is serious talk about future missions, possibly crewed missions, this information, derived mainly from amateur observation, will be invaluable.

A 60–80 mm telescope is of sufficient size to show the polar caps and the better known dark areas. For serious work, telescopes of 150–300 mm aperture using colour filters are preferred. Drawings of the planet should be made with a soft lead pencil on to prepared observation blanks, no more than 50 mm in diameter. As the rotation period of Mars is 24.6 hours, you will need to work fast; after half an hour or so the planet's motion will be obvious. Record your impressions of the planet before consulting a map to identify features; this will eliminate any preconceived ideas of what you would expect to see.

Since the Martian day is about 40 minutes longer than that of the Earth, an observer viewing at the same time each night will see features cross the central meridian 40 minutes later each time. The 40 minute delay equals about 8° of longitude per day, and after six weeks, weather permitting, the observer would have seen all the disc features as they crossed the central meridian.

As the disc is at best only 25 arc seconds in diameter, you will need to use the highest magnification possible. Choose an eyepiece that will give a power of about 1.6 times the diameter of the objective in millimetres. If the atmosphere is steady you may be able to use an even higher power.

A prerequisite for observing detail on Mars is a set of coloured filters. They will improve contrast, bringing otherwise hard-to-detect features into prominence. You can buy sets of high quality glass astronomical filters, but these tend to be expensive. If you own a set of these filters, the type that screw into the bottom of the eyepiece, it is recommended that you hand-hold them over the eyepiece rather than attaching them, as you will need to change filters regularly throughout the observing session. Time spent changing filters can be annoying.

A cheaper alternative is the gelatin filters available from camera stores or optical supply companies. These come in small sheets that can be cut and mounted into 35 mm slide frames. As it is difficult to sketch while holding a filter over the eyepiece, you could cut a hole in a large plastic bottle lid, and glue the filter material over the hole; this way both hands will be free. Take care with gelatin filters, as they are soft and can scratch and mark easily.

Naming features on Mars was a task undertaken by many observers. From Earth the light and dark patches seen were thought to be deserts and seas, thus they were given corresponding names. When spacecraft arrived at the planet they found that the light and dark regions did not correspond to anything much, so a new system of naming had to be adopted. Craters were named after scientists; thus there are two craters called Copernicus, one on the Moon and one on Mars. Large valleys were called *vallis* and are named after towns on Earth or after Mars in other languages—Nirgal Vallis and Kasei Vallis from the Assyrian and Japanese words for Mars respectively. Mountains are named *mons* after the Latin word for mountain and plains are called *planitia* also from Latin.

The surface of Mars shows many different types of terrain, evidence that the planet has been geologically active in the past. Seismometers carried by the Viking landers failed to detect any marsquakes, though the results from the experiments are not conclusive as neither instrument functioned very well. It is expected that if there is still seismic activity on Mars it is fairly minor. Mars has a density of 3.9 $g.cm^{-3}$ meaning that it cannot have a large metal core like the Earth's, though observations of spacecraft orbits have shown that some sort of core does exist. The core is likely to be solid iron sulfide and make up around 16 per cent of the planet's mass. One consequence of having a solid core is that Mars has practically no magnetic field.

A map of Mars shows a number of distinct regions. The southern hemisphere is much more heavily cratered than the northern, indicating that the land in the south is much older. The old surface regions in the south are also, on average, 4 km higher than regions in the northern hemisphere. Some event must have happened in the northern hemisphere causing the reworking of the surface and its lowering by a few kilometres. A second strange feature is the Tharsis bulge near the equator of the planet. This is a huge region, the size of a continent on Earth, which is 10 km higher than its surroundings. It is also the least cratered region, making it the youngest part of the planet's surface.

The largest mountain in the solar system is on Mars. Originally noticed from Earth because of the white clouds which collected around its

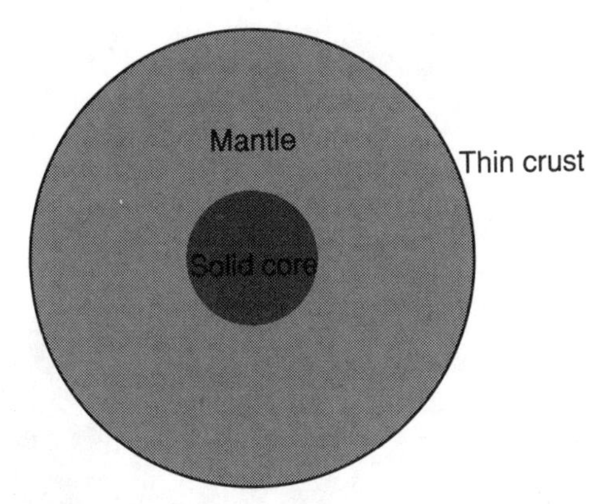


**9.10** Being smaller than the Earth, there is not sufficient internal heat in Mars to keep any of its nickel-iron core liquid. Like the Earth, though, a silicate-rich mantle surrounds the core with a thin rocky crust on top.

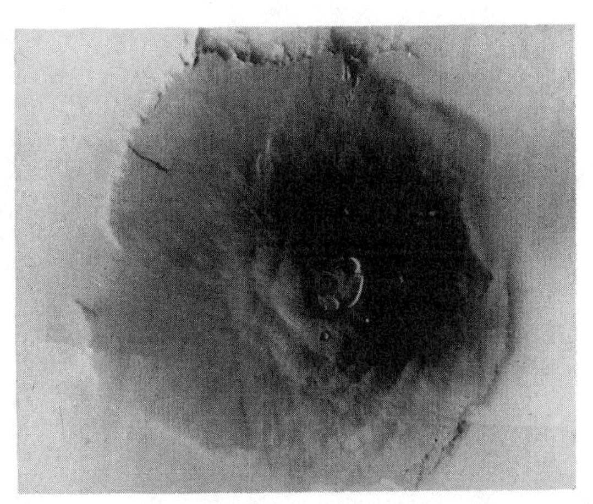

**9.11** This composite photograph of the huge Martian volcano Olympus Mons was taken by the Mariner 9 spacecraft which flew past Mars on 13 November 1971. Olympus Mons is still the largest volcano in the solar system, over 25 km high and 700 km in diameter. The craters on the mountain are obviously volcanic rather than the more common impact craters. PHOTO CREDIT: JPL/NASA

slopes, it was named Nix Olympica (Snows of Olympus). Situated near the Tharsis bulge, Nix Olympica turned out to be an extinct volcano rising 25 km above the surrounding area. The base of this volcano, renamed Olympus Mons, is over 700 km across, making it larger than many countries on Earth. No volcano or mountain on Earth can ever be 25 km high, it could not be supported by the crust, but would just collapse under its own weight. On Mars, where the surface gravity is only two-fifths as much as Earth's, such a mountain can be supported.

As if to balance such a huge mountain, Mars also has the deepest canyon system in the solar system. Also associated with the Tharsis bulge is the Vallis Marinaris, named after Mariner 9 which discovered it. The Vallis Marinaris is so large that it was previously known as the Coprates canal, a canal mapped by Lowell which actually turned out to be a real feature. Vallis Marinaris is just the central part of a series of east-west canyons running for over 4000 km. The individual canyons are around 3 km deep, but Vallis Marinaris is 7 km deep and up to 500 km wide at some places. It appears that Olympus Mons, Vallis Marinaris and the Tharsis bulge are all the result of prolonged tectonic activity in Mars' recent past. Cratering densities show that Olympus Mons is less than 100 million years old, and much of the surrounding area is not much older.

Lowell's canals were non-existent, but had he been a little earlier (4000 million years) he would have seen running water on the surface. The high-resolution photographs of the Martian highlands show numerous examples of the effects of running water. Throughout the highland regions, channels running from high to low are seen, features carved by the erosion effects of running water either from underground springs or from rainfall.

None of these features could be seen by the early astronomers who studied Mars. They concentrated on mapping dark areas which showed seasonal variation, leading them to believe that what they saw were areas of vegetation. Many scientists doubted these explanations and sought alternative solutions to the problem of the light and dark areas. From Earth, Mars appears distinctly reddish in colour; that is why it was named after the Roman god of war. The red colour is caused by the soil which covers the planet. Similar to the iron-rich soils found on many parts of the Earth these soils give the planet its distinctive hue.

As the Viking landers showed, the loose soil is lighter in colour than the rocks underneath it. Winds blowing according to the season can lift the soil, moving it around the planet as dust storms. In some regions the rocks underneath are uncovered, revealing a darker surface; in

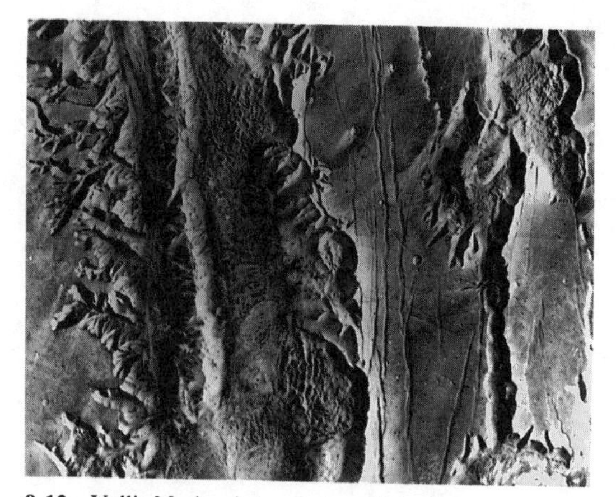

**9.12** Vallis Marinaris is a huge gash in the planet's surface running over 4000 km around the planet, up to 7 km deep and 500 km wide in some places. The valley is the result of prolonged tectonic activity rather than the work of erosion, though erosion has undoubtedly assisted in some of the deepening of the channel and the formation of some of the smaller features associated with it. PHOTO CREDIT: JPL/NASA

**9.13** The compositions of the Martian polar caps, easily visible from Earth, has provided an enigma for many centuries. Here at the north pole of Mars, high cliffs of carbon dioxide ice form as part of the cap melts with the approaching northern summer. PHOTO CREDIT: JPL/NASA

other regions dark rocks are covered by the blown soil, becoming lighter. Because of the regular nature of the winds blowing over the surface, the covering and uncovering of the surface is a regular process leading to the cyclical variation in the colour of some areas, and the assumption that the areas were vegetation.

The view of Mars in an amateur's telescope will reveal the irregular dark, greyish-green surface markings. Over the course of an opposition the dark areas can change in form and intensity as strong winds shift the dust about the planet, covering or exposing them. During the 1988 opposition, observers saw several new features, including four dark areas that had been lost from view since last century.

The most prominent of the markings is known as Syrtis Major, a large, dark, wedge shape found just north of the equator, and visible in even the smallest instruments. Contrasting with the darker area surrounding it, is a region known as Hellas. This lies in the southern hemisphere directly below Syrtis Major. Hellas is a depression and, when covered with dust, it can be very conspicuous. Another feature worth searching for is called the Eye of Mars, or Solis Lacus, a small dark region in the south, ringed by lighter material. A red or orange filter will highlight these dark features, and possibly show others unseen without a filter.

A problem amateur astronomers face when waiting to view Mars near to opposition is the dust storms. Luckily the 1988 opposition was almost free of dust storms, at least until late November some two months after opposition. The southern hemisphere was obscured by dust for two weeks until the storm abated; fortunately it happened many weeks past the best observing period.

The polar caps are another region which can be seen easily from Earth. Through a telescope, the poles of Mars are covered by white caps which grow and shrink according to the season, in the same way as the polar caps of the Earth grow during their winter. It is not obvious that much of the inhabited parts of our northern hemisphere are underneath the Earth's polar cap during their winter, but the snow which covers the ground of Canada, the Soviet Union and much of Europe is part of the seasonal northern polar cap. It is the southern polar cap of Mars which shows the greatest seasonal variation, because of the eccentric Martian orbit. It extends as far north as 55°S during the southern

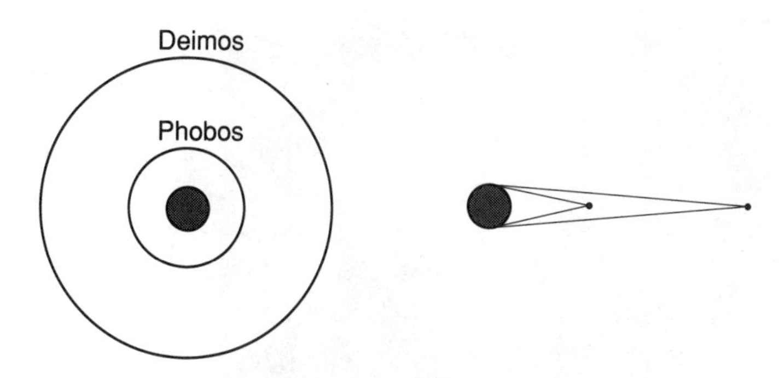


**9.14** Drawn to the same scale here, the orbits of Phobos and Deimos are shown in relation to Mars. Because the moons are so close to the planet, there are places on the surface from which the moons are never visible. To see Deimos an observer must be more than 8° from either pole, while to see Phobos it is necessary to be at least 21° from the poles.

winter, almost disappearing during summer. The seasonal polar caps of Mars are made of frozen carbon dioxide, dry ice. Carbon dioxide will condense out of the Martian atmosphere whenever the temperature is below −150°C. The seasonal caps vary in thickness from a metre near to the permanent caps, to a few centimetres near the edges.

As the seasonal polar caps retreat during spring they reveal a brighter permanent cap which lasts through the summer. Although, like the Earth's polar regions, the Martian south pole receives continuous sunshine through summer, its temperature remains at −150°C so the permanent cap is also probably mostly carbon dioxide ice with a little water.

The southern polar cap shows an interesting pattern as it melts during summer. The area is made of many valleys and mountains; as the sun strikes the slopes the ice there melts, leaving rock, but the ice in the valleys remains, giving a swirled appearance to the polar cap. Because some of it is able to melt away, the permanent cap cannot be more than a few metres deep.

The northern permanent polar cap is much larger than the southern one, never shrinking below 1000 km in size. It too shows the swirl pattern due to valleys and mountains, but there the temperature rises to more than −80°C, much too warm for carbon dioxide to remain solid. The northern cap must therefore be made mainly of water. Observations that the humidity of the Martian atmosphere increases when the northern polar cap shrinks support this theory.

During the 1988 opposition, as the south pole gradually thawed over a period of months, observers noted dark rifts across it. Protrusions and detached areas are commonly seen at the poles, probably caused by uneven melting of the caps because of high or low ground. One such highland area or ridge is named Novissima Thyle, or Novus Mons. As the south pole melts, the ridge can still retain an ice cover although the surrounding low areas are ice free. The same effect happens at the north pole, with a similar ridge named Rima Borealis. A dark collar known as Lowell's Band can sometimes be observed as the poles retreat during their thaw. Green and red filters are best for detecting polar projections and boundaries.

As the polar caps retreat, thin whitish clouds sometimes form. At times the clouds are dense enough to obscure some surface detail, lasting several days before dissipating, but most are tenuous and short lived. It is often difficult to gauge if a white spot is a cloud formation or a bright patch of frost on the surface. Filters can be a tremendous help in identifying cloud type. Yellow or green filters can distinguish surface frost and fogs from the higher level cloud, which is best seen in blue or violet light. If a white area is sharply defined and bright when using a yellow or green filter, it is most probably near or on the surface; conversely blue and violet filters bring the higher clouds into prominence.

Mars has two moons, both discovered in 1877 by Asaph Hall. Named Deimos and Phobos, they appear to be captured minor planets rather than proper moons. The idea that Mars had two moons was not new; in *Gulliver's Travels*, published in 1726, Jonathan Swift has scientists of Laputa state that Mars is accompanied by two satellites. His reasoning was not astronomical, however: as Mars is further from the Sun than the Earth it would need two moons to give illumination at night. Voltaire had followed similar reasoning, while Kepler thought that it must have two as Earth had one moon and Jupi-

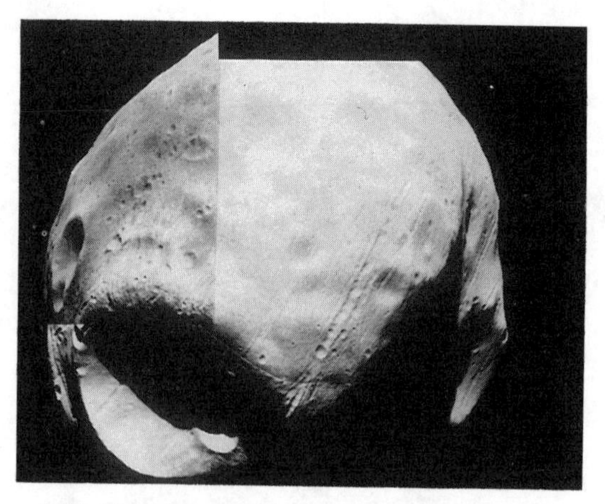

**9.15** This photograph of Phobos, the closer of Mars' two moons was taken from a distance of 600 km by the Viking 1 orbiter in 1978. Phobos is thought to be a good example of what a minor planet would look like close up. Because of the moon's small mass, it can remain irregular in shape, gravitational forces being too weak to mould it into a sphere. Evidence of past collisions and impacts is seen in the heavy cratering which covers the surface of the moon. PHOTO CREDIT: JPL/NASA

ter four, so Mars had to have two to fit the series.

Phobos is the larger of the two moons. Being potato-shaped, it is 27 × 22 × 19 km in size. Phobos orbits Mars at a distance of 9380 km in a period of just 7.7 hours. Because it rotates about the planet much faster than the planet revolves, an observer on Mars would see Phobos rise in the west, scoot across the sky and set in the east about six hours later. Six hours further on the moon would rise again.

Deimos is smaller, 15 × 12 × 11 km, and much further from the planet, $2.3 \times 10^6$ km. Correspondingly, it rotates more slowly, taking 30h18m to circle the planet once. Because of its distance, Deimos cannot be seen from either of the Martian poles; an observer must be within 82° of the equator to see Deimos, and within 69° of it to see Phobos. An observer watching Deimos would see it rise in the east and, because its orbital period is close to that of Mars, it would take 131 hours before it set in the west.

Both Deimos and Phobos look like what we think a minor planet looks like. They are irregularly shaped and pock-marked with craters. Phobos is mostly the same colour while Deimos has spots on its lighter surface. Phobos is only a temporary resident of the Martian system. Because it rotates faster than the planet spins, Phobos' orbit is decaying, bringing it closer and closer to the planet. At the present rate of decay, Phobos will crash into Mars in about 30 million years' time.

With opposition magnitudes of 11.3 and 12.4, both moons are technically within range of a 150 mm telescope; unfortunately, however, the glare from Mars makes it virtually impossible to see them. It is difficult even with a 300 mm instrument.

As Mars diminishes in size and brilliance after each opposition, there is a void left in the amateur's planetary observation program, at least until the next apparition two years later. It is always worth the wait and expectation: nobody can predict what the Red Planet will reveal to us next time.

# 10

# The minor planets

Before all the factors affecting the formation of a planetary system were understood, astronomers were often attracted to patterns they found amongst the planets. Following Kepler's success in calculating the motions of the planets it was hardly surprising that others would follow his lead. Many tried to find out why the planets have the spacing we observe. The best known attempt was made by Johann Titus in 1766. Titus found a surprising relationship between the distances of the planets, discovering that the distance between a planet and the Sun was given by a simple rule:

1 Start with the series 0, 3, 6, 12, 24, ... in which each term is double the one before.
2 Add 4 to each number.
3 Divide the result by 10.
4 You now have the distance to the planet in astronomical units.

Titus included this rule in a translation of a book by Charles Bonnet called *Reflections on Nature* in which Bonnet attempted to show God's handiwork in the order of nature. The rule was later publicised by Johann Bode, who was working at the Berlin Academy of Sciences. He included it in the *Berliner Astronomisches Jahrbuch* he was hired to edit. Published in 1772 without credit to either Bonnet or Titus, the rule became known as Bode's Law. In 1834 Bode finally acknowledged that Titus was the source of the rule.

A glance at Table 10.1 will show a gap between the orbits of Jupiter and Mars. There should have been a planet, but none had been found. In 1781, when Uranus was discovered it was found to fit into the pattern quite well: the Titus-Bode rule predicted 19.6 AU and Uranus was found at 19.2 AU, further strengthening the reliance on the rule. Bode took this as proof that the law was correct. His confidence was matched by the Hungarian astronomer Franz von Zach, who called a meeting of his astronomer friends in Lilienthal, Germany on 21 September 1800. The meeting was held at the home of the town's chief magistrate, Johann Schröter, an avid amateur astronomer who owned a reflecting telescope built by William Herschel of 8 m focal length, the largest telescope in Europe.

The six astronomers who met with von Zach devised a plan to find the missing planet. Since they expected to find the planet near to the ecliptic they divided up that area of the sky into 24 sections, each of one hour right ascension. They then gave these sections to astronomers who would help them with the search. Von Zach named the group the Lilienthal Detectives. Letters went out to all astronomers notifying them of their assignments and the arrangements for the search.

One letter went to the observatory in Palermo, Sicily, but it was too late. Working at the observatory was the priest and mathematician Giuseppe Piazzi. Piazzi had started the observatory in 1780 and in 1788 went to England to buy the best instruments. While there he visited George Airy, the Astronomer Royal, and William Herschel. While observing with Herschel, from whom he had bought some equipment, he had fallen from a ladder and broken his arm, demonstrating the dangers involved in astronomy. In 1790, equipped with his new instruments, Piazzi began the task he had set himself, measuring the exact positions of stars, nearly 10 000 of them.

Eleven years later, on 1 January 1801 the first day of the nineteenth century, he was still at it when he noticed an object in the constellation

**Table 10.1 Titus-Bode rule**

| Planet | Predicted distance | Actual distance |
|---|---|---|
| Mercury | $\frac{0+4}{10}=0.4$ | 0.39 |
| Venus | $\frac{3+4}{10}=0.7$ | 0.72 |
| Earth | $\frac{6+4}{10}=1.0$ | 1.00 |
| Mars | $\frac{12+4}{10}=1.6$ | 1.52 |
| ? | $\frac{24+4}{10}=2.8$ | |
| Jupiter | $\frac{48+4}{10}=5.2$ | 5.20 |
| Saturn | $\frac{96+4}{10}=10.2$ | 9.54 |
| Uranus | $\frac{192+4}{10}=19.6$ | 19.20 |

of Taurus which wasn't in the records he was checking. Further observations over the next few nights showed that the object was moving and so was a member of the solar system. To his colleagues Piazzi described the object as a comet, but in a letter of 24 January 1801 Piazzi said '. . . it has occurred to me several times that it might be something better than a comet'. It was several months before Piazzi's letters got to the astronomers he had advised of his discovery and by then the object had moved too close to the Sun in the sky to be observed. Bode immediately suspected that the object was the planet that the Detectives had planned to find; von Zach was delighted. The philosopher Georg Hegel wasn't amused. In 1801, before word of the discovery had reached the world, he had published *Disertatio Philosophica de Orbitis Planetarum* in which he 'proved' by logic that only seven planets could exist and the Titus-Bode rule was a figment.

The fact that Piazzi's object had only been observed for a brief time before being lost made it difficult to calculate its orbit and hence to predict where it would be when it reappeared. Enter the mathematician Carl Gauss. In autumn 1801 he published a method by which the orbit of an object could be calculated from only three positional measurements and later in the year he delivered his orbital calculations and positional predictions to von Zach. On 31 December 1801, von Zach found the object again, just half a degree from Gauss' prediction. Gauss also confirmed that the planet's distance from the Sun was 2.77 AU, close to the 2.8 predicted by Bode's Law.

Given the choice, Piazzi chose the name Ceres Ferdinanea, Ceres of Ferdinand. Ceres was the Roman goddess of agriculture and the patron goddess of Sicily, while Ferdinand was the ruler of Sicily. Because of its unwieldy length, and its political nature, it was quickly shortened to Ceres.

On 28 March 1802, Wilhelm Olbers, one of the Lilienthal Detectives, discovered another object. Olbers knew the area of sky in which he was looking as he had been searching there for Ceres earlier in the year. In fact he had found that object on 1 January 1802, independently of von Zach, so he immediately noticed the strange object. After its orbit was calculated, Olbers named the object Pallas, after the goddess of medicine. This created a problem, the Titus-Bode rule predicted one planet at 2.8 AU, now there were two. Neither planet was a normal size, estimates ranged from 260 and 177 km to 993 and 540 km for Ceres and Pallas respectively. Herschel suggested that these objects were a new class of solar system member and suggested the name 'asteroid' because they appeared as stars through the telescope. The name stuck, but the preferred name is minor planet, as proposed by Piazzi in 1802, as they really have nothing to do with stars.

Olbers suggested an idea to explain the two objects. Perhaps they were the fragments of a large planet that, for one reason or another, had broken up. He identified the two regions of sky where the orbits of Ceres and Pallas came closest together and suggested that searches be mounted there to identify other fragments of the catastrophe. On 2 September 1804 Carl Harding, Schröter's assistant and another of the Detectives, discovered Juno, the third minor planet. On 29 March 1807 Olbers, taking his own advice, discovered Vesta. From then on, despite their searching, the Detectives could not find another planet and eventually they stopped looking.

It wasn't until 1845 that another minor planet was discovered, again by a German amateur. He named it Astræa. Since that time not a year has gone by without more minor planets being discovered, many by amateurs. The thousandth minor planet, Piazzia, was found in 1923, the two thousandth, Herschel, in 1973 and the three thousandth, Leonardo, in 1984. Today over

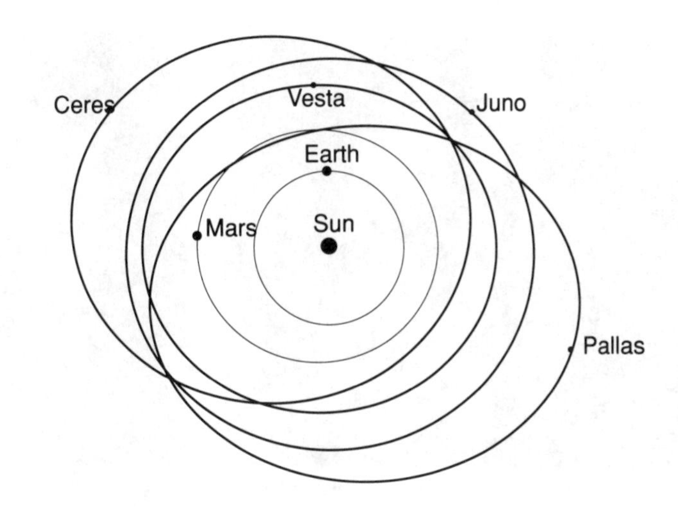


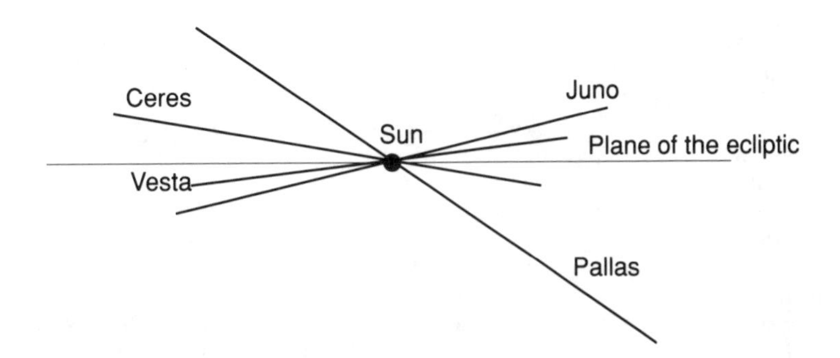


**10.1** The orbits of the four largest minor planets are here shown with the orbits of the Earth and Mars. All the orbits have large eccentricities compared with those of the planets, and the inclinations of their orbits are also much greater than those of the planets.

3700 minor planets are known. Minor planets are fairly easy to find, most discoveries now being accidental, the object straying across a photographic plate being exposed for some other purpose. Often the discoveries annoy professional astronomers, spoiling their photographs. To be officially counted and entered in the list of minor planets, an object must be observed long enough for its orbit to be calculated and its motion predicted for years to come.

Though minor planets are generally cursed by the amateur who does regular novae or supernovae patrol work, they make an interesting study in their own right. As there is great prestige associated with the discovery of a nova or supernova, the reporting of such a find is often made in haste to secure the amateur fame as the discoverer. On many occasions faint minor planets in line of sight with a distant galaxy have been embarrassingly reported as supernovae by the inexperienced observer, and brighter ones have set off alarms for galactic novae. It is a good idea to be patient and wait for an hour or two to see if your find has moved.

The minor planets are often neglected by amateur astronomers, but they are fun to track down and observe. There are many minor planets that attain magnitudes brighter than 10 at opposition and therefore modest instruments can be used. For telescopes in the 80–150 mm range, there are several hundred minor planets which have magnitudes from 10 to 12. As a rule, the best observing window is two weeks either side of an opposition; outside that time the brightness decreases sharply.

Seen from Earth, Vesta is the brightest minor planet. Second only to Ceres in size, it is the only one that can sometimes be observed with the unaided eye, though those with very keen eyesight could possibly see Pallas with a magni-

**10.2** These two pictures show the minor planet Vesta in the constellation of Virgo on 17 April 1985. Vesta is the brightest object in either photograph. The photo on the right was taken four hours and 40 minutes after the left-hand photo and shows the movement of the minor planet in that time. PHOTO CREDIT: STEVEN QUIRK

tude of 6.5 at a favourable opposition. Reaching opposition about every two years, Vesta's magnitude varies, depending on the distance between Earth and Vesta. During the 1986 opposition it reached magnitude 5.3 and was easily seen with the unaided eye but at an unfavourable opposition it can remain hidden at magnitude 7.

Most ephemerides list the right ascension and declination of the four largest and brightest minor planets: Ceres, Pallas, Juno and Vesta. These positions may be plotted on a good star atlas. Some ephemerides will only give positions every few days or weeks apart, but it is easy to interpolate between dates using an atlas. If you have never tried to find an minor planet against the stellar background, it may be easier to wait until one passes close to a bright star.

To make a positive identification of any asteroid you should sketch the surrounding field stars as accurately as possible, then follow up a day or so later with another sketch, comparing the field with your first drawing. The point of light that has moved is the minor planet. When making sketches it is unnecessary to draw every single star, especially if you are looking for a magnitude 8 minor planet in a field that contains many fainter stars. Just draw the main stars near the suspected position down to about magnitude 9.

Minor planets are easy to track down with an SLR camera fitted with a standard 50 mm lens. Using film rated at about 400 ISO, they can be detected to about magnitude 11 in a five-minute exposure, assuming dark skies. The camera will need to be fixed to an equatorial mounted telescope with motor drive. Just for fun, the camera could be pointed in the general direction of the ecliptic and an exposure taken, then compared with another of the same region several nights later, to see if any of the 'stars' have moved. If you point the camera to a known position of a minor planet, stop the lens down to the smallest aperture to reduce sky fog, and take an hour-long exposure. You may find a small elongated trail on the photograph.

Occasionally a minor planet can occult a star, causing a momentary drop in magnitude. The accurate timing of these events, using the simple technique described in the section on lunar occultations, is all that is necessary; the data will help in determining the size and shape of a minor planet. *Sky and Telescope* magazine publishes predictions for these occultations. Because of parallax, an occultation cannot be observed worldwide, but only over narrow stretches of the globe. The predictions are only approximate and are usually refined shortly before the actual occultation.

Minor planet occultations, like lunar and planetary occultations, call for team work and co-ordination. If your interest leans in this direction, it is recommended that you join a local group that has members active in this area.

Many professional astronomers no longer

bother with minor planets, so the field is open to amateurs with large enough telescopes and the necessary calculation skills. The responsibility for keeping track of all the minor planets belongs to the International Astronomical Union Minor Planet Centres in Leningrad and Cambridge, Massechusetts. When an astronomer discovers a minor planet one of the centres is notified and a temporary number is assigned. Its position and predictions of its motion are then circulated. Only two years later when the minor planet is observed after another orbit, is a permanent number assigned. Along with the permanent number comes the name.

Originally minor planets were named after Greek and Roman goddesses. When these ran out, any feminine name was allowed. When a masculine name was applied it was given the feminine Latin ending, Piazzia for example. Nowadays the requirement for feminine names has been dropped altogether and minor planet names come from a huge variety of people, places and even companies. Some astronomers even sell the naming rights to raise money for research. Many amateur astronomers also have their names immortalised, a reward for many cold nights spent observing.

The Titus-Bode rule which led to the discovery of the minor planets held a major place in astronomy. Indeed, although no-one could explain why it worked, its influence was so great that when it came time to calculate the position of Neptune, both mathematicians working on the problem started with the prediction that it would lie at 38.8 AU. When Neptune was discovered near the predicted position it seemed another triumph for the rule, but when the orbit of the planet was worked on it was only 30.06 AU from the Sun, an error of 29 per cent over the actual value. When Pluto was discovered in 1930 it turned out to be at 39 AU, nowhere near the 77.2 AU predicted. With this new evidence against it, the Titus-Bode rule returned to being what it had been originally: a curiosity.

The problem with the minor planets is that they are too minor. Ceres turned out to be the largest minor planet with a diameter of 1000 km, though Vesta is brighter as it has a higher albedo. It is estimated that 99 per cent of the minor planets bigger than 100 km and 50 per cent of those larger than 10 km are known. Added up the total mass of the minor planets is just 5 per cent of the mass of our Moon, or one-two thousandth that of the Earth, too small for them ever to have been part of a planet.

It is not surprising that more small minor planets exist than large ones; an estimate of the distribution of sizes gives us information about the sizes and densities of craters on other worlds. Consider 10 km minor planets. We would expect there to be 1000 1 km minor planets for each 10 km planet as a 10 km body has 1000 times the mass of a 1 km one. The observed distribution doesn't quite fit this rule, however; there are fewer small minor planets than there should be.

Determining the size of a minor planet is not a simple task, as all are too small to show up as anything but a point of light in a telescope. Other methods must therefore be used. One method which often uses the services of amateur astronomers involves the passage of a minor planet in front of a star. These occultations are observed by a group of astronomers spread along a north-south line, so each observer sees the star hidden behind a different piece of the minor planet. By exactly timing when each observer saw the star disappear and reappear, the size and shape of the minor planet is determined. Unfortunately such events are rare, and only a few dozen minor planets have been measured in this way.

A second method of finding out about minor planets involves measuring their brightness over a period of time as they rotate. By watching the variation of the brightness of a minor planet over an extended period, dips and bumps will be seen in its light curve. This gives astronomers some idea of the shape and surface markings of the body.

Most of the minor planets are concentrated in the region between the orbits of Mars and Jupiter at an average distance of 2.2 to 3.3 AU. Scientists estimate that 100 000 objects larger than 1 km can be found in that region. Though 100 000 might sound like a lot, the huge volume of space they fill means that on average they are millions of kilometres from each other. For this reason, there is virtually no danger of a spacecraft accidently colliding with a minor planet; indeed, the trajectories of spacecraft so far have never passed close enough to any minor planet to allow pictures or other measurements to be taken. By carefully calculating the path of the craft, the United States' probe, Galileo, will pass close to two minor planets, Gaspara and

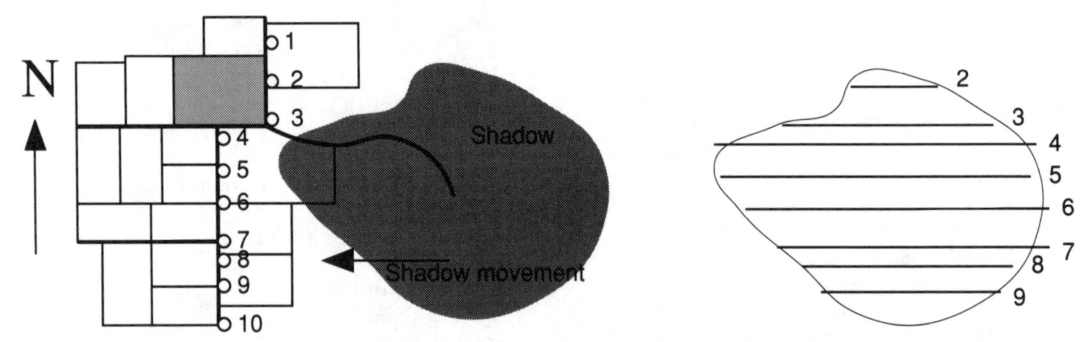


**10.3** One of the most important observations a group of amateur astronomers can make is the occultation of a star by a minor planet. By placing observers at different positions across the expected path of the minor planet's shadow and then accurately recording the times of disappearance and reappearance of the star for each observer, it is possible to determine the shape of the shadow cast by the minor planet and hence the shape of the minor planet itself.

**10.4** By carefully measuring the amount of light received from a minor planet over a period of time and noting its variations, the period of rotation and sometimes the shape of a minor planet can be determined.

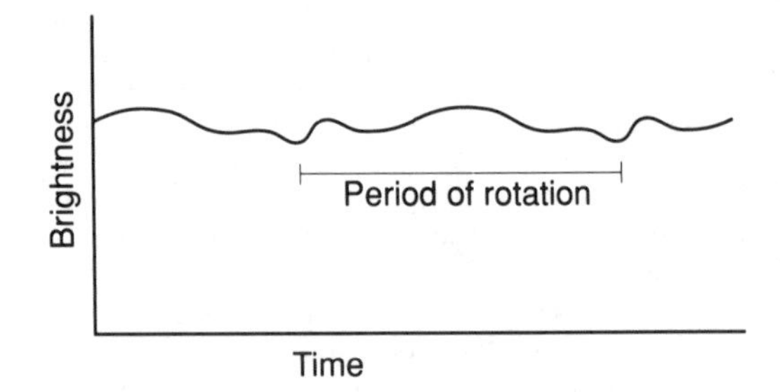


Ida, on its six-year voyage from Earth to Jupiter. The craft will pass Gaspara in October 1991 and travel close to Ida in August 1993, before arriving at Jupiter in December 1995.

The orbits of the minor planets are fairly stable, having inclinations between 0° and 30° with a peak at 3°, and eccentricities of 0 to 0.5 with the majority being around 0.15. With these stable configurations collisions between the minor planets are uncommon, only one per 100 000 years or so, but the results of these collisions, as we will see, do affect the Earth.

The orbits of the asteroids in the main belt are not evenly distributed. Some orbital distances have many asteroids, while others have none at all, as shown in Figure 10.5. The gaps in the belt are called *resonance gaps, or Kirkwood gaps*, after the astronomer who discovered them. The gaps in the belt are caused by Jupiter's gravitation. At any one time the influence of Jupiter's gravity is small—not enough to disturb the orbit of a minor planet—but if the effect can be enhanced in some way then the minor planet will be moved from its orbit. Such an enhancement happens if a minor planet orbits the Sun in exactly half the time it takes Jupiter to do so, so that for every two revolutions the minor planet makes, Jupiter makes one.

Consider an asteroid circling the Sun at a distance of 3.3 AU. Its orbital period is six years and Jupiter's is twelve. This means that every twelve years the minor planet will be disturbed by Jupiter in exactly the same place in its orbit. The cumulative effects of Jupiter's gravitational tugs will be enough, over a relatively short time, to move the minor planet into a different orbit. The effect is known as *resonance*. The resonance at 3.3 AU is called the 2:1 resonance, as the minor planet makes two revolutions to Jupiter's one. Other resonances also exist: a 4:1 resonance at 2.0 AU, a 3:1 resonance at 2.5 AU and less important resonances of 5:2 and 3:2 at 2.84 and 4.0 AU respectively. We will see resonances again in the rings of Saturn.

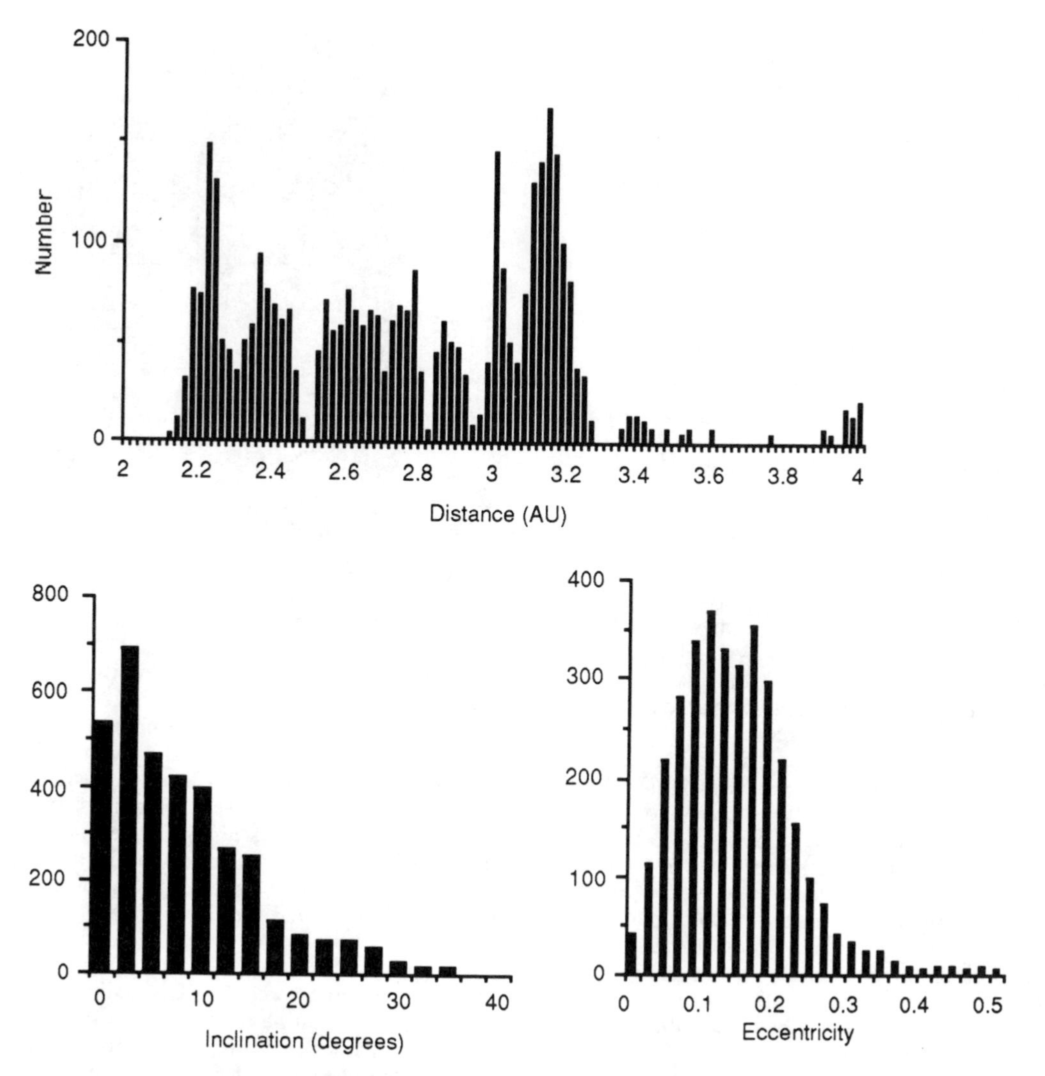


**10.5** These three graphs show the distribution of minor planet orbits. The first graphs the number of minor planets with various orbital radii. As you can see there are a number of gaps where no minor planets are found, and other regions where there are a large number. Thus clumping together of orbits is partially due to resonance effects with Jupiter's orbit.

The lower graphs show the distribution of inclinations and eccentricities of minor planets' orbits.

The minor planets exhibit patterns other than resonance gaps. A minor planet family is defined as a group of objects with similar orbits. Such groups may well have had a common origin, or have become grouped due to some other phenomenon. Around half the known asteroids belong to families, three families accounting for nearly 10 per cent of them. In the case of these three families, the Eos, Themis and Kronos families, the minor planets of which they consist are probably the result of fairly recent collisions between larger asteroids.

Beyond the outer limit of the main minor planet belt, the distribution of asteroids falls to almost nothing, with most objects there confined to a few orbital periods with wide gaps in be-

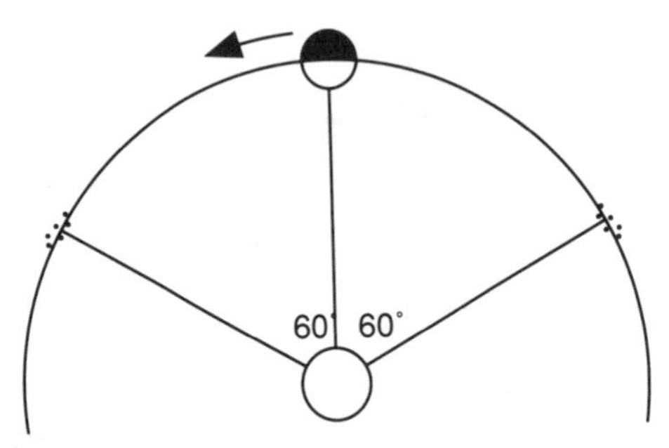


**10.6** There are two points, known as Lagrangean points, 60° in front and 60° behind a planet in its orbit about the Sun in which an object can travel and remain stable. The positions which correspond to these two Lagrangean points in the orbit of Jupiter contain a small family of asteroids which stay in the stable regions.

**10.7** This small meteorite shows pitting and melting caused by its fiery plunge through our atmosphere. PHOTO CREDIT: DAVID REIDY, UNIVERSITY OF WOLLONGONG COLLECTION

tween. Two particular groups of asteroids outside the main group are of interest. These groups have the same orbit as Jupiter, but are 60° in front and 60° behind the planet. These two points are stable regions known as *Lagrangian points* after the French mathematician who calculated their existence. The regions around the Lagrangian points in which a minor planet can have a stable orbit are large and several dozen minor planets are found in each. The first minor planet discovered there was 624 Hektor and all the others have also taken names from the heroes of Homer's *Iliad*. Collectively they are known as the *Trojan asteroids*. The composition of the Trojan asteroids is quite different from the compositions of those in the main belt; they appear to be made from material left over from when Jupiter formed.

If we have never sampled an asteroid, how do we know their composition? Simple—the solar system provides a 'home delivery' service. Throughout recorded history there have been tales of stones falling from the sky, although until the last century such stories were classed with fairy tales and dragons, by serious scientists. In 1769 three 'thunderstones' were submitted to the French Academy of Sciences, but although chemical tests showed interesting results, the story of the origin of the stones, 'they fell from the sky', was rejected. It was only in 1803 after a fall at L'Aigle near Paris, that the European scientific community accepted the origin of these stones, now called *meteorites*.

Meteorites are constantly reaching the Earth. Every year several falls are observed and most are recovered. Most of these meteorites are stone or metallic masses around 1 kg in mass. More rarely a larger object will strike the Earth, although in most cases these large objects break up in the atmosphere scattering smaller particles over a wide area. Even more rarely a large meteorite will survive its passage through the atmosphere and strike the Earth, causing a crater like those on the Moon.

The question arises, will another large meteorite hit the Earth? Yes. There is no doubt that the Earth will again be hit by a meteorite large enough to cause significant damage given a sufficiently long period of time. Early in 1989 a body over 1 km long passed within a million kilometres of the Earth, so the possibility definitely exists, but in any one year the chance is remote. The probability of an object large enough to make a crater 10 km in diameter striking the Earth is one in a million in any given year, so there is really no need to worry. Even if we were aware of such an object hurtling towards the Earth, there is nothing we could do about it, even with our present technology.

The meteorites which have been recovered show quite a range of characteristics, suggesting that they originated in different environments. The traditional classification of meteorites

**10.8** This small slice of nickel-iron meteorite shows the classic crystal pattern. The pattern is caused by the boundaries between crystals of iron and nickel within the meteorite. The pattern is enhanced by applying a weak acid to the surface to etch the metals slightly. PHOTO CREDIT: DAVID REIDY, UNIVERSITY OF WOLLONGONG COLLECTION

**10.9** This stony meteorite shows small, brown-coloured pebbles within a larger mass of dark material. About 85 per cent of all meteorites are thought to be stony types. PHOTO CREDIT: DAVID REIDY, UNIVERSITY OF WOLLONGONG COLLECTION

depends upon their bulk composition. Three classes are generally used: iron, stone and stony-iron.

Iron meteorites are nearly pure nickel-iron and have a density of 7 $g.cm^{-3}$, corresponding to the material we think makes up the Earth's core. Most of the iron and nickel on the Earth occurs in the form of oxides rather than the pure metal found in the meteorites. It has been hypothesised that the recovery of such meteorites played an important part in the transition of cultures from the bronze age to the iron age. Stony meteorites look much the same as terrestrial stones, while the third group is the combination of stone and iron as their name suggests. The stony meteorites are made from materials similar to the ingredients which make up the Earth's crust.

Another method of classifying meteorites is according to the type of body from which they originated. Primitive meteorites are ones which contain material which has been little altered since the formation of the solar system—85 per cent of all meteorites are of this type. All primitive meteorites are stony types, as the nickel-iron type need to have been inside a planet for the metal to have become liquid and formed the meteorite. The meteorites which show evidence of at one time being part of a planet are called *differentiated meteorites*. Differentiated meteorites show that some parts of the solar system must have been broken up in the process of planetary formation and reformed into another planet.

The iron meteorites show the greatest amount of processing. They must have come from a body with sufficient internal heat for the iron and nickel to melt and begin to form a core. The related stony-iron meteorites have the same nickel-iron compounds, but contain other rock-forming minerals as well. These meteorites, which account for only 4 per cent of falls are thought to be from the boundary between the core of the parent body and its mantle.

Radioactive techniques have been used to calculate the age of meteorites. Radioactive dating depends upon carefully measuring the amounts of two related radioactive elements and working out how much of one has turned into the other—for example, atoms of uranium will, over a very long period, turn into atoms of lead. Measuring the amounts of the two compounds will then give the age of the rock. A number of different pairs of compounds can be used on a

**10.10** This is a slice of a stony iron meteorite which fell near Imilac in Chile. The metal part of the meteorite is as shiny as stainless steel being made from an iron-nickel alloy. In the metal are large crystals of olivine, a brown-coloured mineral found in large quantities on Earth. PHOTO CREDIT: DAVID REIDY, UNIVERSITY OF WOLLONGONG COLLECTION

**10.11** A small part of the Murchison meteorite, a carbonaceous condurite which fell near Murchison, Victoria in September 1969. Scientific study of the fall was swift and showed the presence of amino acids and other organic materials which are the basis for life on Earth. The fact that the meteorite was recovered quickly without time for terrestrial contamination to have occurred meant that these chemicals could only be of extra-terrestrial origin. This one meteorite revolutionised scientific thinking about the possible origins of life on Earth. PHOTO CREDIT: DAVID REIDY, UNIVERSITY OF WOLLONGONG COLLECTION

sample to get more reliable results. The ages measured by most methods will be the time since the meteorite became solid, since from that point there is no chance of the material having been lost.

The most primitive meteorites of all are the carbonaceous meteorites. These meteorites are rich in carbon and volatile compounds such as water which combine with other minerals to form clays. They are less dense than other meteorites and generally fairly fragile. An interesting point about some carbonaceous meteorites is that they show evidence of being altered by liquid water while part of their parent bodies. Perhaps the most interesting aspect of the carbonaceous meteorites is that they contain up to 3 per cent complex carbon compounds.

Carbon compounds in which carbon joins with hydrogen are called organic compounds, and such substances play an important part in life on Earth. On Earth, non-organic carbon compounds are rare, showing up in carbonate rocks and in coal and oil; even so, these were once part of living organisms. Though associated with life on Earth, the carbon in meteorites does not suggest that there was life in the solar nebula or elsewhere, as many other chemical reactions could give rise to the existence of the compounds. Most of the carbon compounds are best described as tars, but they also contain compounds of importance to life: components for proteins and nucleic acids. These compounds were first recognised with certainty in the Murcheson meteorite which fell into Australia in 1969. This meteorite was found to contain sixteen different amino acids, only five of which are common on Earth. Also interesting was that both left- and right-handed versions of the acids were found in almost equal numbers; on Earth it is rare to find right-handed versions at all.

It is clear from their different natures that meteorites must have come from different bodies. The primitive meteorites must have come from relatively small bodies that formed directly from the solar nebula. The bodies must have been small, otherwise there would have been too much heat from gravitational contraction and from radioactive heating and the primitive bodies would have been altered. Therefore the parents of these meteorites could have been no more than a few hundred kilometres in diameter. These small parents must have also

**10.12** These two small meteorites are tektites, small objects which appear to be either terrestrial or lunar in origin and were almost certainly formed by material thrown up by a large meteorite impact falling back to Earth through the atmosphere. PHOTO CREDIT: DAVID REIDY, UNIVERSITY OF WOLLONGONG COLLECTION

contained organic chemicals and liquid water. The chemical composition of the primitive meteorites shows that there must have been quite a number of parent bodies.

The differentiated meteors must have come from other bodies, ones in which the internal heat was enough to allow the chemicals in the original mixture to separate out. Again, these bodies were limited in size to a few hundred kilometres, so it is not clear why some bodies retained heat and differentiated out while others remained primitive.

Ninety per cent of all meteorites fall into one of the two previous groupings, but a small number come from larger bodies, ones which were active at least $1.5 \times 10^9$ years ago. These meteorites must be the result of collisions between asteroids and a planet which threw material up from the planet and into solar orbit. One small class of meteorites, the SNC type, has been linked to Mars while a number of other meteorites have come from lunar impacts.

# 11

# Jupiter

Long before it was known to be the largest of the planets, ancient astronomers named Jupiter after the most important of the gods. Their insight was remarkable, because Jupiter is neither the brightest of the planets nor the most obvious from Earth; both those qualities go to Venus as it graces the morning or evening skies. With its retinue of companions, however, Jupiter certainly appears now to be the king of the planets. In our night sky Jupiter comes in second in the brightness stakes, but the detail which is visible on the planet's surface with even a small telescope makes it one of the favourite objects for the amateur.

Jupiter is the first of the gas giant planets, the closest body to the Sun to retain its original hydrogen and helium after the new Sun started shining. Jupiter was able to retain its gases because of two factors. Firstly, its size meant that its gravitational hold on the gas was strong and secondly, because of its distance from the Sun, the solar wind was not strong enough to blow them away. There is so much hydrogen and helium in Jupiter that the surface of whatever solid there is at the centre of the planet will remain forever hidden from view beneath thousands of kilometres of swirling clouds and gases. When we look at Jupiter through the telescope we see only the tops of those clouds. Unlike the clouds of Venus though, which present a blank, white view, the clouds of Jupiter are a mass of colour: reds, browns and oranges.

The majority of Jupiter's atmosphere is hydrogen, with a decent amount of helium mixed in. As hydrogen and helium were the most abundant elements in the early solar system, it is hardly surprising that they make up the bulk of the largest planet. It is not hydrogen or helium, however, which creates the colours we see, but other, more complex chemicals, the result of chemical reactions within the atmosphere.

The first gas to be identified in Jupiter's atmosphere was methane, a compound made from one carbon and four hydrogen atoms, closely followed by ammonia, consisting of one nitrogen and three hydrogen atoms. The clouds we see at the top of Jupiter's atmosphere are made of crystals of ammonia ice. These ammonia clouds are white to pale yellow in colour and can be easily seen as the lightest coloured objects in the Jovian atmosphere. These clouds are patchy, like the clouds on Earth, so in many places they part, giving a view of the layers of gas below.

The clouds below are not white; they are red or brown, indicating that the chemicals making them up are quite different from the ammonia clouds. Despite the visits of a number of spacecraft, the exact composition of these clouds is still uncertain. It could be that they are ammonium hydrosulfide, or other sulfur compounds, many of which would make brown-coloured clouds. Other compounds such as phosphine may be responsible for the clouds we see, but only further exploration of Jupiter by spacecraft will reveal the answers.

Because it is a gaseous planet, Jupiter's rotation is a complex matter. Unlike the Earth, which as a solid rotates at a constant rate, the clouds of Jupiter are free to rotate about the planet at any speed which local conditions can support. Near the poles, the clouds of Jupiter rotate about the planet once in 9h55m41s, on average. Closer to the equator a jet stream effect speeds the clouds so that one revolution takes only 9h50m30s. Studies of radio waves emitted from deep within the planet show that

**Table 11.1 Jupiter's atmosphere**

| Gas | | Fraction |
|---|---|---|
| Hydrogen | $H_2$ | 86.1% |
| Helium | He | 13.8% |
| Methane | $CH_4$ | 0.09% |
| Ammonia | $NH_3$ | 0.02% |
| Ethane | $C_2H_6$ | 0.004% |
| Ethyne | $C_2H_2$ | 0.00008% |
| Phosphine | $PH_3$ | 0.00004% |

the core of the planet rotates every 9h55m30s, making Jupiter the fastest rotating body in the solar system.

To distinguish the clouds on Jupiter and their different rates of rotation, the planet's features are grouped into two systems: System I for the clouds near to the equator, with their rapid rotation, and System II for the clouds further from the equator with their slower periods.* A planet made only of clouds may seem boring, but Jupiter's turbulent atmosphere is far from dull. The complex weather patterns, combined with the different colours of the clouds, mean that even a small telescope can reveal much activity and structure in the Jovian atmosphere. The close-up pictures taken of Jupiter, first by Pioneers 10 and 11 in December 1973 and December 1974, and later by Voyagers 1 and 2 in March and July 1979, show that on a small scale Jupiter is a constantly churning mass of atmospheric turbulence. From Earth these small structures aren't seen, though a number of much more stable, permanent features are prominent.

The first thing observers of Jupiter notice is the alternating bands of light and dark clouds which encircle the planet. These bands are remarkably constant, having been seen over the 400 years of telescopic observations of Jupiter. The bands indicate weather cells in the Jovian atmosphere, the light colours being caused by the light ammonia ice clouds which float high in the atmosphere, the darker bands by the coloured clouds below showing through. The most interesting observations are made along the edges where two bands meet. There turbulence between the bands can cause wisps of cloud to extend over the darker regions, or holes to appear in the lighter places, allowing the clouds beneath to show through.

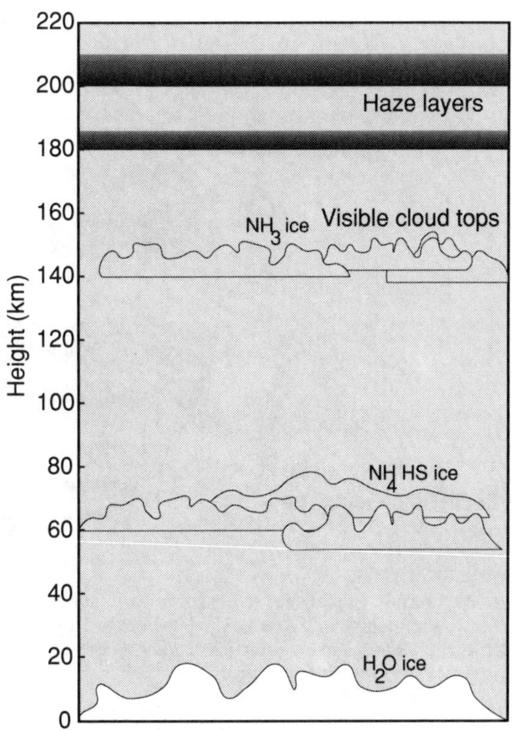


**11.1** Jupiter's atmosphere is made of layers of cloud and haze. The haze layers towards the top of the atmosphere obscure some features in the clouds below. The upper layer of $NH_3$ clouds is responsible for the lighter coloured zones seen encircling the planet. Where these zones are missing, the darker coloured clouds below show through as dark belts. Below the $NH_4HS$ clouds is a final layer of clouds formed of water ice like the clouds of Earth. The pressure at the base of the ice clouds is around 1 megapascal, ten times the pressure at the Earth's surface.

The bands of clouds are caused by the rapid rotation of the planet. The weather patterns on Jupiter are controlled by two sources of energy, the radiation received from the Sun and the internal heat of Jupiter. Jupiter radiates more energy than it receives from the Sun—twice as much—however this energy is released not as light we can see, but as infrared radiation. Calculations show that from the energy received by Jupiter from the Sun, the planet should have an average temperature of −166°C, but it is 20° warmer than this. How does Jupiter produce this energy?

* Periods measured with respect to the rotation of the planet's core are called System III, but these are of little use in observational astronomy.

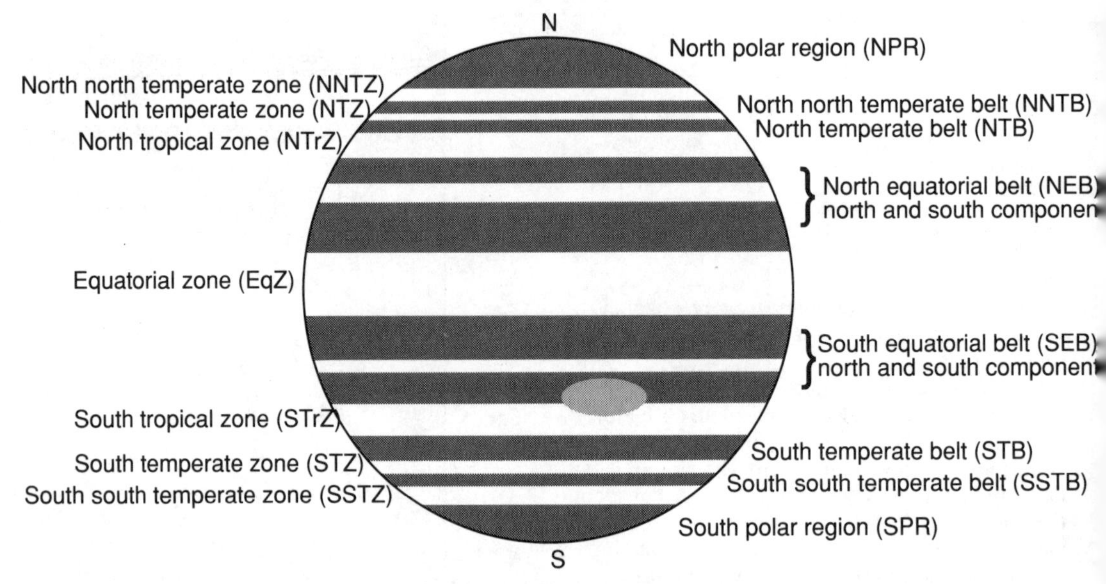


**11.2** Because of their permanence, astronomers have given names to the bands of clouds in Jupiter's atmosphere. The darker bands are known as belts while the lighter bands are called zones. Not all the belts and zones will be visible at all times in amateur telescopes, as the contrast between adjacent bands can vary considerably.

Jupiter's energy production cannot be nuclear like the Sun's. Massive though it is, Jupiter is almost a hundred times too small to form a star. The energy instead comes from gravitation. When an object falls to Earth under gravity it gains energy as it falls, energy which it then loses when it strikes the ground. In the case of Jupiter the whole planet contracts and the energy gained by the particles falling towards the centre is released as heat. Jupiter is still undergoing the gravitational contraction which began when it started to form from the solar nebula. In the early days of its formation Jupiter would have been releasing enough energy to glow a dull red. Nowadays Jupiter contracts at a much slower rate, around 1 mm per year.

Given that we can only see the tops of the clouds of Jupiter and that there is no chance of ever landing a probe on the surface of the planet or even sending one to penetrate very far into the atmosphere because of the enormous pressures found there, the investigation of the structure of Jupiter has been a largely theoretical affair. The problems are compounded when it is realised that we know little about the materials which make up the planet when they are at the incredible temperatures and pressures found in

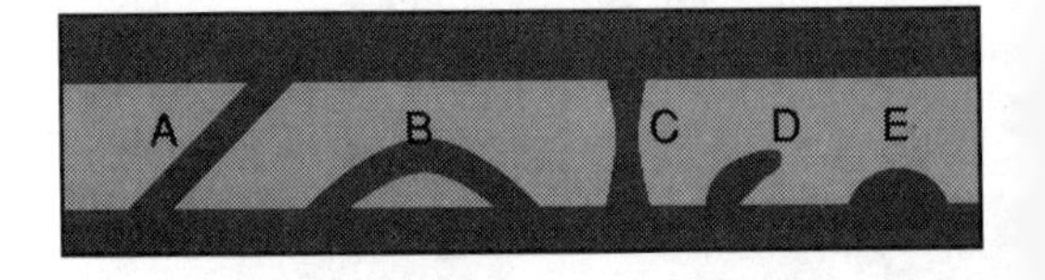


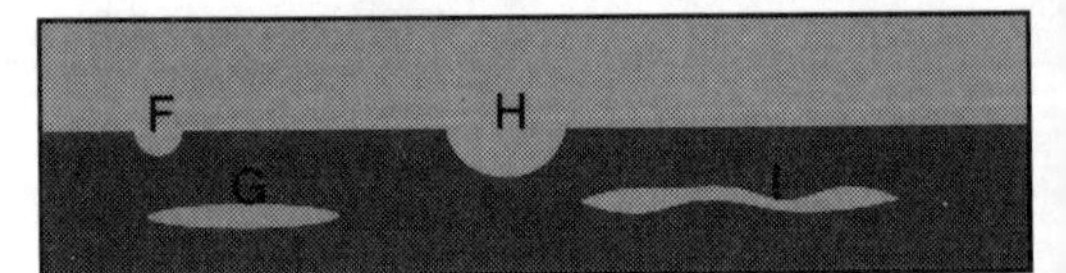


**11.3** Features are often seen in Jupiter's belts and zones, the more common ones have names as shown here.

| | | | |
|---|---|---|---|
| A | Feston | F | Notch |
| B | Loop festoon | G | Streak |
| C | Column | H | Bay |
| D, E | Projections | I | Rift |

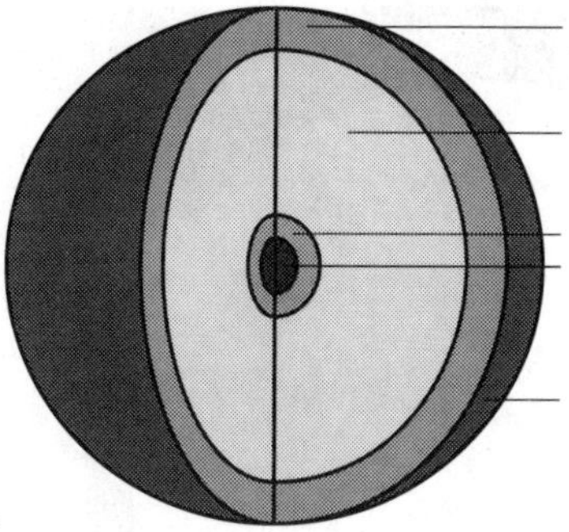

**11.4** This view to the centre of Jupiter is based upon current knowledge gained from observations made by spacecraft of the planet's electrical and magnetic properties. From Earth only the tops of the clouds are ever visible. At Jupiter's centre is a hot rocky core, about twice the size of the Earth; around this are ices made solid by the enormous pressure of the gases above them. Around the ice is a layer of hydrogen so compressed that it behaves more like a metal than a gas. Finally there is a layer of gaseous hydrogen crowned by the clouds we see from Earth.

Jupiter. For Jupiter to exhibit the properties of heat and radio emission we see, the temperature at its centre must be 25 000°C, and the pressure must be at least five thousand million tonnes per square metre.

We know that under extreme conditions such as that found in Jupiter, hydrogen stops behaving like a gas and begins to act more as a metal. Calculations suggest that this happens about 25 per cent of the way from the surface to the core. But because of the difficulty of reproducing such conditions in the laboratory, the exact conditions under which the transition occurs are uncertain. Just above the layer where hydrogen turns metallic, it will be a liquid; still higher it will take on its familiar gaseous form, making the atmosphere we observe.

At the centre of the planet is a solid core. This is known from the observations of the regular rotation of Jupiter's magnetic field and radio emission. Observations of the planet's density also show that a heavy core must be present. Jupiter's core is between ten and fifteen times the mass of the Earth and between two and two and a half times its size. Jupiter's core is probably made from the same sort of materials that make the other rocky planets: silicates and metals, with perhaps some ices and heavy elements.

It is unlikely that we will ever find a planet much larger in size than Jupiter; calculations have shown that although planets can be heavier, they won't be larger in diameter. Adding more mass to Jupiter would increase its gravitational field and lead to a lessening in size as the gases became more compressed. If even more mass was added, taking Jupiter to a hundred times its present mass, compression of the gases would cause temperatures at the centre to rise until they were high enough for nuclear reactions to start and a star to form.

Jupiter is an ideal subject for a small telescope, even at conjunction the disc is over 30 seconds of arc in diameter, larger than Mars at its best. At opposition Jupiter's 50 arc second disc is smaller than Venus at inferior conjunction, but a 50 mm telescope will show the bands and binoculars, the four largest moons.

The opposition dates for the next twelve years are shown in Figure 11.6. When Jupiter is at perihelic opposition it is about 587 million km from Earth, and magnitude −2.5. At aphelic opposition, it will be 661 million km from Earth and magnitude −2.0. As the synodic period is 399 days, oppositions occur about one month later each year, the most favourable every twelve years. The sidereal period of Jupiter is 11.86 years, and the planet will be seen to occupy each hour of right ascension for about six months. There is really no such thing as an unfavourable opposition, nor do you need to wait for the close approaches to do useful work.

Two things will be immediately apparent when observing the planet's disc: the distinct flattening of the poles and the parallel markings. Because Jupiter is a gaseous body rotating quickly, it bulges very noticeably at the equator; the pole-to-pole diameter being 9500 km smaller than the equatorial diameter of 142 800 km. Each band of Jupiter's clouds contrasts in intensity with its neighbour, making them easy to see. Jupiter's axial tilt is only 3° from perpendicular, so there are not any seasons as we know them, and as the equator is very close to being edge on, the belts and zones always appear as straight lines. Figure 11.2 shows the various belts and zones. Do not expect Jupiter to look exactly like the diagram; the boundaries are not always so regular.

Drawings of the changing disc features make interesting comparisons from one observation

**11.5** Because of the polar flattening of Jupiter, an observing blank must be an oval rather than a circle. This observing blank is suitable for whole-disc observations using a small to medium sized telescope.

to the next. Because of the polar oblateness, sketches should be made on a prepared blank that matches Jupiter's shape. It is not easy to draw an accurate ellipse this small and it is recommended that you photocopy the example in Figure 11.5 for your sketches.

The main difficulty in drawing the gas giant is its fast rotation. Unless you work quickly you will continually lose detail around the limb while new features become visible from the opposite limb; drawings of the whole disc should be finished within ten minutes. An interesting variation on full global sketches is to choose a small region that shows detail, and draw this in a rectangular box.

Of all the Jovian phenomena, the study of the belts and zones is probably the most interesting, as there can be very complex detail. The south and north equatorial belts are the most obvious features in small instruments, the two being separated by the equatorial zone. The temperate banding can be less clear, and the dusky north and south polar regions appear lighter than the equatorial belts. The amount of detail visible on the disc will depend to a large degree on the observing conditions and telescope aperture. A large amateur instrument will show a wealth of small detail in the boundaries between belts and zones. The colour of Jupiter will also depend on instrument size and the observer's colour perception. Generally in medium to large sized telescopes (150 to 300 mm) the zones will look white to cream, and the belts brown to reddish-brown. It is worth experimenting with colour filters; red or orange will generally enhance dark markings.

The importance of amateur observations was highlighted in 1979 when a disturbance in the south equatorial belt was found. Once notified of the position of the disturbance, the Voyager mission controllers were able to study the disturbance with the approaching Voyager 2. With the Galileo mission due to arrive at Jupiter in 1995, the amateur will again have the opportunity to make useful observations that could assist in unveiling some of the gas giant's mysteries.

After the Voyager probes returned incredible close-up images of Jupiter, astronomers were able to measure wind speeds across the planet by noting the movement of various features. It appears that the speed varies between bands and zones, and it is at the boundaries where there is a substantial change in wind speed that the amateur will note most of the small-scale detail. The irregularities in the belt system range from bright and dark circles and ovals to wisps, projections and bridges in the zones. Figure 11.3 shows some examples of the features that can be observed.

The timing of a prominent feature across the central meridian (the imaginary line passing through the two poles) allows the longitude of the area to be established; this in turn will provide useful information on Jupiter's winds. It is easy after a little practice to time transits to an accuracy of a minute or two; the polar flattening helps in the determination of the central meridian. Tables of the longitude of the central meridian for both systems I and II are available in the *Astronomical Almanac* and in publications from the larger astronomical groups. It is a simple matter to calculate the longitude of any particular feature crossing the central meridian from these tables.

In 1939 dark features appeared in the south temperate zone. These gradually changed into three white ovals measuring 100 000 km in width. These ovals could still be observed in 1989, but had shrunk to less than 10 per cent of their former size. The longest-lived and most remarkable feature of Jupiter's atmosphere

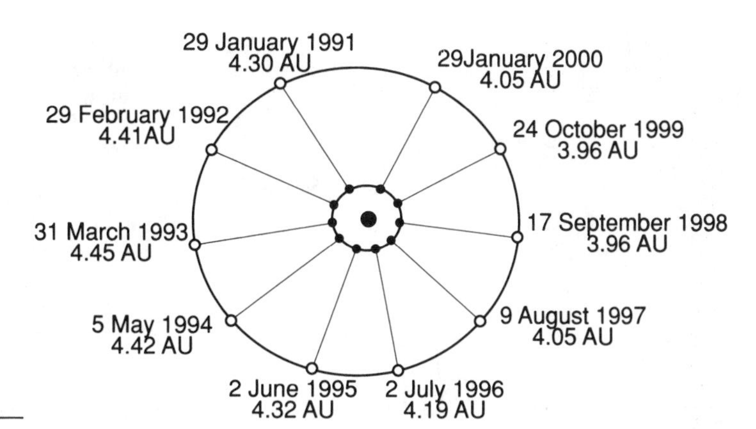


**11.6** Oppositions of Jupiter until 2000. The distance between the Earth and Jupiter is given for each.

is the famous Great Red Spot (GRS): a tremendous swirling storm that has raged for at least 300 years. The GRS has been observed since the seventeenth century, but it was not until 1878 that it attracted much attention after undergoing a dramatic change in intensity. It is difficult to imagine a storm of such gigantic proportions, ellipsoid in shape, and at present about 30 000 km in length.

The GRS is located at about −25° latitude in the south tropical zone, splitting it in two, its northern extremity protruding into the south equatorial belt. In 1878 the colour gradually changed from a pinkish hue to brick-red, and four years later the spot faded to a light grey colour. Since that time the GRS has been seen with varying degrees of intensity and colour, but nothing like the brick-red described last century. Since the late 1800s the GRS has been steadily shrinking and is now about half its former size. The GRS can occasionally fade to invisibility, but its presence can always be indirectly detected by the Red Spot Hollow: a large depression in the south equatorial belt where the GRS's northern boundary lies. During the past decade the GRS has displayed a pinkish hue, rimmed with a darker band, allowing the full ellipse to be traced. It can sometimes be hard to find with small instruments, but an 80 mm refractor should have no difficulty, although the subtle colour will only be visible in larger amateur telescopes.

Most features on Jupiter are transitory and last only a few days or weeks, though some will remain for months and on rare occasions, for years. All features show some degree of change throughout their lives. A spot known as the South Tropical Disturbance was discovered at the turn of the century, and it lasted almost 40 years. The disturbance travelled faster than the mean System II period, overtaking the Great Red Spot on many occasions during its life.

Jupiter was the first planet found to be a radio transmitter. In 1955 it was observed transmitting radio signals at a frequency of 20 MHz. These broadcasts don't indicate the presence of life on the planet; they are instead the result of electrons moving in Jupiter's magnetic field. Indeed it was these transmissions which first showed that Jupiter had a magnetic field. Some of the broadcasts are so intense that Jupiter is often the brightest radio object in the sky.

Jupiter's magnetic field, or *magnetosphere*, spreads out from the planet into space over an immense distance. In the direction towards the Sun, Jupiter's field extends three million kilometres; in the direction away from the Sun the tail of the magnetic field reaches to the orbit of Saturn. If it were possible to see Jupiter's magnetic field from Earth, the planet would appear to be the same size as the Moon appears in the sky. The magnetosphere of Jupiter has a major influence on the satellites of the planet.

One of Voyager's major discoveries at Jupiter was the thin ring encircling the planet. Within a few days of the discovery by the spacecraft, astronomers on Earth were also able to find and observe the ring—it helps if you know where to look. Jupiter's ring was found 128.3 thousand km from the planet's centre, 56.9 thousand km above the cloud tops. The ring is about 3000 km wide and 30 km thick. It is very tenuous blocking only a tiny fraction of the light passing through it, much less than a clear sheet of glass.

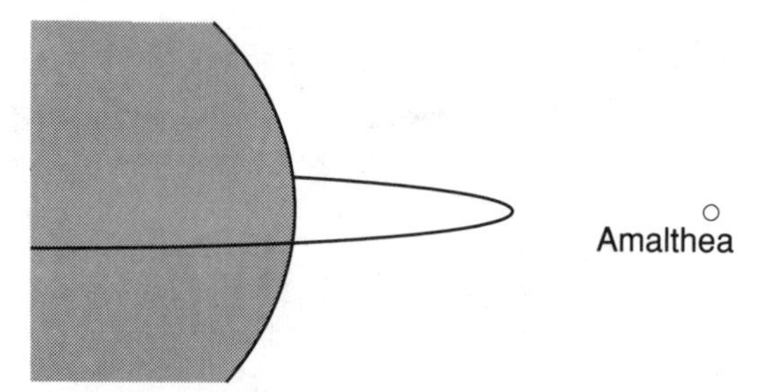


**11.7** One of the most unexpected features found by Voyager I at Jupiter was a thin ring circling the planet just 54 000 km from the cloud tops. The ring is 5000 km wide and very thin. The ring is so tenuous that it absorbs only one-millionth of the light passing through it, making it thousands of times clearer than window glass.

Jupiter's ring is made from tiny grains of silicate material. There are no large particles in the ring indicating that it is a very old structure, all the larger particles having worn themselves down by collisions. Because of the continual collisions, many of the particles appear to have acquired a static electricity charge. This makes them mutually repel and form a very insubstantial cloud of charged atoms around the main ring.

On 7 January 1610, while observing Jupiter with his first telescope Galileo saw 'three little stars, small but very bright, near the planet'. The next day he looked again and wrote, 'I turned again to look at the same part of the heavens, I found a very different state of things, for there were three little stars all west of Jupiter, and nearer together than on the previous night . . . There are three stars in the heavens moving about Jupiter, as Venus and Mercury round the Sun.'

A few nights later Galileo noticed a fourth star moving similarly to the other three. At the time the ideas of Copernicus and Ptolomy were in dispute. Ptolomy's geocentric system in which everything moved about the Earth was the officially accepted view, but Copernicus' heliocentric system was being propounded by many scientists. Galileo's observations of the four 'stars' clearly showed that not everything orbited the Earth. Galileo continued his observations, mapping out the motions of the four objects night after night. A short time later astronomer Simon Meyer suggested the names Io, Europa, Ganymede and Callisto as names for the satellites. These names have stuck and the group is now collectively known as the Galilean satellites.

It was quickly reasoned that the dots could not be stars. Although they were much too small to be seen as anything but dots, their motion clearly didn't match the motion of the other stars, but was more like that of the Earth's Moon, so scientists deduced that that was what they were: moons. Studies of these four new bodies fascinated scientists. They were able to observe many different phenomena within the Jovian system, occultations of one moon by another, eclipses of the satellites and other mutual interactions.

Jupiter's Galilean moons revolve very close to the plane of the planet's equator, and as Jupiter's axial tilt is only about 3° we see the equator and satellite motion nearly edge on. Each moon shuttles back and forth in front of and behind the planet, staying in an almost straight line.

There are accounts of people with very keen eyesight who claim to have seen the Galilean satellites with the unaided eye. Since it is only Jupiter's excessive brilliance that prevents us from catching glimpses of the satellites, you could try searching for them with Jupiter hidden behind a distant wall or post. The task will be easier if you first view Jupiter in a telescope, or consult an almanac, to ensure that a satellite is a reasonable distance from the disc.

Of the Jovian satellite family, only the Galilean satellites are of any real interest; the others are too small and faint for amateur instruments. The most obvious feature of the Galilean satellites is their rapid orbital period, which is noticeable from one night to the next. Io whirls around Jupiter in 1.769 days, Europa a little more slowly at 3.551 days, Ganymede at 7.155 days, and Callisto, the slowest, at 16.689 days.

Drawings made of the moons over a week or two will provide an interesting insight into their motion; each satellite can be identified by their maximum angular separation from the planet. When Ganymede and Callisto are at their extremes on either side of the planet, the distance between the two is about 16 minutes of arc, or about half the lunar or solar diameter, but under moderate magnification Jupiter's disc and moons will be within the same field. Figure 11.9 shows the daily movement of the satellites.

The Galilean satellites provide an impressive display of events that can be studied almost any

**11.8** This photograph shows Jupiter and the four Galilean satellites at 19:10 on 22 October 1984. The planet is over-exposed to allow the moons to be seen. The moons are, from left to right: Callisto, Io, Europa, Ganymede. PHOTO CREDIT: STEVEN QUIRK

**Table 11.2 The satellites of Jupiter**

| Name | Discoverer and year | Diameter (km) | Mass (kg) | Orbital Distance (km) | Period (days) | Eccentricity | Inclination (°) |
|---|---|---|---|---|---|---|---|
| Metis | Voyager 1979 | <40 | — | 127 800 | 0.249 | 0 | 0. |
| Adrastea | Voyager 1979 | 40 | — | 129 000 | .300 | 0 | 0 |
| Amalthea | Barnard 1892 | 240 | — | 181 300 | 0.498 | 0.0028 | 0.4 |
| Thebe | Voyager 1979 | 80 | — | 221 700 | 0.675 | — | 1.25 |
| Io | Galileo 1610 | 3650 | $8.92 \times 10^{22}$ | 421 600 | 1.769 | 0 | 0 |
| Europa | Galileo 1610 | 3120 | $4.86 \times 10^{22}$ | 670 900 | 3.551 | 0.0003 | 0.02 |
| Ganymede | Galileo 1610 | 5280 | $1.48 \times 10^{23}$ | 1 070 000 | 7.155 | 0.0015 | 0.09 |
| Calisto | Galileo 1610 | 4870 | $1.06 \times 10^{23}$ | 1 180 000 | 16.689 | 0.0075 | 0.43 |
| Leda | Kowal 1974 | 10 | — | 11 094 000 | 239 | 0.148 | 27 |
| Himalia | Perrine 1904 | 170 | — | 11 470 000 | 250.6 | 0.158 | 28 |
| Lysithea | Nicholson 1938 | 24 | — | 11 710 000 | 260 | 0.13 | 29 |
| Elara | Perrine 1905 | 80 | — | 11 740 000 | 260.1 | 0.207 | 28 |
| Ananke | Nicholson 1951 | 20 | — | 20 700 000 | 617 | 0.17 | 147 |
| Carme | Nicholson 1938 | 30 | — | 22 350 000 | 692 | 0.21 | 163 |
| Pasiphae | Melotte 1908 | 36 | — | 23 300 000 | 735 | 0.38 | 148 |
| Sinope | Nicholson 1914 | 28 | — | 23 700 000 | 758 | 0.28 | 153 |

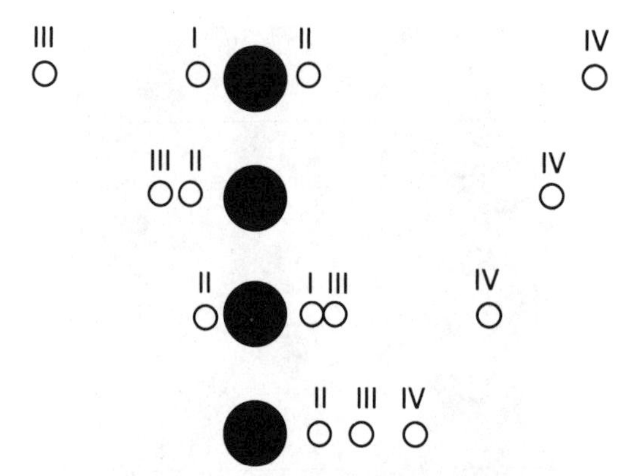


**11.9** The movement of the four Galilean satellites in their orbits around Jupiter makes them a fascinating set of objects to watch. Here we see a number of observations made over a period of days, showing the changing relationships between the positions of the moons. The standard method of labelling the moons with roman numerals has been used: I—Io, II—Europa, III—Ganymede, IV—Callisto.

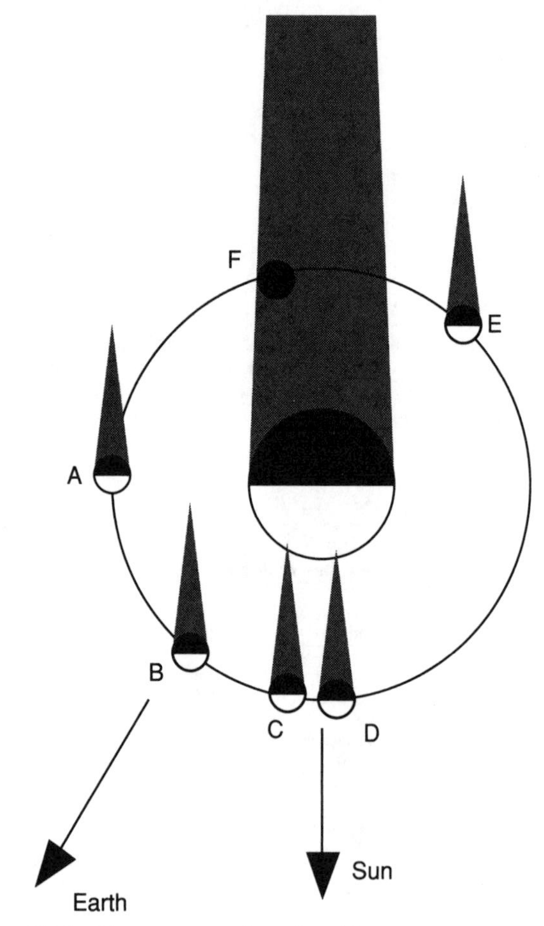


**11.10** There are a number of satellite phenomena which can be observed from Earth. These are:
A moon seen away from planet;
B moon seen against planet;
C moon and shadow seen against planet;
D shadow seen against planet, Moon seen away from limb;
E moon hidden by planet;
F moon hidden by planet's shadow.

night that the planet is visible. The events range from satellite eclipses to occultations, transits, and shadow transits. The timing of these occurrences has become an important aspect of amateur work which helps professional astronomers calculate the precise orbital elements of the satellites. An eclipse occurs every time a moon passes into the shadow cone of the planet. From our viewpoint, except during the brief period of opposition when eclipses cannot be seen, the shadow cone will be found on one side or other of the planet's limb.

After opposition, with Jupiter in the evening sky, a satellite will stay visible until it is occulted by the planet, and reappear from eclipse at some distance from the opposite limb. Observations before opposition, with Jupiter in the morning sky, will have the satellite vanishing into eclipse before reaching the planet, and emerging from occultation on the opposite side. Near quadrature, the shadow cone will be at its maximum slant from the planet. During eastern quadrature, because of their distance, both Ganymede and Callisto can be eclipsed, only to reappear briefly before being occulted by the planet; at western quadrature the sequence is reversed. Callisto is sometimes excluded from these events, resulting from its distance and the tilt of Jupiter's axis.

As a satellite approaches the disc from the eastern side, it will pass across or transit the planet. Before opposition, the satellite's dark shadow can be seen distinctly projected against the planet some time before the actual transit. The reverse happens after opposition: when the satellite leaves the disc on the west side, its

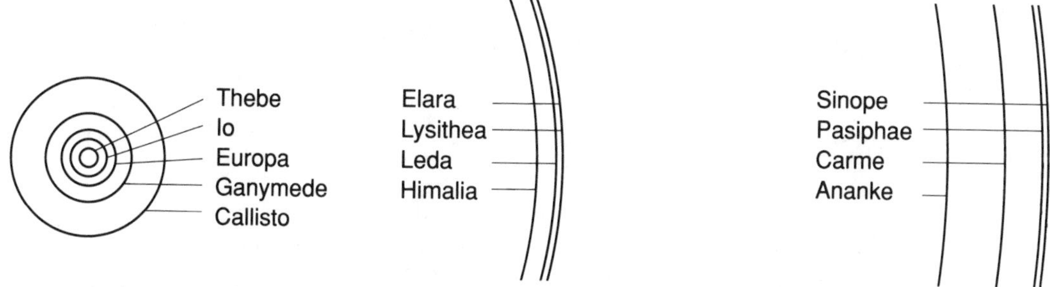


**11.11** The orbits of Jupiter's moons, here shown to scale, fall neatly into families. The outermost four moons are small bodies which orbit Jupiter in the retrograde direction and are probably captured minor planets. The next four are small but all travel around Jupiter in the expected direction, anticlockwise when viewed from above the north pole; these moons probably formed with Jupiter. The inner group of moons contains the four Galilean moons: Io, Europa, Ganymede and Callisto, and four smaller moons, of which only one, Thebe, is shown here.

shadow lags behind. As each satellite crosses the disc, it can be seen with varying degrees of visibility, although it is sometimes difficult against a light zone in small instruments.

It is possible to see Jupiter without any moons on either side. Although not all that common, it will happen when all four satellites are simultaneously undergoing a transit, occultation or eclipse. This line-up will occur on 2 January 1991 and 27 August 1997. These two events are ideally suited to observation with the planet at opposition on both occasions. When the Earth is directly in the plane of the orbits of the Galileans, every five to six years, another series of interesting events happens: the moons begin to eclipse and occult each other.

The satellites of Jupiter have also played an important part in the history of science, for it was using them that the first accurate calculations of the speed of light were obtained. It was the Danish astronomer Ole Roemer who, in the late 1600s, first noticed that the period of one of Jupiter's moons, Io, varied over the year. By careful calculation, he found that the period of the moon was shorter when Jupiter and the Earth were approaching and longer when they were receding. Roemer concluded that the difference was due to light having a finite speed and it taking less time to travel the shorter distance between two identical positions as the Earth approached and longer for the increasing distance as it receded. Further calculations by the astronomer based on this and other phenomena of Jupiter's satellites led to a calculation of the speed of light.

Though the speed of light can be measured much more accurately in the laboratory, the study of the satellites of Jupiter is now even more intense. Following Galileo's discovery of the first four satellites, other astronomers continued to observe the planet and find more moons. The first, Amalthea in 1892, was followed by nine more up until the late 1970s. The discovery of these satellites was of little more than academic interest to many observers, as even the largest of the Jovian satellites appears as nothing more than a point of light in all but the largest telescopes. It wasn't until spacecraft began visiting Jupiter that the diversity and uniqueness of the satellites became apparent.

Scientists had supposed that the moons of Jupiter would be cold and devoid of anything of interest, apart from meteorite craters and associated features. Little did they realise that the Jovian satellite system would display most of the processes which led to the formation of the planets, and that the study of the satellites would lead to further knowledge of the origin and formation of the whole solar system.

A glance at Figure 11.11 will show that Jupiter's family of moons is divided into three groups. The outermost group contains four small moons, all orbiting the planet in the retrograde direction. These small bodies are probably minor planets captured by Jupiter's gravitational field. None of them has been seen close enough for any details to be discerned. Closer to the planet is another group of four moons. These also are small in size, though Himalia is a reasonable 170 km. These moons

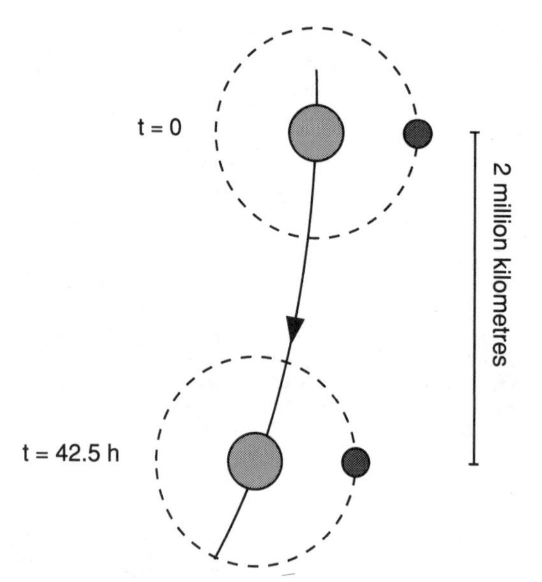


**11.12** In the 1600s Dutch astronomer Ole Roemer used the study of Jupiter's moon Io to calculate the speed of light. Roemer noticed that as Jupiter and Io approached the Earth the period of the moon was shorter than when the two were receding. The reason is clear from the diagram above. When Jupiter and Io are moving towards the Earth, in the 42.5 hours it takes Io to make one revolution, the pair would move two million km closer; this shorter distance would mean that the light would arrive six seconds earlier than would be expected. The effect is small over one revolution, but when many very careful observations were made Roemer was able to calculate a fairly accurate number for the velocity of light, 303 000 km $s^{-1}$.

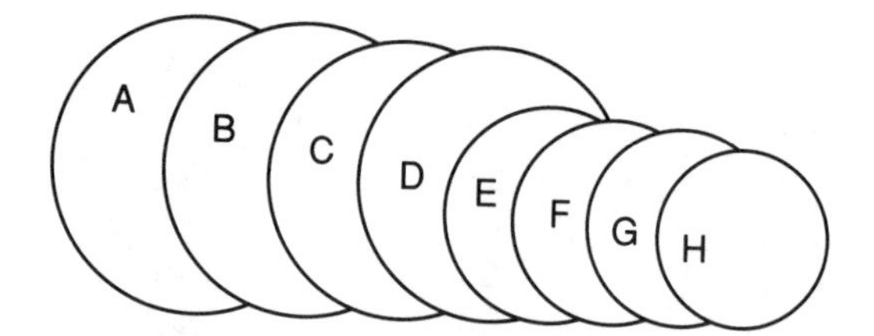


**11.13** The relative sizes of some of the solar system's moons and the planet Mercury; diameters are given for each.

| | | | | | |
|---|---|---|---|---|---|
| A | Ganymede | 2631 km | E | Triton | 1900 km |
| B | Titan | 2575 km | F | Io | 1815 km |
| C | Mercury | 2435 km | G | Moon | 1738 km |
| D | Callisto | 2400 km | H | Europa | 1569 km |

are probably also a group of captured minor planets as, though all have orbits in the correct direction, their orbital inclinations are quite steep, around 28°.

It is the group of moons closest to Jupiter which is the most fascinating. This group contains the four Galilean satellites and four smaller bodies. The four small bodies, of which Amalthea is the largest at 240 km, are closest to Jupiter. Unlike the small moons further from the planet, all four of these moons have orbits in the regular direction and inclinations around 0°, indicating that these moons formed with the planet rather than being captured afterwards. Two of these moons, Metis and Adrastea, act as shepherds, keeping the particles in Jupiter's thin ring in place.

The four Galilean satellites were amongst the major targets for the two Voyager spacecraft at Jupiter. Before the arrival of the spacecraft at Jupiter, scientists debated what they would find when they got there. Like the satellites and planets of the inner solar system, it was expected that the surfaces of the satellites would have suffered extensive bombardment by meteorites throughout their formation. The question was: would these craters be preserved?

The composition of the moons would answer that question. Ice was the favoured material for the surface of the satellites. On Earth large amounts of ice can flow slowly—glaciers are such a phenomenon, the ice in them flowing slowly down valleys. In the outer solar system the temperature is too low for ice to flow; its properties resemble those of rock. On the Jovian satellites, however, the temperature is a little warmer than in the further reaches of the solar system, so it was conjectured that the ice there would be able to flow, and hence would eventually collapse and fill in any craters.

Callisto is the furthest of the Galilean satellites from Jupiter, orbiting the planet once every sixteen days and sixteen hours. Callisto's diameter of 4840 km makes it almost the same size as the planet Mercury, but whereas Mercury has a density of 5.5 g.$cm^{-3}$, being made of rock, Callisto's density is just 1.8 g.$cm^{-3}$ suggesting that it is composed of equal parts of rock and ice. The temperature on Callisto varies from −120°C during the day to −170°C at night, making conditions ideal for a surface of ice.

The ice on the surface of Callisto is very easy

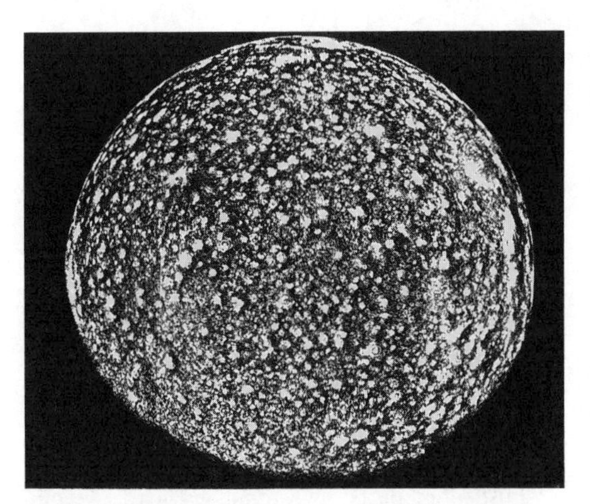

**11.14** This picture is made from nine images taken by Voyager II on 7 July 1979. The surface of Callisto is completely covered by impact craters. The most cratered object in the solar system, the surface of this moon must be very old. The surface of Callisto is made of ice, but the ice is so cold that it acts like rock and has preserved the evidence of impacts for 4000 million years. Near the upper right of the moon can just be made out the Valhalla basin, a circular feature 600 km across formed by a huge impact after most of the cratering on the moon had finished. PHOTO CREDIT: JPL/NASA

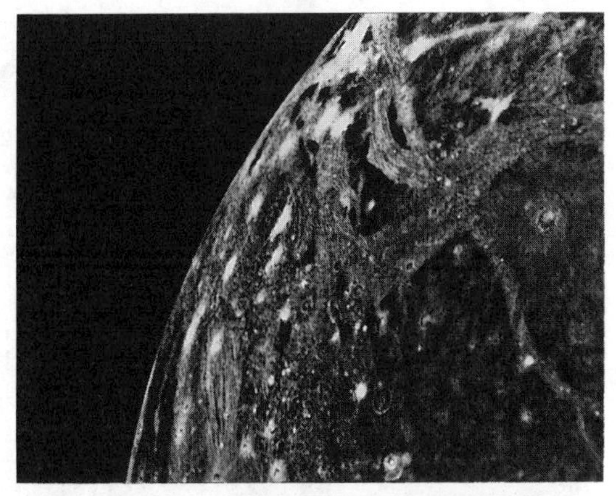

**11.15** Ganymede shows a mixture of ancient and recent terrains. The ancient terrain seen at the lower left of this picture shows a great range of impact craters. The different sizes and overlapping of the craters indicates that the cratering took place over an extended period of time. The more recent terrain of Ganymede is seen near to the limb of the moon where bright bands cut across the surface of the moon. These bands appear to be made of long parallel channels 10 to 15 km apart and about 1 km deep. These formations were probably made after a great impact late in the moon's history which displaced the crust and led to deformations in the ice.

to spot; although the surface is dark in colour, an albedo of 0.18, the presence of ice is immediately revealed in the spectrum of light reflected from the moon where it shows in the infrared region. The dark colour of Callisto's surface could either be due to the original material forming the moon not separating to a rocky core and icy crust, or to overcoating with some foreign material.

The first of these explanations is unlikely. A body the size of Callisto would have generated sufficient heat during its formation to melt some of the ice and allow the heavier rocky particles to make their way towards the centre. Even if there had been insufficient initial heating, the temperature of the ice now is such that large blocks of rock would still slowly sink towards the middle.

Callisto's surface shows the result of heavy meteoritic impacts, being as heavily cratered as the highlands of our own Moon. This is only possible if the surface of the moon has remained undisturbed for many thousands of millions of years. It is this that explains the dark coating on the satellite: it must be covered with the accumulated dust from millions of tiny micrometeorites. Over 4000 million years the dust would build up on the surface, mixing with the ice and giving the dark colour seen today.

Although the cratering density is as high, the craters marking Callisto's surface are quite different from lunar craters. Instead of being deep bowl-shaped depressions, the craters are flattened into shallow circles with low walls. This deformation is caused by the slow flowing of the ice on the surface, as had been predicted. The amount of cratering leads to the conclusion that the surface is indeed very old, similar amounts of cratering having occurred on our Moon 4000 million years ago.

A few places on the surface are much brighter indicating the impact of a large body in recent times. These events allow new, clean ice from below the surface to flow out, but there are no features to compare with the large mare regions of our Moon or the small flooded plains of Mercury, indicating the lack of any really large impacts near the end of cratering. The largest of

**11.16** The surface of Europa, here seen by Voyager 2 at a distance of 241 000 km, is light in colour and covered by curving darker lines like cracks in the surface. These fractures are possibly the result of the surface cracking as the moon expanded when the ice froze. The surface of the moon is almost devoid of meteorite craters indicating that it must be quite young. The smallest features visible on this picture are about 4 km across. Look carefully near the terminator: the shadows cast by the features show clearly the small relief of the Europan surface.
PHOTO CREDIT: JPL/NASA

the clean ice flows is a circular feature called Valhalla over 600 km in diameter with shock features spreading out over a further 500 km. The clean, bright centre and the concentric circles of ridges around it give Valhalla the look of a bullseye.

The largest satellite of Jupiter is Ganymede. Indeed Ganymede is the largest satellite in the solar system*. Ganymede is the next closest satellite to Jupiter after Callisto. It has a similar density, 1.9 g.cm$^{-3}$, and is just 410 km bigger. It, too, clearly shows from spectroscopy that its surface is ice covered. With all these similarities it could be expected that Ganymede and Callisto would be twins, yet even a superficial look at the two moons shows that they are very different from each other.

Half of Ganymede's surface does resemble Callisto. It has the same dark, heavily cratered regions, even down to similar bullseye patterns from recent impacts, but other parts of the surface are quite different. In amongst the craters are large, lighter coloured areas with significantly fewer craters, indicating a younger surface.

In places throughout the lighter coloured areas, large regions of grooves can be seen, patterns of parallel ridges and valleys crossing and intersecting with each other. Similar features are seen on both Earth and Venus. The process of formation of these ridges is not completely understood but, unlike the ridges on Earth and Venus which form when the crust is compressed, the systems on Ganymede appear to be due to stretching. Tension in the moon's surface has caused some parts to drop, forming valleys, whilst other parts have remained high. Accompanying this activity would have been the leakage of water from beneath the surface leading to the lighter colour.

The other striking feature of Ganymede is the extremely bright impact craters and their rays which dot the surface. These bright features indicate that almost pure water must have come from beneath the surface, and in fairly recent times. This suggests the presence of clean water beneath Ganymede's surface, something Callisto appears to lack.

The next moon towards Jupiter is something of a puzzle. Europa has a density of 3 g.cm$^{-3}$ so instead of being half ice and rock, it must contain only around 10 per cent ice. Yet despite the lack of ice, the surface of Europa is the brightest of the four large satellites with an albedo of 0.70. This and its reflected spectrum indicate that the surface must be made of almost pure ice. This leads to the conclusion that Europa must be fairly well differentiated, with all the rock at the centre and a thin covering of water ice over the top.

What makes this apparently greater differentiation difficult to fathom is why the surface of a smaller moon, 2000 km smaller than Ganymede, should be better differentiated than the larger moons nearby. The light-coloured crust also suggests something is happening on the moon to keep the ice uncontaminated by dust. The idea that the surface of Europa is slowly being reformed is supported by the lack of impact craters on the moon. The crater densities are roughly the same as the density on the Earth, indicating that the surface we are seeing is quite recent.

The lack of craters is made up for by the

* For many years it was thought that Saturn's Titan was the largest satellite, but the discovery of an atmosphere around it makes it slightly smaller than Ganymede.

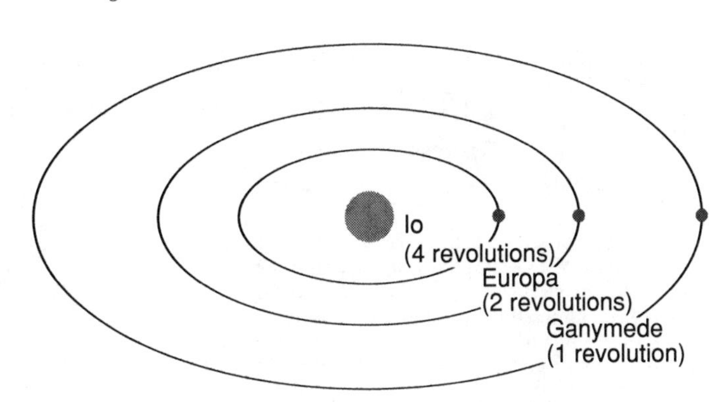


**11.17** Resonances occur between the orbits of three of the Galilean satellites. For every one revolution of Ganymede, Europa orbits Jupiter twice and Io orbits Jupiter four times. This resonance of orbits causes stresses in the crusts of all three moons, leading to many of the structures we see on their surfaces.

presence of long ridges of light material and streaks of dark material covering the surface. The ridges, only a few hundred metres high, stretch for thousands of kilometres across the face of the moon. With no features more than a few hundred metres high, Europa turns out to be the smoothest object in the solar system. The reason for this flatness is that beneath the relatively thin solid crust, an ocean of liquid water probably exists and this, of course, would not have the structural strength to support very tall features.

Why has this moon retained oceans of liquid water when the larger objects further from the planet have frozen solid? The answer appears to be the presence of Ganymede. A look at Table 11.2 will show that the orbital period of Ganymede is exactly twice that of Europa, so the pair of moons is locked into synchronous orbits about Jupiter. The deformation of Europa by the tidal forces of Ganymede every second turn around Jupiter is enough to provide the heating necessary to maintain liquid oceans beneath the icy crust.

Another case of synchronous orbits within the Jovian system exists between Europa and the last of the Galilean satellites, Io. On Io the energy provided by the synchronous orbit, and the resulting stretching and compressing of the moon, makes Io the most active object in the solar system. Io has a density of 3.6 $g.cm^{-3}$ and a diameter of 3650 km, making it similar to our Moon in both size and density; this suggests that the composition of Io is similar to that of our Moon.

The photographs which the Voyager spacecraft returned from Io were one of the biggest surprises of the mission. Instead of the rocky, grey terrain of the other moons, Io is red, orange, yellow, brown, black and white, an object which resembles a pizza more than a moon. The origin of the colours is not completely certain. Sulfur is one material which has the required range of colours, taking hues from yellow to black depending upon the way it cools from a liquid, and certainly sulphur compounds have been found both around Io and in the charged particles around Jupiter, but there are other chemicals which can provide the same colour range. The solution will only be revealed by further study.

Io's surface is completely devoid of impact craters, indicating that the surface is very young. The rate at which the moon is expected to be impacted upon and the lack of such impact markings leads scientists to believe that Io's crust must be formed at the rate of a few centimetres per century—extraordinarily fast.

Though Io lacks impact features, it doesn't lack craters. However its craters are volcanic in origin. Some of the volcanoes associated with the craters are over 9 km high. The number of volcanic features on Io is remarkable, but what was more astounding was the discovery of still-active volcanoes which were erupting as the spacecraft flew by. This was the first time volcanism had been seen anywhere other than on Earth.

It is not molten rock, however, which drives the volcanoes of Io, but sulfur and its compounds. The volcanoes appear to originate from rifts in the moon's surface where the material is thrown 300 km into space above the moon before it rains back down on the surface. These plume eruptions as they are known, throw 100 000 tonnes of material upwards each second, easily enough to account for the resurfacing of the moon. In other parts of the surface, lakes of

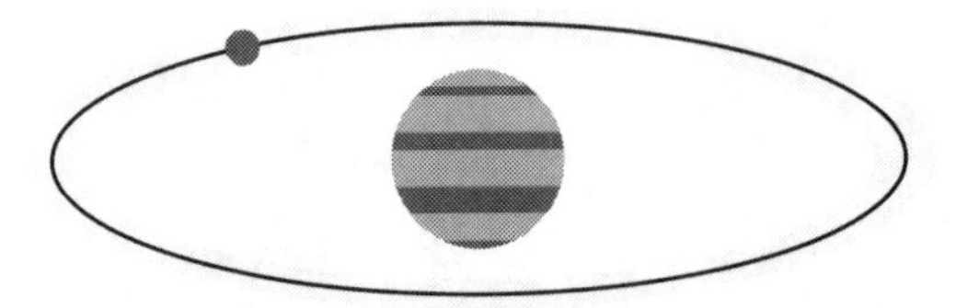

**11.18** The low escape velocity of Io means that the spectacular sulfur volcanoes which dot its surface are able to throw material clear of the moon's gravitational field and into orbit around Jupiter. This material forms a doughnut-shaped ring around Jupiter known as the Io plasma torus.

liquid sulfur can be seen, indicating temperatures over +100°C.

The energy to power all the volcanic activity on Io comes from its resonant orbit with Europa and tidal heating from Jupiter. Io is almost the same distance from Jupiter as our Moon is from Earth, yet Jupiter is 300 times as massive as Earth, and its stronger tidal force has caused a large bulge to form on the side of the moon facing the planet. Normally this deformation would not lead to heat as the moon would always keep the bulge pointed towards Jupiter. Io's orbit is, however, distorted by Europa and Ganymede. Io makes two orbits for every one of Europa's, and four for every orbit of Ganymede.

It is this resonance of Io's orbit with Europa and Ganymede which provides the heat. Because the gravitational tugs of Europa and Ganymede always happen at the same point in its orbit, Io is unable to travel around Jupiter is a circle; the tugs distort it into an ellipse. This results in the bulge on Io shifting back and forth with respect to Jupiter. This twisting leads to strains in the moon, which in turn leads to heating.

All this energy, around a hundred million megajoules per second, has driven all the water and other light compounds from Io. Sulfur and its compounds are all that remain. Io's core has been stretched and strained so much that it is almost entirely liquid; only a thin crust, which is being constantly recycled, is able to form. The changes are so marked that the maps drawn from the Voyager visits will be different from those drawn after Galileo arrives in 1995.

As if to reciprocate, Io has quite an effect on Jupiter via the planet's magnetic field. Moving within the Jovian magnetic field, Io acts like the wire in a generator, creating quite high currents of electrons and ions which cycle from the moon to the planet and back again. The currents generated are in excess of three million amps and cause heating in Jupiter's atmosphere at the point where they reach the planet, creating vast auroral displays. Another effect on Jupiter comes from Io's volcanism. Some of the huge amounts of sulfur thrown into space does not fall back to Io's surface; instead it forms a tenuous doughnut-shaped ring of sulfur ions around the planet. This ring, called the *Io plasma torus*, surrounds Jupiter and contributes most of the heavy particles in the planet's magnetosphere.

On 17 October 1989 the United States' space shuttle Atlantis lifted off from the Kennedy spaceport with the Galileo spacecraft aboard. Galileo, which had been delayed since the explosion of the space shuttle Challenger in January 1986, and by law suits attempting to block its launch because of nuclear materials it carries for power, was finally on its way to Jupiter. To get to Jupiter using the minimum amount of fuel, the path Galileo is following is the most complex since Voyager 2's grand tour of the solar system, taking it past Venus once and Earth twice before it is finally headed towards its final destination.

At Jupiter, Galileo's mission is to explore the planet and its four largest moons. Galileo, unlike the Pioneer and Voyager missions, is not a flyby. The craft will enter an orbit about Jupiter, allowing it to study the planet's atmosphere over an extended period. It is envisioned that the craft will orbit Jupiter about ten times before using the gravitational effects of the large moons to move it into a complicated path of encounters with Jupiter's satellites. The craft carries cameras to send back pictures of the planet and its moons, spectrometers to analyse the chemicals found both in visible and ultraviolet light and numerous other instruments to analyse magnetic fields and other phenomena at Jupiter and along its journey to the planet.

In addition to the main spacecraft, Galileo includes a small probe designed to enter Jupiter's atmosphere. To achieve this task the probe will separate from the main craft five months before Galileo reaches Jupiter in December 1995. As the probe reaches Jupiter it will be travelling at around 48 kilometres per second. Using the drag of Jupiter's upper atmosphere, and a heat shield, the craft will decrease speed until it is slow enough to release the shield and open a parachute. Over the next hour and a

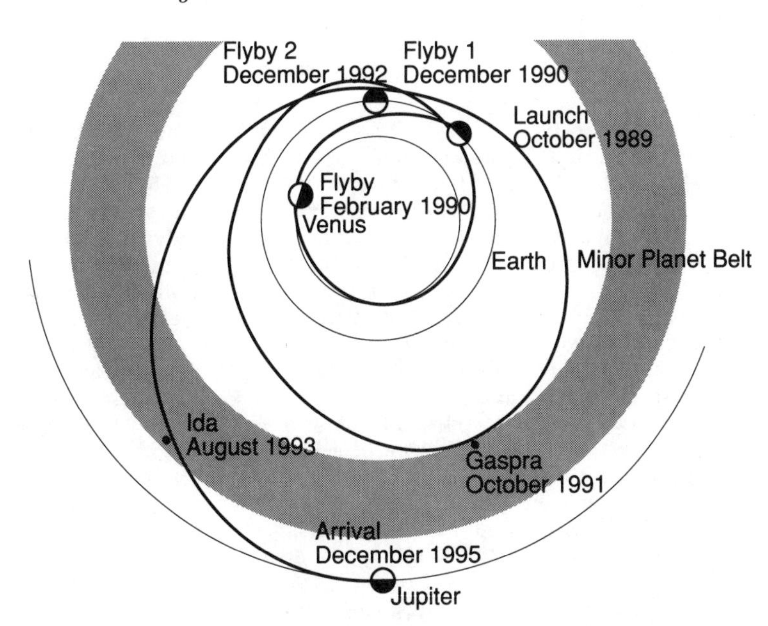


**11.19** After its launch on 17 October 1989, the United States' spacecraft Galileo began to follow a complex path which will have it reach Jupiter in December 1995. Galileo flew past Venus on 16 February 1990 and will pass the Earth twice, in December 1990 and 1992. After its first passage past the Earth, Galileo will travel into the region between Mars and Jupiter and pass close to the minor planet Gaspara before passing just 300 km from the Earth on the final stage of its flight. Before reaching Jupiter, however, Galileo will pass another minor planet, Ida, arriving at Jupiter two years later.

quarter the probe will drift through the clouds and gases, taking readings and making analyses of what it finds. The probe should pass through three layers of clouds on its way down: the high ammonia, the lower ammonia hydrosulfide and finally perhaps water vapour clouds. Below the clouds the probe will stop working as it is crushed when the pressure of the atmosphere rises to about 25 times that of Earth.

Even if Jupiter were a moonless world, it would still provide the amateur with countless hours of enjoyable observation. However, the study of Jovian satellite phenomenon is a worthy pursuit that is not only fun, but of immense scientific value; the Galileo project, when it arrives at Jupiter in 1995, will be assisted by amateur timings.

*Voyager 2's journey*

On 20 August 1977, Voyager 2 was carried from the Earth by a Titan-Centaur rocket on a journey that would write nearly everything we know about the outer solar system. Voyager 2's place in history came about a once-in-175-year opportunity, to send a spacecraft past all the gas giant planets of the solar system in a reasonable time, provided that the craft could survive more than twelve years at temperatures of −240°C and miss the hazards along the way.

The discovery of the possibility of such a 'grand tour' came from the work of a Ph.D. student, Gary Flandro, who was working at JPL in a summer job in 1965. He and Michael Minovice found that by approaching a planet in the right way, the speed of a spacecraft could be increased by the gravitational field of the planet, thus giving a huge boost in speed. In particular they found that Jupiter's gravity could be used to speed a spacecraft enough to visit the outer planets with little expenditure of fuel, and that the next time would be in 1977. In 1965 no one had walked on the Moon and only one spacecraft, Mariner 4, had successfully reached an outer planet, Mars. How could a spacecraft be ready in just twelve years to take advantage of this rare opportunity?

The obstacles to the journey were immense. At the distance from the Sun of the outer planets, solar energy could not be used to power the spacecraft, and direct continual instructions from Earth would be out of the question. The time delay was too long, so the spacecraft would have to handle its own emergencies. Given that the space age was just eight years old, how could a craft be made which would last at least twelve years? NASA and JPL having finally got approval for the mission, the United States Congress cancelled the project in December 1971, saying that $1000 million was too much. In 1972 they relented and approved $360 million allowing flights to just Jupiter and Saturn for two probes patterned after the successful Mariner probes to Mars and Venus. This plan needed spacecraft which could last four years, not twelve, a much easier task. The four-planet tour was dead.

Well, nearly dead. In 1975 the Mariner-Jupiter-Saturn project was given the name Voyager. The aims for the mission were to obtain data from the two planets and close-up pictures of their moons, particularly Io and Titan. It was still possible to fly the grand tour, but not if images were wanted of Io and Titan. One spacecraft had to be sacrificed. If Voyager 1 flew close to Titan it would obtain nice views, but be headed in the wrong direction to continue further. On the other hand, having taken the pictures, Voyager 2 could fly past, neglecting Titan, and travel on to Uranus. In this way the central goals of the mission could be accomplished and Voyager 2 still travel on, but only if Voyager 1 succeeded at Saturn. Which it did, brilliantly. The grand tour, cancelled twice, was really on.

Well, nearly on, Voyager 2 was getting old. Eight months after launch a power surge broke one of Voyager 2's radio receivers and damaged the other one. Instead of being able to tune into broadcasts from Earth over a wide range of frequencies, the tuner could accept signals over a range of just 192 hertz. Just after its encounter with Saturn, the platform holding the cameras jammed, making it impossible to move the cameras and causing some data to be lost. Of course, as we know, these problems were overcome.

The problem of the tuner was easy to solve, in principle; just make sure that the transmitter on Earth broadcast at the right frequency for the spacecraft. Unfortunately there were a number of other effects to be taken into account. Both the spacecraft and the Earth were moving, which introduced a Doppler shift in the frequency which had to be taken into account. The temperature of the receiver also altered the frequency at which it worked, so that had to be allowed for. The action of using the thrusters also altered the temperature, again making corrections necessary. Besides, no one was exactly sure of the temperatures the spacecraft would encounter at Neptune or elsewhere.

The scan platform problem had rectified itself to some extent as lubricant seeped back into the system allowing movement again, but only at low speeds, not the high rates needed to compensate for the motion of the spacecraft. Not wanting to lose any data at Uranus or Neptune, engineers came up with a way to turn the spacecraft to compensate for its motion, in the way a television camera operator pans the camera to capture fast moving action. Using the small hydrazine thrusters and a new program on the onboard computer, Voyager was able to turn to follow a planet at both very slow and fairly high speeds. With these problems fixed Voyager was set for its visits to Uranus and Neptune.

# 12

# Saturn

Saturn has been described as the jewel of the solar system, its system of rings making it the focus of attention for artists and scientists alike. Even though rings have been found around all three other gas giant planets, Saturn is still *the* planet with the rings.

Although about 20 per cent smaller in diameter than Jupiter and almost twice the distance from the Sun, Saturn still presents a disc size suitable for study with a small telescope. Saturn's diameter (excluding rings) during opposition is 21 seconds of arc, or two-fifths of Jupiter's largest diameter, and fifteen seconds of arc at conjunction, similar to Mars at a poor opposition. The maximum opposition magnitude of Saturn is about −0.4, but this may drop about half a magnitude if the ring system is seen edge on.

Aside from the magnificent rings, the most obvious feature of the planet in a telescope is the polar flattening. Saturn's rapid axial rotation period is 10 hours 14 minutes at the equator, a little slower than that of Jupiter. The fast spin rate and low density lead to a much greater equatorial bulge than Jupiter's. The equatorial diameter exceeds the polar diameter by 11 per cent, and it is most obvious when the rings are presented edge on.

The synodic period of Saturn is 378 days, so oppositions happen about two weeks later each year. The slow sidereal period of 29.46 years means that Saturn will reside in each zodiacal constellation for about two and a half years, long enough to distort a constellation and confuse the untrained observer into treating it as a fixed star. Figure 12.1 shows the coming oppositions until the turn of the century.

Galileo was the first person to see the rings of Saturn, but because of the poor quality of his telescope, or perhaps because no one had ever imagined that rings might circle a planet, he never identified them as such. Galileo originally thought they might be handles or moons on either side of the planet, maybe a triple planet. After a few years of observing the planet Galileo was unable to see the rings at all and began to dismiss them as early mistakes, but then they reappeared again.

It was not until 1659, almost 50 years later that the Dutch astronomer Christian Huyghens realised that Saturn was surrounded by 'a thin flat ring which nowhere touches the planet'. He said that the ring encircled the equator of the planet and that it must be very thin. This explained the disappearance of the rings that Galileo saw. During part of its journey around the Sun, during the Saturnian spring and autumn, the rings of the planet appear edge on to the Earth and hence are invisible. This continues for a few months and then the rings slowly become visible again as thin bands. Galileo had originally seen the rings when they were near their best; a few years later they were edge on and he was unable to see them at all.

In 1675 Jean Dominique Cassini was observing Saturn's rings when he noticed that there were two rings, the inner and outer separated by more than 3000 km through which the planet could be seen. This division is known today as the *Cassini division*. Later, other observers found other gaps in the rings, and other rings.

Study over 300 years has revealed much about the rings of Saturn. One of the first things deduced was that the rings must be made of small particles, not large solid sheets. If the rings were sheets, then the laws of planetary motion would quickly break them up; particles on the inner edge of the rings would need to move much faster than particles further out. No material

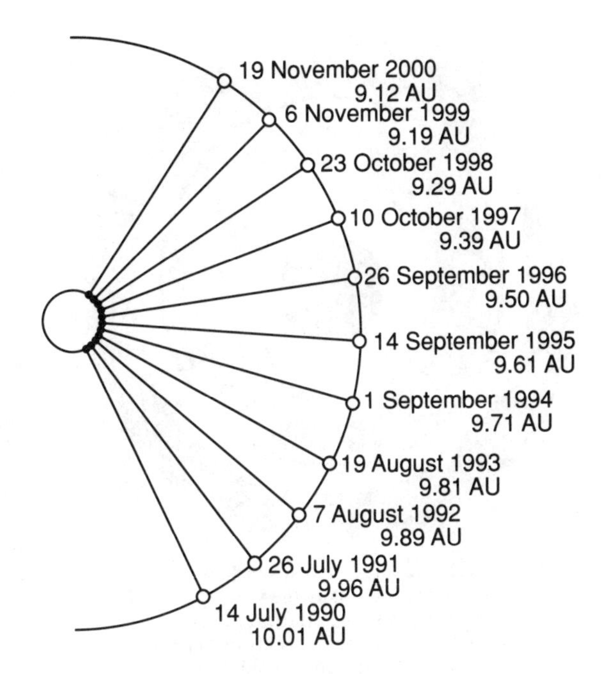


**12.1** Oppositions of Saturn until 2000. The distance between the Earth and Saturn is given for each.

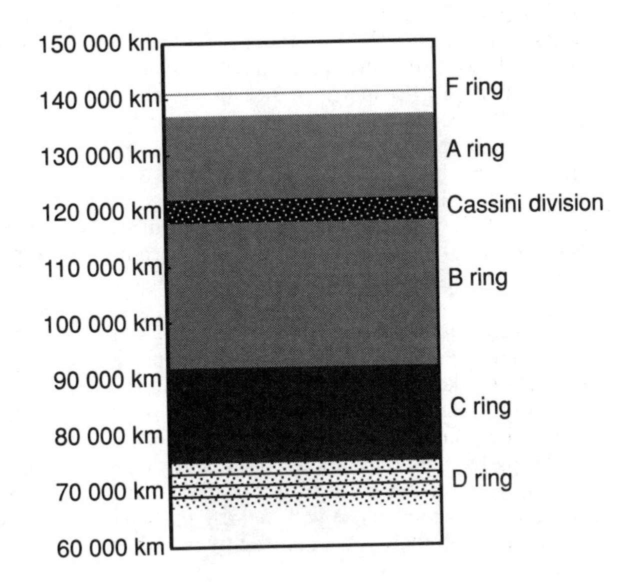


**12.2** The structure of Saturn's rings in terms of their distance from the centre of the planet. The G and E rings do not appear in this diagram. They are 170 000 and 230 000 km from the planet's centre respectively.

**12.3** An eccentric ring is one in which all the particles have orbits of exactly the same eccentricity which are aligned with each other. If the orbits were not so aligned the ring would quickly disintegrate, as collisions between the particles would push them into new orbits away from the ring. How such a ring remains stable and how it forms are still mysteries.

would be able to withstand the tidal forces generated by such an arrangement and would soon crumble.

Observations from Earth in the early 1970s showed that the rings of Saturn are made primarily of water ice. The first radar observations in 1973 showed that the ice is made of chunks around a tenth of a metre in size up to a few metres. The same observations also revealed that, though thin, the rings had a reasonable thickness and had not collapsed to being only one particle thick. When Pioneer 11 arrived at Saturn in 1979 it was able to reveal much more about the rings, but the real information came with the arrivals of Voyagers 1 and 2 in 1980 and 1981.

The Voyagers confirmed much of what was already known. The rings are made of icy particles, almost like snowballs. Each particle in the rings follows an almost perfectly circular orbit around the planet. If the particles didn't follow circular paths, then they would continually move closer to and further away from the planet; the collisions this would cause with nearby particles would be such that any eccentricity in orbits would quickly be absorbed. Similar arguments explain why the particles in the rings are in a flat band; a particle with an inclined orbit would have to pass through the main ring plane twice per orbit, and again collisions would soon restrict it to the plane. All these features had been expected, but there were many things revealed which had not been predicted: narrow gaps, waves, radial markings, eccentric rings, kinky rings, even braided rings.

Naming the rings presented a problem. As three rings could already be seen from Earth, they had already been named. The outermost ring is the A ring, the next the B ring and the faint ring close to the planet the C ring. Voyager discovered another ring closer still to the planet, called the D ring, and another outside the A ring, the F ring. Further still from the planet is the thin G ring and then the diffuse E ring just inside the orbit of the moon Enceladus. Therefore the sequence from the planet is D, C, B, Cassini division, A, F, G, E.

It is the A and B rings which are easily visible from Earth. The B ring starts 32 000 km above Saturn's cloud tops where the ring particles, spread thinly throughout the C and D rings, suddenly bunch together. It is here at the innermost edge of the B ring that the ring system is brightest. Throughout its 25 000 km span, the particles making the B ring are so closely packed that the ring appears opaque. The particles making the ring are between 100 mm and a few metres in size and fairly light in colour. From Earth, no gaps in the B ring can be seen though the cameras of the Voyager spacecraft showed that the B ring is, in fact, made from thousands of tiny ringlets so close together that from the distance of Earth they seem to combine to form one large, flat ring.

Outside the B ring is the only gap in the rings that can be seen clearly from Earth, the Cassini division. Although this region of the rings appears as an empty 3500 km gap, the region contains many small ringlets. There are some small gaps, though even these contain the occasional ring—one gap contains a rare eccentric ring. In an eccentric ring the particles do not follow a circular path around the planet, but follow a standard elliptical orbit. As it was pointed out earlier, it would be expected that the orbit would be destroyed by collisions between particles as they moved closer and further from the planet. This doesn't happen in an eccentric ring because all the particles have their

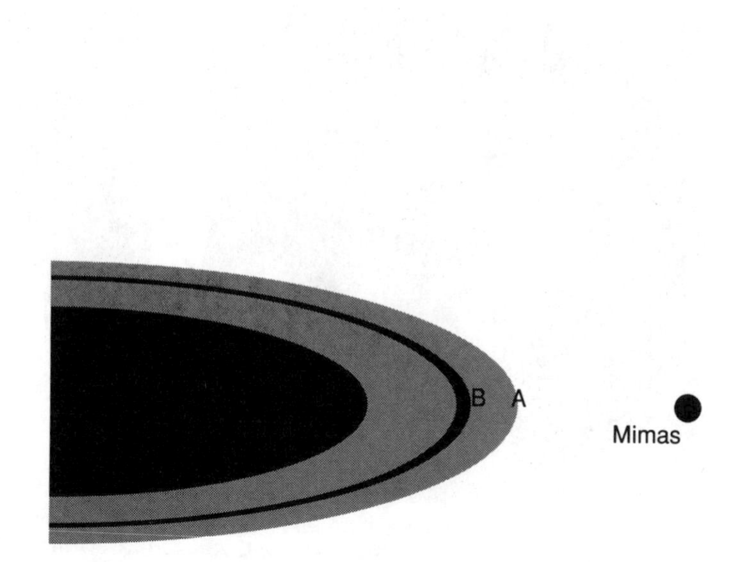


**12.4** The satellite Mimas is responsible for both the outer sharp edge of Saturn's rings and the gap known as the Cassini division. Particles orbiting Saturn at the distance of the outer edge of the A ring (A in the diagram) complete three revolutions for every two of Mimas. Particles which move further out than this experience an accelerating force which makes them move back into the main part of the ring, thus the edge is kept clean. The Cassini division results from the particles having a two to one resonance with Mimas: for every one revolution of Mimas a ring particle in the Cassini division would revolve twice. This strong resonance moves the particles into either the A ring or B ring, leaving the division relatively free.

orbits aligned, so that they reach their furthest point from the planet at exactly the same place, eliminating collisions. Exactly how such a ring could form and remain stable is unknown.

Marking the outer edge of the Cassini division, 62 000 km from the planet, is the beginning of the A ring. The A ring is not as bright as the B ring, nor is it as opaque. The particles forming the ring are approximately the same size and colour as those of the B ring, but there are fewer of them. Like the B ring the A ring consists of thousands of tiny ringlets, each in a circular orbit about the planet. Within the A ring is a 360 km wide gap called the *Encke division*; this gap can be seen from Earth only with large telescopes and excellent seeing conditions. Within the Encke division the Voyagers found two kinky rings. A kinky ring is one where the ring particles do not appear to form a simple thin ring, but clump together and stray from the expected path. Like eccentric rings, the processes which form and maintain a kinky ring are still unknown, but it may involve the presence of a small (a few tens of kilometres) moon orbiting within the rings.

Closer to the planet than the B ring are two more rings, only one of which can be glimpsed from Earth, the crepe or C ring. The C ring begins 14 000 km above the surface of the planet. The particles making the C ring are fairly well spread out so the planet can easily be seen through the ring. Within the C ring are more gaps, one of which contains another eccentric ring. Below the C ring is a small collection of ringlets, known as the D ring; these contain so few particles that they cannot be seen from Earth.

The boundary of Saturn's ring system is sharp. It finishes abruptly 77 000 km from the planet's surface. There is no gradual fading of the rings and no stray particles drifting away. They end precisely. Four thousand kilometres beyond the outer edge of the A ring, however, is the F ring, discovered by Voyager 1. Unlike the broad bands closer to the planet, the F ring is a single ringlet 100 km wide. Like the two rings in gaps, this ring is also eccentric.

Unlike the two other eccentric rings, however, the F ring also shows places where it splits into a number of smaller ringlets which then twine about each other in a pattern similar to braiding. Initially scientists were at a loss to explain such phenomena, but with the discovery of two small moons, one on either side of the F ring, an explanation was worked out. The two satellites, 1980S28 and 1980S27, or Prometheus and Pandora, are called shepherd satellites. Their role is to keep the particles of the F ring confined to the small band they occupy; it is these satellites too which generate the fine structure in the F ring. The braiding is explained by gravitational interactions between the two

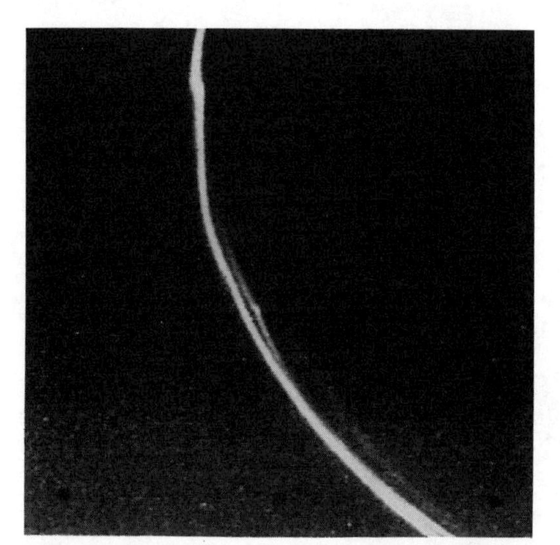

**12.5** Initially the braided structure of Saturn's F ring puzzled scientists and led to media claims that the laws of physics had been violated, but the explanation was simple once the presence of two small shepherd moons was detected. The interaction of these two small moons, Prometheus and Pandora, causes particles in the F ring to break into strands which twist about each other. Up to five separate strands have been found in the F ring. PHOTO CREDIT: JPL/NASA

**12.6** A photograph of some spokes in Saturn's B ring. When seen in reflected light the spokes appear dark, as seen here; when seen in scattered light against the rings, they appear bright indicating that they are made from small particles of dust. PHOTO CREDIT: JPL/NASA

shepherd moons when they pass close to each other, which could cause the ring to split into many strands.

We have seen the effects of resonant orbits on the moons of Jupiter. The resonant effects of satellites also have an influence on the orbits of the particles in Saturn's rings. It is Mimas, the closest of the large satellites, which is responsible for the Cassini division; particles orbiting at the inner edge of that division have a period exactly half that of the moon. Resonance between the orbits of the particles and Mimas therefore moves the particles out of that orbit, causing the gap. It is also Mimas which maintains the sharp outer edge of the ring system. Particles at the outer edge of the A ring have a period exactly two-thirds of the period of Mimas; again resonance removes any particles which stray beyond the edge. Other satellites create resonances within the rings too, but they lead to only small gaps as the moons are either too small or too far from the rings and so do not have simple resonances.

Within the rings the Voyager spacecraft were able to see much fine structure. One feature which puzzled scientists was the existence of spokes in the rings, lines which stretched across the width of the ring and which maintained their structure over time. It was thought that no radial structures could exist in the rings as the differing rotation rates between the inner edge and outer edge would destroy any straight feature in minutes, yet some of these spokes lasted for over ten hours. The spokes were found to be clouds of dust which rise out of the rings and appear dark against the lighter ring material. It is thought that the dust is expelled from the ring by some electrostatic process. It is this electrostatic force which keeps the particles in a straight line even though they would normally have different rotation periods. The clouds form over just a few minutes and can last for many hours before dispersing.

The Cassini division can be easily seen in small telescopes of 75–80 mm aperture, while the crepe ring needs about 150–200 mm of aperture and the hard-to-detect Encke's division 200–250 mm. The difficulty in observing these features depends on the tilt of the rings; when they are open to their maximum extent the task will obviously be much simpler. It is best to search for Encke's division at the ansæ, or ends, of the rings, instead of near where the rings cross the disc. Of interest in any size telescope is the shadow cast by the globe on to the rings, and the shadow of the rings on the globe.

Saturn's axis is tilted at 26.45° to the perpendicular, and as the rings are in line with the plane of the planet's equator and maintain a fixed orientation in space, we see them open and close during Saturn's 29.46 year orbit. Since 1980 the northern face of the rings has been presented to the Earth. They reached maximum inclination in 1988, and have been gradually closing since then. In 1995 the rings will again be seen edge on. As they reopen we will be afforded a view of the southern face until the next edge on occurrence fifteen years later. When edge on the rings are very hard to detect with small telescopes and may be lost for several months or more. They may even disappear in large instruments for a few days, and reappear as a thin line across the face of the globe, or as exceedingly thin needle-like projections on either side of the planet.

The interval between successive edge on appearances is not a constant 15 years, but alternates between about 13.75 and 15.75. The reason for this is that Saturn's orbital speed is greater during perihelion (when we see the southern face of the rings) than at aphelion (when the northern face is visible). When we view the southern face of the rings, a portion of the northern hemisphere is hidden, and when the northern face is visible the rings cover part of the southern hemisphere. Only for one or two years, when we see the ring plane almost edge on, do we get a full view of the planetary cloud detail.

When the rings are seen edge on, it is common for us to pass through their plane three times. This will happen during the 1995–96 presentation. In May 1995, four months before opposition, the Earth will cross from the north to the south face of the rings. In August, one month from opposition, we pass from the south to the north face. During November the rings will be edge on with respect to the Sun, and will be difficult to see as the illumination on the thin system will be minimal. In February 1996 we again pass from the north to the south face and this aspect will remain for the next 13.75 years.

On rare occasions during Saturn's journey across the sky, it will occult, or pass in front of a star. Planetary occultations are of great value to astronomers. Not only do they provide information about the structure of a planet's atmosphere, but they can be used as a tool of discovery; such was the case when a star was dimmed as the unseen rings of Uranus passed over it in 1977, and those of Neptune in 1984, even before Voyager 2 rendezvoused with those planets.

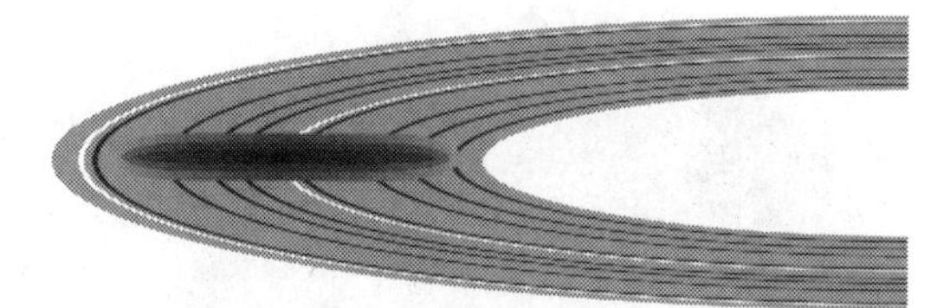

**12.7** Spokes appear as radial lines stretching across Saturn's rings. The spokes are clouds of dust which rise from the rings, obscuring the structure beneath, and are held in place by magnetic fields. They remain for a few hours before dispersing.

Stars of magnitude 8 and fainter are overwhelmed by the Saturnian system's brightness, and are useless for amateur observation. Only a handful of occultations by stars of sufficient brightness have occurred this century, and these proved (before Voyager) that the rings had more gaps than could be seen visually.

The brightest star so far occulted by Saturn was 28 Sagittarii. The event happened on 2–3 July 1989 and amateurs worldwide observed the event. The aim of the combined effort was to time any fluctuations in magnitude that the star underwent as it passed through the rings. Any changes in the star's brightness outside the known rings could also prove the existence of any undiscovered ring material.

From the east coast of Australia, amateurs were only able to witness one half of the event, as the disappearance of 28 Sagittarii behind the rings and globe happened during daylight hours. From my vantage point in the Blue Mountains west of Sydney, the sky looked excellent, especially just after months of the worst weather we had experienced in many years. Having set up our telescopes before sunset, we awaited our first glimpse of Saturn in the twilight.

Just fourteen minutes before 28 Sagittarii was to emerge from behind the globe, cloud began to form directly over the area that Saturn occupied! It could only happen with one of the rarest astronomical events of this century. Murphy's Law struck again. All was not completely lost,

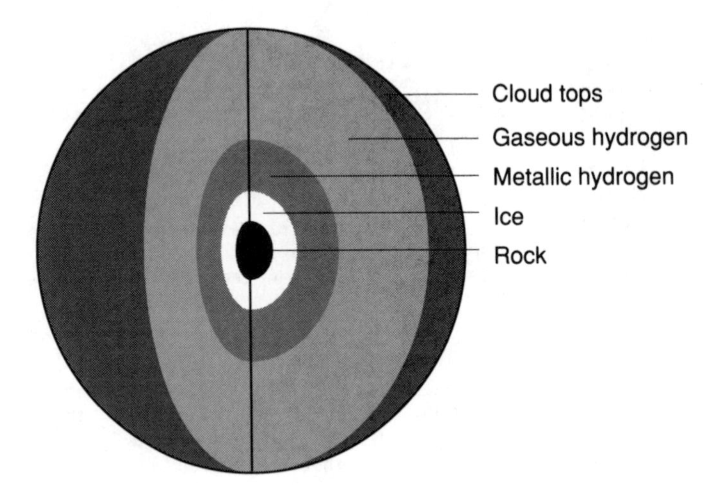


**12.8** The core of Saturn is probably a rocky ball a little larger than the Earth, covered by a layer of ice 8000 km thick. Above this is a layer of hydrogen so compressed that it behaves like a metal rather than a gas, and above this a layer of gaseous hydrogen and helium. Because of the low temperature of Saturn the helium mixed with the hydrogen is able to form droplets and condense towards the centre of the planet. This generates the heat we see powering some of the weather of Saturn.

**Table 12.1 Saturn's atmosphere**

| Gas | | Fraction |
|---|---|---|
| Hydrogen | $H_2$ | 92.4% |
| Helium | He | 7.4% |
| Methane | $CH_4$ | 0.2% |
| Ammonia | $NH_3$ | 0.02% |
| Ethane | $C_2H_6$ | 0.0008% |
| Phosphine | $PH_3$ | 0.0003% |
| Ethyne | $C_2H_2$ | 0.00001% |

as we could just make out Saturn's feeble glow penetrating the cloud. Fifteen minutes after emersion a 60-second break in the cloud revealed the star three-quarters of the distance from the globe to the rings. Thirty minutes later, a twenty-second break showed the star shining brilliantly through the Cassini division. Our last view saw the star some distance from Ring A, forming a right-angle triangle with two of Saturn's satellites, Rhea and Dione.

While we only glimpsed a few minutes of the event, we were not disappointed, although the cloud had prevented any serious attempt at data collection. From an aesthetic view, the sight of the bright orange star in and near the rings was splendid. In Europe, after the main event, the star was occulted by Saturn's largest moon, Titan. Several hundred observers on the centre line saw a flash at mid-occultation, apparently caused as light from 28 Sagittarii was refracted around Titan by its atmosphere.

With all the interest in the rings of Saturn it is easy to forget that there is a planet below them. Saturn is the second largest planet in the solar system and like Jupiter, it is made predominantly of gases, mainly hydrogen and helium, but in different proportions. Studies suggest that Saturn has a similar structure to Jupiter, though due to its smaller mass Saturn's metallic hydrogen layer is smaller than Jupiter's while its rocky core and ice layers are larger.

Also like Jupiter, Saturn emits more energy than it receives from the Sun—about 1.7 times as much—but the mechanism by which it makes this energy is different. Because Saturn is smaller than Jupiter it never reached the red hot stage Jupiter did, and its energy derives from this fact. In Jupiter's interior temperatures were high enough for the hydrogen and helium liquids to combine in a mixture; in Saturn this was not possible, so helium formed little droplets within the hydrogen liquid. The helium droplets are denser than the hydrogen and so they fall towards the planet's centre. The energy released by this fall is what gives Saturn its energy today. Until recently this was just a theory, but the Voyager observations supported it by finding that the amount of helium in the upper atmosphere of the planet was smaller than would have been expected if it were not falling to the planet's core.

The study of the atmosphere of Saturn has been going on for as long as the study of Jupiter. Unlike Jupiter, with its riot of colours and bands, Saturn's clouds are much more subdued due to two factors: the lower gravity and lower temperature. Because of its lower gravity, Saturn's atmosphere is much more extended than Jupiter's; hence the cloud layer is thicker and less likely to allow gaps to form.

The lower temperature means that for a given height in the atmosphere the composition of gases will be different. On Jupiter, where the temperature is higher, the cloud layers appear higher in the atmosphere and are thinner. On Saturn, though, the clouds start deeper in the atmosphere and extend further upwards. All this leads to a rather featureless disc. This doesn't mean that there is nothing to see; there is some banding similar to Jupiter's. However it is not easily seen, and there are no features similar to the Great Red Spot scooting around in the atmosphere.

Saturn, like the other gas giant planets, rotates fairly quickly, its core completing one revolution in 10h39m24s. Because of its gaseous nature, its surface clouds rotate with different periods, those at the equator (System I) rotating once in 10h14m while nearer to the poles (System II) the rotation period is 26 minutes longer. The cloud tops of Saturn give the planet an equatorial diameter of 120 000 km and a polar diameter of 106 900 km, so the flattening due to its high-speed rotation is clearly visible.

Measurements of the rotation of Saturn's interior came, of course, from studies of its magnetic field by visiting spacecraft. Saturn's magnetic field is over 500 times stronger than the Earth's, but ten times weaker than Jupiter's. Though stronger than Earth's field at the centre, the larger size of Saturn means that at the visible surface the magnetic field is slightly weaker than the field at the Earth's surface. Like the field of Jupiter, though, its north magnetic pole is close to the south pole of the planet. A feature of Saturn's field which sets it apart from the other planets of the solar system is that the magnetic field and the axis of rotation of the planet almost coincide. They differ by only 1°.

To the amateur astronomer, the disc of Saturn presents a system of belts and zones similar to Jupiter's, but much less obvious because of a high-level obscuring haze in the atmosphere. A wide, bright zone bordered by the equatorial belts and the dark polar regions is easily observed in small telescopes, while the dusky temperate belts and zones generally need a larger instrument. The polar regions and the equatorial zones do occasionally fluctuate in brightness; either of the poles can brighten and the equatorial zone may darken.

Detail in the belts and zones is infrequent and

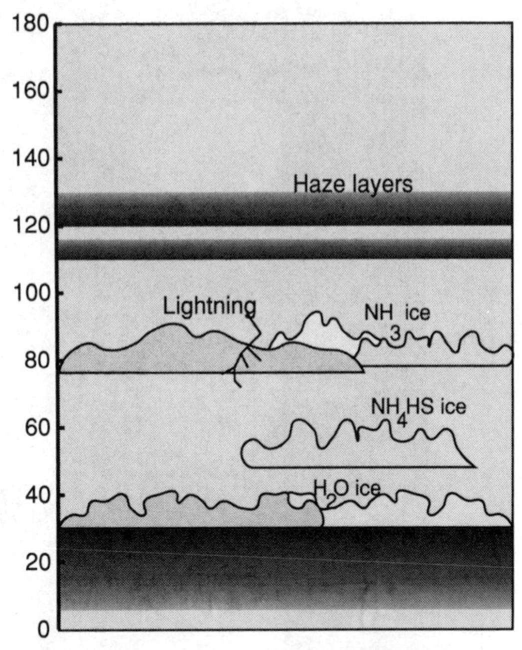


**12.9** The structure of Saturn's atmosphere is similar to that of Jupiter, but because of the planet's lower gravity ammonia ($NH_3$) is able to condense over a greater range of heights giving thicker clouds and less visible banding. The pressure at the zero level of this diagram is three megapascals, about 30 times the pressure at the Earth's surface.

short lived, and observations of any features are important to our knowledge of the atmosphere. Disturbances in the belts usually manifest themselves as a bulge or festoon which may link adjacent belts. Within the zones, white or occasionally dusky spots may appear.

The best known feature is the Great White Spot (GWS) which has a lifetime of only a few months. Discovered in 1876 in the equatorial zone, the GWS reappeared in the years 1903, 1933 and 1960, and astronomers predict a return some time in the early 1990s. The GWS may only be visible for several months, but in October 1969 a spot was observed in the south temperate regions that broke all records; it lasted for sixteen months, and measured 8000 km by 6000 km, with the long dimension aligned north-south.

Observational techniques for Saturn are similar to those for Jupiter, and any spot or marking should be timed as it crosses the central meri-

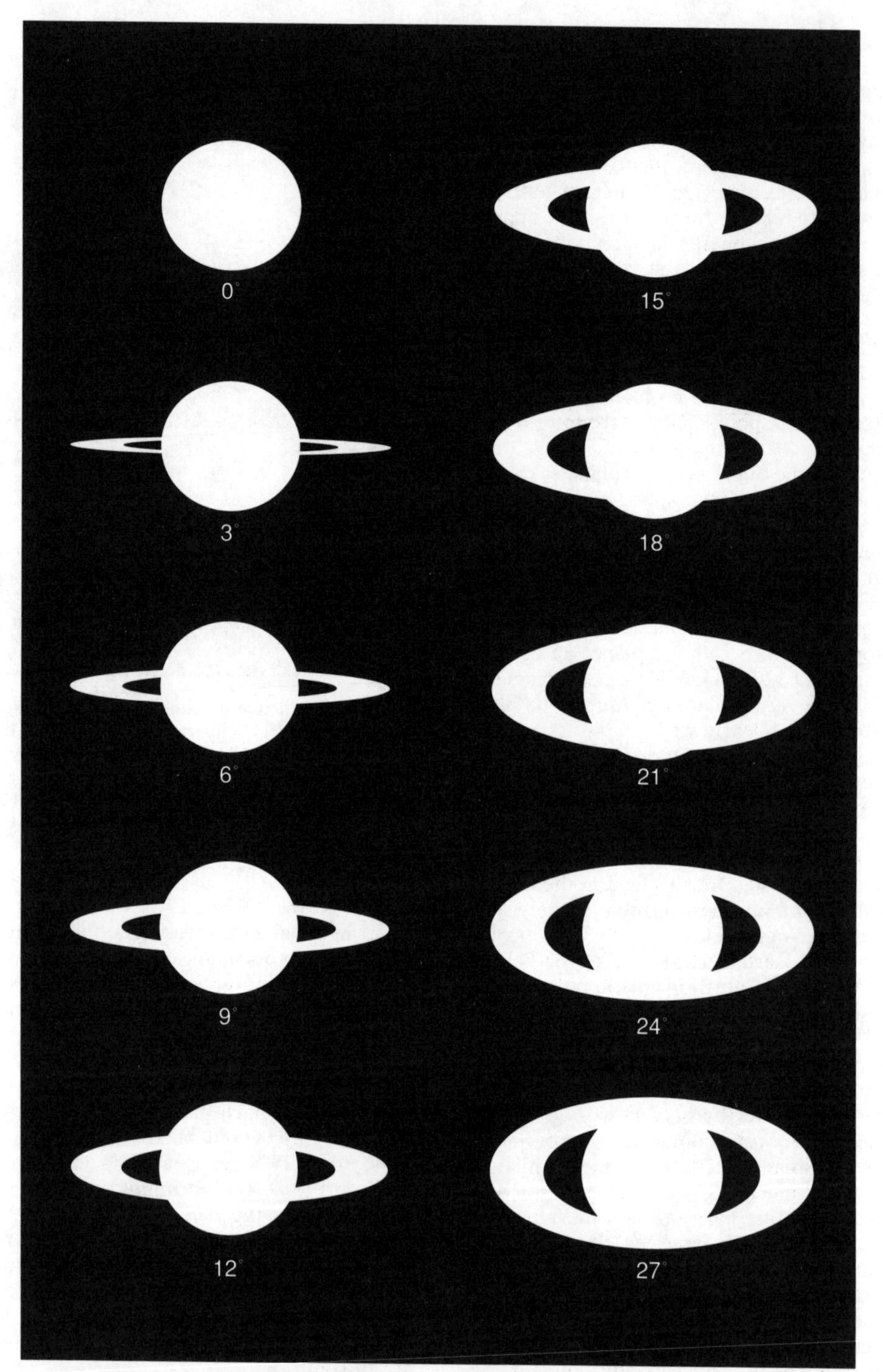


**12.10** Observing Saturn can be difficult without the correctly shaped blank. This set of blanks in 3° steps provides you with the correct shape for any apparition. Just select the blank which most closely matches the planet's shape. The blanks are a suitable size for most observations which can be made with an amateur telescope.

**Table 12.2 The satellites of Saturn**

| Name | Year of discovery | Diameter (km) | Mass (kg) | Orbital distance (km) | Period (days) | Eccentricity |
|---|---|---|---|---|---|---|
| Unnamed | 1985 Voyager | 15 | $3 \times 10^{15}$ | 118 200 | 0.48 | — |
| Unnamed | 1985 Voyager | 15 | $3 \times 10^{15}$ | 133 600 | 0.58 | — |
| Atlas | 1980 Voyager | 40 × 20 | — | 137 670 | 0.602 | — |
| Prometheus | 1980 Voyager | 110 × 90 × 70 | — | 139 350 | 0.613 | — |
| Pandora | 1980 Voyager | 110 × 90 × 70 | — | 141 700 | 0.6285 | — |
| Janus | 1980 Cruikshank | 140 × 120 × 100 | — | 151 472 | 0.6947 | 0 |
| Epimetheus | 1976 Fountain and Larson | 220 × 200 × 160 | — | 151 422 | 0.6944 | — |
| Mimas | 1789 Herschel | 392 | $3.7 \times 10^{19}$ | 185 540 | 0.94242 | 0.0201 |
| Enceladus | 1789 Herschel | 500 | $8.4 \times 10^{19}$ | 238 040 | 1.3702 | 0.0044 |
| Tethys | 1684 Cassini | 1060 | $7.55 \times 10^{20}$ | 294 670 | 1.8878 | 0 |
| Telesto | 1980 Voyager | 34 × 28 × 26 | — | 294 670 | 1.8878 | — |
| Calypso | 1980 Voyager | 34 × 22 × 22 | — | 294 670 | 1.8878 | — |
| Dione | 1684 Cassini | 1120 | $1.05 \times 10^{21}$ | 377 420 | 2.7369 | 0.0022 |
| Helene | 1980 Laques and Lecachaux | 36 × 32 × 30 | — | 378 060 | 2.7391 | — |
| Rhea | 1672 Cassini | 1530 | $2.49 \times 10^{21}$ | 527 100 | 4.5175 | 0.0010 |
| Titan | 1655 Huyghens | 5150 | $1.35 \times 10^{23}$ | 1 221 860 | 15.9454 | 0.0289 |
| Hyperion | 1848 Bond | 410 × 260 × 220 | — | 1 481 000 | 21.2766 | 0.1042 |
| Iapetus | 1671 Cassini | 1460 | $1.88 \times 10^{21}$ | 3 560 800 | 79.3308 | 0.0283 |
| Phoebe | 1898 Pickering | 220 | — | 12 954 000 | 550.45 | 0.1591 |

Note the relationships between Tethys, Calypso and Telesto. All three moons share the same orbit, the two smaller moons occupying the Lagrangian points 60° ahead of and behind Tethys. Janus and Epimetheus also share an orbit alternately approaching and receding from each other but never colliding.

dian. Tables of the longitude of the central meridian are available in the *Astronomical Almanac* and similar publications. Drawings of Saturn are useful, but unless any unusual features or markings are visible it is probably not worth the effort. Figure 12.10 provides a series of blanks that can be used for your drawings. They can be photocopied and enlarged to a convenient size; a 50 mm globe diameter is recommended.

The use of red and orange coloured filters is sometimes helpful in improving the contrast of belts and features. Always alternate between unfiltered and filtered observations as it may be that the unfiltered view is best at that time.

Perhaps it is not surprising that the planet with the best ring system in the solar system would also have the largest number of moons. To date, 21 moons have been found, ranging in

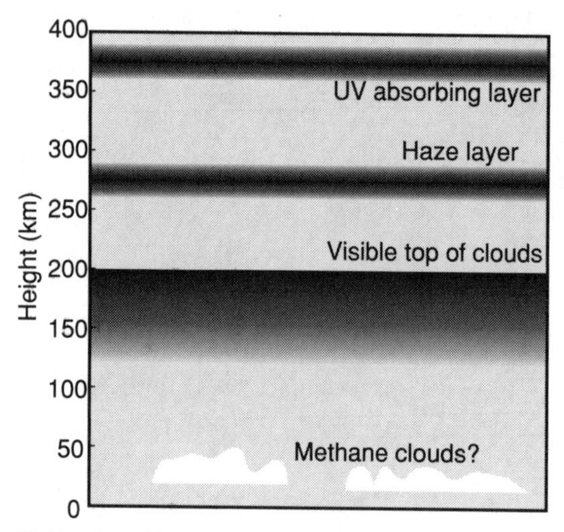


**12.11** A profile of the layers and composition of Titan's atmosphere. Notice that because of the lower mass of the moon, and hence its gravity, the atmosphere extends much further into space than the atmospheres of many of the planets. The pressure at the base of the atmosphere is close to that at the base of the Earth's atmosphere, and like the Earth the atmosphere is largely made of nitrogen.

size from the little ring guardians only a few tens of kilometres in size to Titan, over 5000 km in diameter. The satellites of Saturn again provide an opportunity to view a solar system in miniature, many of the moons providing surprises for scientists who thought they'd seen it all at Jupiter.

Saturn's largest satellite, Titan, was for a long time thought to be the largest moon in the solar system. In the 1940s astronomer Gerard Kuiper carried out observations of many of the satellites in the solar system, recording their spectrum to determine the composition of their surface. At Titan he found traces of methane, but at the temperature found on Titan methane would still be a gas; the only conclusion was that Titan had an atmosphere. This was not expected. Moons such as Ganymede and Callisto at Jupiter are similar in size to Titan, but they have no atmosphere, so what was special about this moon?

The Voyager 1 encounter with Titan in 1980 was awaited with high hopes that the information returned would solve the mystery. The pictures, though, were disappointing. Titan presented a blank orange disc, the surface completely hidden by smog layers in the atmosphere. As Voyager 2 would do at Neptune, Voyager 1 passed behind Titan on 11 November 1980, the signal from its transmitter passing through the moon's atmosphere before being blocked by the moon itself. This passage allowed scientists to make many measurements of the structure of the atmosphere. The results showed that the surface pressure at Titan is 150 kPa, 50 per cent higher than the pressure on Earth. Because of the low surface gravity, that meant that there was ten times as much gas in the atmosphere of Titan that there is on Earth.

It was the composition of Titan's atmosphere which was surprising. The methane detected in the 1940s was only a minor part; most of the atmosphere—99 per cent is nitrogen and argon. In addition to these two major gases, many organic compounds were also found in trace amounts. Most of the chemistry in Titan's atmosphere takes place because ultraviolet light from the Sun breaks molecules apart, which then reform as new compounds. It is thought that similar reactions took place in the atmosphere of the early Earth, creating some of the chemicals necessary for life to begin.

The haze which blankets Titan, hiding its surface, is complete. No photographs of the moon showed any detail at all—not even enough structure in the clouds to allow a determination of their rotational period. The haze is even more featureless than the clouds of Venus. Given the impossibility of seeing it, scientists have developed models of what they think the surface of Titan might look like. Most of the surface would be water ice, rock hard due to the low temperature. This terrain may be crossed with rivers of hydrocarbons emptying into seas of ethane beneath which would be a layer of heavier organic chemicals, an oil company's paradise.

Clearly more research is needed at Titan. A mission is currently being planned by NASA and the European Space Agency. Called Cassini, the craft will orbit Saturn, taking photographs of the planet and its moons. The craft will also include a probe to descend through Titan's atmosphere to sample its structure and return data from the surface.

If the pictures returned from Titan were disappointing, then the images of the other moons of Saturn certainly weren't. After Titan, Saturn has six medium satellites and a large number of smaller ones. The largest is Rhea with a dia-

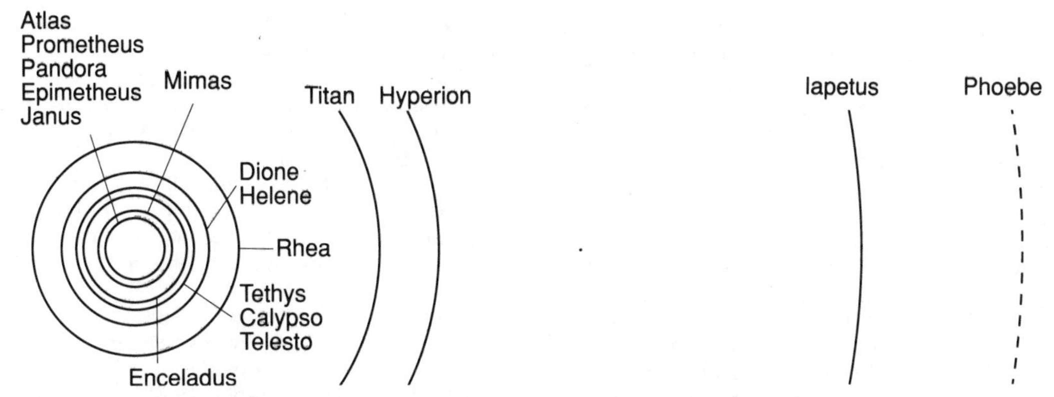


**12.12** Unlike Jupiter's family of moons, Saturn's does not fall nicely into groups. Shown here to scale, except for Phoebe, most of Saturn's satellites are close to the planet and the ring system. Unlike other satellites in the solar system, some of Saturn's moons share orbits. Tethys shares its orbit with two small satellites, Calypso and Telesto, which occupy the Lagrangean points 60° in front of and behind Tethys. Dione also shares its orbit with a small moon, Helene, at the Lagrangean point 60° behind Dione.

meter of just over 1500 km. Rhea has a density of 1.3 $g.cm^{-3}$ lower than the large satellites of Jupiter. The satellite appears to be primarily made of water ice and pictures of the surface show that it is uniformly and heavily cratered. Though it has been cratered, the surface of the moon is bright, with an albedo of 0.6 showing that there has been little dust to darken its surface over the past 4000 million years. On the hemisphere facing away from the direction of travel, bright streaks cross the surface. Their exact nature is unclear as they were only seen from a great distance, but they are possibly remnants from some ancient event which made fresh ice flow from beneath the surface.

Dione is the next medium satellite. Like Rhea, its surface is heavily cratered and it too shows the white streaks on its trailing hemisphere. Unlike Rhea, though, it shows evidence of internal activity with parts of its surface covered by floods of ice from the interior covering older craters. In the same orbit as Dione is the small moon Helene. This object occupies the Lagrangian point 60° behind the moon; in that position the smaller body is stable and in no danger of colliding with Dione.

Tethys also shared in heavy cratering, and it too shows evidence of internal activity. A notable feature of this moon is a huge fissure system called the Ithaca Chasma. This valley is over 100 km wide and stretches three-quarters of the way around the moon. It may have been caused by the expansion of the interior of the moon as it cooled, but such an expansion would have increased the volume of the moon by 20 per cent. Tethys too has companions in its orbit: Calypso and Telesto occupy the Lagrangian points in front of and behind the moon.

The last of the 'normal' satellites is Mimas. It is as heavily cratered as the oldest parts of our Moon and shows little evidence of any internal activity. Its most obvious feature is a huge crater on the leading hemisphere. The crater is over 150 km in diameter, one-third the diameter of the moon. Had the object which caused it been much larger it would have split the moon in two.

Cassini was a man who spent much of his life studying Saturn, its rings and moons. In 1671, working at the Paris Observatory, he noticed that the moon he had discovered and named Iapetus was a little strange. The brightness of the satellite varied over a remarkable range; it appeared that the moon must have one very dark hemisphere and one very light one. Indeed measurements of the albedo showed 0.5 for the bright, leading hemisphere but only 0.03 for the darker, trailing half.

The dark material is reddish in colour, similar to the dark material thought to cover Pluto and some comets and similar to the material in the Murcheson meteorite. The material appears to be organic in nature, perhaps made from the decomposition of methane by ultraviolet light and its subsequent reformation into longer organic molecules, a sort of tar.

The origin of the material is another mystery. Is it part of the moon and the lighter part a coating, or is the moon light with a dark coating? The mass and size of Iapetus identify it as an icy moon, so we would expect it to be light in colour. Voyager's images confirm this, the light hemisphere showing all the normal features of a cratered icy moon, so it is the black material which is the coating. It seems that it must be external, for it has formed on the trailing hemisphere and not the leading. If it were of internal origin then there is no reason why it would be aligned with the moon's orbital motion.

If the material is of external origin, then where did it come from? None of the other satellites of the Saturnian system has such markings, so it seems unlikely that the material would have originated outside the system. It is difficult to find somewhere within the system of rings and moons for the material to have originated. Phoebe and Titan are the two moons closest to Iapetus; as they are uncoated, one of them may have been the maker of the material. It could be that Titan's haze layer hides the answer to the puzzle. Only further exploration of the system will tell.

Enceladus, too, is a problem. Its surface is much too bright, reflecting light almost as well as a mirror, with an albedo of close to 1. In addition, just inside the orbit of the moon is a very broad diffuse band of particles forming the E ring of Saturn. The particles making the ring are tiny, much too small to remain in the ring for long as radiation pressure from the Sun would quickly blow them away, so either something must be replenishing them or the ring is very young. Either way it seems that Enceladus must have something to do with the ring.

Studies of the surface of Enceladus show that it must have been active in the last few hundred million years as most of the cratering on its surface has been completely erased. The mechanism for this activity remains unexplained. There is no tidal heating such as that associated with Io at Jupiter, nor is any other simple explanation apparent. The only suggestion is that very recently, in the past ten to twenty thousand years, Enceladus suffered an enormous impact which punched all the way through to its liquid water core. This would have caused a fountain into space, providing the material for the E ring, and led to the surface being completely reworked. Again, this satellite will repay much more study.

Only some of Saturn's moons are suitable for amateur viewing. Titan, the largest moon in the system, at 8.3 magnitude is easily visible in a 50 mm telescope or binoculars. A 200 mm telescope will show at least five or six moons, depending on their nearness to Saturn's glare, the tilt of the rings, and the limiting magnitude of the observing site.

In order from the planet, those satellites within reach of amateur instruments are: Mimas (magnitude 12.9), Enceladus (11.7), Tethys (10.2), Dione (10.4), Rhea (9.7), Titan (8.3), Hyperion (14.2) and Iapetus (variable between 10.0 to 12.0). The moons out to Rhea are at distances comparable with those of the Galilean satellites. A large gap separates them from Titan and Hyperion, and a still larger gap leads to the outermost observable moon, Iapetus.

Mimas and Enceladus generally need telescopes in the 200–300 mm range as they are close to Saturn's glare. Tethys, Dione, Rhea and Titan can be all visible in 80 mm telescopes. Hyperion, because of its faintness, will require a large instrument. As the magnitude of Iapetus varies between eastern and western elongation, 12th and 10th respectively, a small telescope may detect it when brightest.

Most of the satellites orbit on a similar plane to the rings, and two years before and after the edge on appearance of the rings, they will undergo eclipses, occultations and transits of both moon and shadow. The sight of the moons shuttling across the ring plane in similar fashion to Jupiter's moons is a sight that should be eagerly awaited.

Identifying Titan is simple because of its brightness, but the identification of the other moons is more difficult. If the rings are open the moons will be randomly scattered across the entire field, and are easily confused with background stars. An interesting exercise is to note all points of light that can be seen around Saturn, taking care to draw their positions exactly. Follow this up each night for about a week and you will soon learn which 'stars' belong to Saturn.

Astronomy magazines and almanacs also publish diagrams and tables of the satellite positions. The tables provide dates of elongation of each satellite; it is then a simple matter to calculate the time from the last elongation to the observation time. The interval in hours or days can then be plotted from the elongation point on the diagram, enabling you to find with accuracy the current position of each satellite. For the slow-moving outer moons, Titan, Hyperion and Iapetus, the times of eastern and western elongation, with inferior and superior conjunctions are given. For the faster inner moons, only the eastern elongations are supplied.

The Saturnian system has much to offer the casual observer. An 80 mm telescope is perfectly adequate for excellent views, and it is positively a magnificent sight to introduce your friends to the marvels of astronomy. On the serious side, there is much work that may be done by the amateur, but this calls for much patience, and telescopes of at least 200 mm aperture.

# 13

# Uranus

*On Tuesday the 13th of March, between ten and eleven in the evening, while I was examining the small stars in the neighbourhood of H Geminorum, I perceived one that appeared visibly larger than the rest: being struck with its uncommon magnitude, I compared it to H Geminorum and the small star in the quartile between Auriga and Gemini, and finding it so much larger than either of them, suspected it to be a comet.*

It was with these words that on 26 April 1781 William Herschel announced to the Royal Society in London the discovery of what would be a new planet. For thousands of years, people had observed the five wandering planets. They were given names, worshipped by many peoples and tracked methodically throughout the centuries. The idea that there may have been more than five planets never entered the collective imaginations of people, so it came as a bit of a surprise when another was discovered.

We might expect that a new planet would be found by the combined efforts of the world's astronomers searching diligently night after night, so it is more surprising still that this discovery was made by an English amateur astronomer using a homemade telescope. The story of William Herschel and his sister Caroline and how they found a planet is fascinating.

William (Friedrich Wilhelm) Herschel was born 15 November 1738, the third child of Isaac and Anna Herschel in Hanover, Germany. Isaac was an oboist in the Hanoverian Guards military band, and saw to it that his four sons and two daughters received training not only in music, but in cultural and scientific matters also. William was a fine musician and entered his father's band as oboist and violinist in 1753.

In 1756 at the beginning of the Seven Years' War the Hanoverian guards and their band were posted to England for a few months, but were recalled as war at home seemed imminent. After the battle of Hastenback in 1757 at which the Hanoverian forces were defeated, William decided that the military life was not for him, and with his father's help secured a discharge from the band. It was just in time, as in September of 1757 the guards were forced to surrender and interned in a military camp for the next two years.

Late in 1757 William and his brother Jacob left for England to find careers as musicians. William copied manuscripts while Jacob taught. In 1759 Jacob returned home while William stayed in England giving lessons when possible and composing music of his own. By 1762 he was sending money back home to help support his family. In 1766, as organs were being installed in large numbers in English churches, William secured a position as organist and choir director in the city of Bath. Because of his skill and reputation, students flocked to William for lessons and his combined income gave him a good living.

In 1767 Isaac, William's father, died, and his musically talented daughter, Caroline, found herself trapped as the house cook and cleaning lady. William tried to bring her to England but only succeeded five years later when he was able to give his mother sufficient funds to hire a full-time servant. Caroline was then able to join her brother in England. Going from managing one household to another did not worry Caroline as she was now free to study her music once more and help her brother in his studies. From his study of music Herschel was led to the mathematics behind harmonies, and in turn to the study of maths for its own sake. Maths led to optics and optics led, in 1773, to the study of astronomy. William bought some lenses and assembled some small telescopes, but the views he obtained were disappointing as the telescopes had to be small to avoid the problems of chromatic aberration. To make larger telescopes required larger lenses; this in turn made the telescopes enormously long. Herschel built one, 10 m in length, but eventually abandoned it as unwieldy.

Herschel then turned his attention to reflecting telescopes, trying first to find a suitable metal alloy from which to fashion a mirror. At the same time one of his other brothers, Alexander, visited contributing his engineering skills to the problem. Eventually a suitable alloy, 71 per cent copper and 29 per cent tin was found and, using the kitchen as a foundry, William began casting mirrors using moulds he made from horse dung.

Flushed with success at his telescope making, Herschel devoted more and more time to astronomy, cutting back on his music teaching. With the poor state of English weather, every clear night was devoted to the study of the skies. No time was to be lost and William, working daily at telescope making and nightly at observing, rarely stopped for meals, Caroline feeding him at the telescope. Caroline wrote: 'If it had not been for the intervention of a cloudy or moonlight night, I know not when my brother (or I either) should have got any sleep.' By 1781 Herschel had completed a survey of the sky in which he catalogued every star of magnitude 4 or brighter. This review, and reports of the quality of his telescopes, generated much interest in English scientific circles. Not content to rest Herschel began a second all-sky survey (he would later complete another two) using a 16 cm reflector built after the first survey.

It was on the evening of 13 March 1781 as he was observing near the star H Geminorum that he observed what he described as a 'curious either nebulous star or perhaps a comet'. Comets had been known for a long time; astronomers such as Halley and Newton had calculated their orbits and found them to be shaped like a parabola. However, similar calculations on Herschel's object did not fit; the object was not comet-like.

It was Neville Maskelyne, the Astronomer Royal, who first publicly suggested that the new object might be a planet. After observing the object himself for a few nights, Maskelyne wrote to Dr William Watson, Herschel's friend who had first alerted him to the discovery: '[Its motion] convinces me that it is a comet or a new planet, but very different from any comet I ever . . . saw.' Later he wrote to Herschel: 'It is likely to be a regular planet moving in an orbit nearly circular round the Sun.' Mathematical confirmation of these suspicions came separately from Swedish astronomer Anders Johan Lexell and from the French mathematicians Jean Bochart de Saron and Pierre Laplace in the summer of 1781. Laplace and André Méchain calculated the first elliptical orbit for the planet in 1783. Six months after the discovery most astronomers acknowledged Herschel's 'comet' as a large, new planet. Its calculated orbit placed it almost twice as far from the Sun as Saturn and Herschel's observations gave it an estimated diameter of 54 700 km, four and a half times that of Earth. Herschel had doubled the size of the solar system.

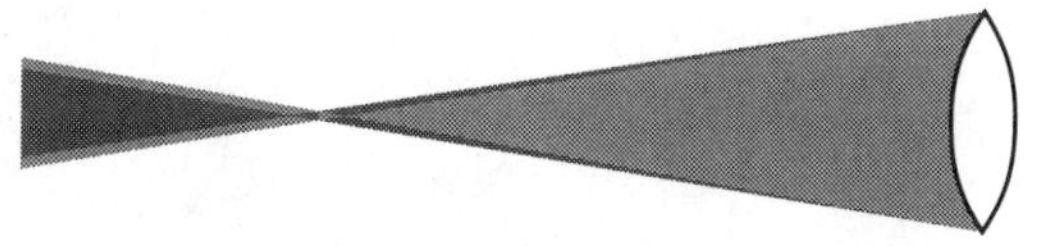

**13.1** Because a lens acts like a prism in splitting light into its component colours, the images formed by the red light from an object and the blue light from an object are at slightly different distances from the lens. To overcome this difficulty, before the invention of the achromatic lens astronomers had to use very long focal lengths, up to 10 m in some cases.

The Royal Society, England's foremost scientific establishment, gave Herschel its Copley Medal in 1781 and honorary membership. Herschel's achievement in discovering the new planet by recognising its disc as being different from the stars stunned the astronomical community. Many would not believe the claims he made about the quality of his telescopes, but when Neville Maskelyne viewed through some of the instruments and declared them to be superior to those of the Royal Observatory, these critics were soon silenced.

In the summer of 1782, King George III granted Herschel a royal pension of £200 per year to continue his work; in recognition of her contribution, Caroline was given £50 per year, making her the first woman to be a professional astronomer.

By October of 1781, with all the world now acknowledging the planet, the need for a name became urgent. English astronomers, aware of

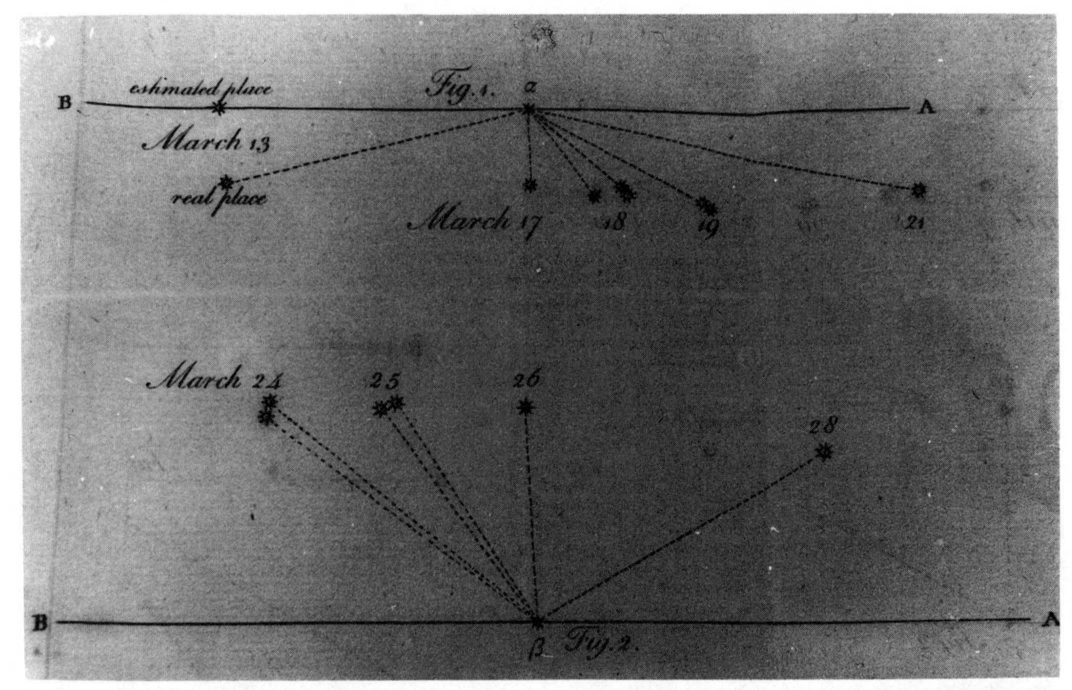


**13.2** William Herschel drew these two charts to show the position of Uranus over a two-week period following his discovery of the planet. In the top chart he has taken positions measured from the star α Geminorum for the dates 13 March to 21 March 1781. In the lower chart, bearings are taken form β Geminorum. Notice that for 13 March both the estimated place of the discovery and the actual place calculated from the later observations are shown.

the role the French played in helping verify the planet's orbit, were concerned that they would take the honour of naming the object upon themselves. President of the Royal Society, Joseph Banks, wrote to Herschel making just such a point and asking him to name it, but Herschel ignored the request. Indeed the French were busy inventing names, but quite generously suggested that the planet should be named Herschel.

Another suggestion came from Johann Bode (of Bode's Law) editor of the *Berlin Astronomical Yearbook*. He suggested Uranus. His reasoning was simple: as Jupiter was the father of Mars, Venus and Mercury, and Saturn was Jupiter's father, and Uranus was Saturn's father, it made sense to keep all the planets part of one family, albeit mythological. These two proposals, along with others for the name, Neptune, Astræa and others, were discussed and debated, yet still Herschel remained silent.

After Herschel's pension came through, with its condition that he live close enough to Windsor Castle that he might show the King and his family the heavens whenever they wished, he wrote to the King suggesting the name Georgium Sidus (George's Star) in his honour. The name was instantly unpopular everywhere except England because of its political flavour, its length and because the object wasn't a star. This left everyone to go their own way: Herschel, then Uranus in France, Uranus in Germany and, for 60 years, the Georgian Planet in Britain. It wasn't until 1847 that the *Nautical Almanac* finally used the name Uranus.

Herschel moved to Slough, within sight of Windsor Castle, where he spent the rest of his life. He gave up music and devoted himself full time to astronomy. He was 42 when he found Uranus, but for the next 40 years Herschel led the way in astronomy. Herschel married a widow, Mary Pitt, in 1788 and in 1792 their son, John, was born. From a young age John helped his father and Aunt Caroline in their work. It would be John who would carry on his father's work after William's death.

Herschel died on 25 August 1822, just before his 84th birthday. In his time observing, Herschel had discovered one planet, the sixth and seventh moons of Saturn, two moons of Uranus, the existence of double star systems and the motion of the Sun through space. He had also completed four catalogues of the stars of the heavens. Caroline survived her brother by 26 years, having contributed throughout his observing career, while John would take one of his father's telescopes to the Cape of Good Hope to survey the southern stars, making his own contribution to astronomy.

Herschel did much to dispel the opinion that Uranus was discovered by accident. He was at the time engaged in his second sky survey, looking at stars down to magnitude 8, ten times fainter than Uranus. If anyone was to find the planet, this man with his homemade telescope and determination would do it. As Herschel later explained: 'it was that night its turn to be discovered'.

To the unaided eye Uranus presents itself as a faint star slightly brighter than 6th magnitude, and the practiced observer should have no difficulty in finding it under dark-sky conditions. From suburban skies, where the limiting magnitude may be around 4 at best, binoculars or a finderscope will be needed.

Uranus is easy to find. Simply look up the right ascension and declination in an almanac and plot the position on a star atlas, noting the constellation in which it resides. A planisphere will establish the location of the constellation and, if it is above the horizon, it is then a simple matter to 'star hop' from known brighter stars using binoculars or a low-power finderscope. Should you have an equatorial mount correctly polar aligned, you can simply use the right ascension and declination co-ordinates to dial it up with setting circles.

Uranus' motion against the background stars is slow, about 4° per year. At the time of writing the planet is in the constellation Sagittarius, and will move into Capricornus in 1995. It will reside there until 2003 before moving into Aquarius. Uranus' orbital plane is inclined to the ecliptic by only 46 minutes and 22 seconds of arc, so it never strays far from the zodiac. With an orbital period of 84.01 years, the planet has only completed about two and a half revolutions around the Sun since William Herschel discovered it.

An instrument from 60 to 80 mm aperture using about 50x magnification will resolve the star-like point into a small, pale green disc. Increasing power in small telescopes will do little to help see detail, and will only lead to a bigger but much fainter image; larger apertures fare little better. The greenish colouration of Uranus results from the absorption of red light by small amounts of methane gas in the atmosphere, and most of the light reflected is in the blue-green part of the spectrum.

Little detail can be observed on the disc in amateur telescopes. Under ideal conditions with instruments in the 200–250 mm range, the dusky poles may be seen separated by a brighter equatorial region; when one of the poles is directed towards Earth, the disc will seem featureless. A high-level obscuring haze similar to Saturn's exists in the upper atmosphere, effectively masking detail in the cloud belts below.

With a disc size just under four seconds of arc, it is not worth trying to photograph the planet through a telescope. A more rewarding pursuit is to point a single-lens reflex camera with a standard lens, mounted piggyback on an equatorial telescope with motor drive, in the general direction of Uranus. As Uranus and Neptune are situated close together for some years to come, 1° apart in 1993 and gradually moving out to 12° in 2000, you will be able to capture both of them on the same frame.

An exposure time of three or four minutes, with a 50 mm *f* 2.0 lens, using fast black and white or colour film, is enough to fill your photographs with an abundance of detail. You should experiment with different films and *f* stops to give the most satisfying picture. When working at a fast *f* stop, you may find the stars around the edge of the print appear elongated; the reason for this is that most lenses aren't perfect around the outer edge, a simple solution is to stop the lens down. A change from *f* 2.0 to *f* 4.0 will call for an exposure of double the length to capture the same detail, but you will be rewarded with a sharper, more pleasing result. Figure 13.3 shows the type of work that can be achieved with a standard SLR camera.

Over the next few years, Jupiter, Mars and the minor planet Ceres will all pass near to Uranus and Neptune, and photographs taken a few days or weeks apart will show their motion against the stellar background. In 1986, the planets Saturn, Uranus, Neptune and minor

**13.3** Uranus is seen here over-exposed with two of its moons, Titania and Oberon, seen close to the left side of the planet. This photograph was taken on 12 July 1985 with a 250 mm reflecting telescope using a 2 minute exposure at prime focus on 400 ISO film. It is a good example of what an amateur astronomer can achieve with patience and perseverance. PHOTO CREDIT: STEVEN QUIRK

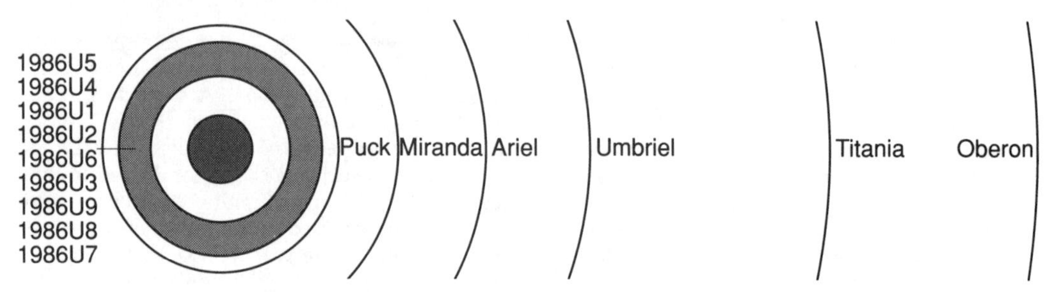


**13.4** From the discovery of Uranus to the visit of Voyager 2 in 1986, five moons were known to orbit the planet. During the encounter of Voyager 2 with Uranus another ten moons were discovered, most lying in a band from 49 000 to 75 000 km from the planet's centre. All of the new moons are small, with Puck at 170 km the only one larger than 100 km.

planet Vesta were all within a circle less than 15° in diameter, making a very picturesque photograph against the rich Milky Way region of Sagittarius.

After discovering the planet, astronomers were not content and the search for moons began. On 11 January 1787, six years after the planet's discovery, William Herschel found not one but two moons. This time he was using his 48 cm reflecting telescope. Two more satellites were discovered by William Lassell, an English amateur astronomer, much in the manner of Herschel himself, in 1851.

It was William's son, John, who in 1852 took on the task of naming the moons. The first two he named Titania and Oberon, not from Greek mythology, but from Shakespeare's *A Midsummer Night's Dream*. Lassell's two moons were called Umbriel and Ariel, from Alexander Pope's *The Rape of the Lock*. When Gerard P. Kuiper discovered a fifth moon in 1948, Shakespeare's *The Tempest* provided the name of Miranda.

Of the fifteen known satellites, only the two found by William Herschel are visible to the amateur. The three next brightest lie within the orbits of Titania and Oberon and are lost to the planet's glare, except in the largest of telescopes or by photographic methods. The ten moons found during the Voyager 2 encounter are beyond the scope of any Earth-based techniques.

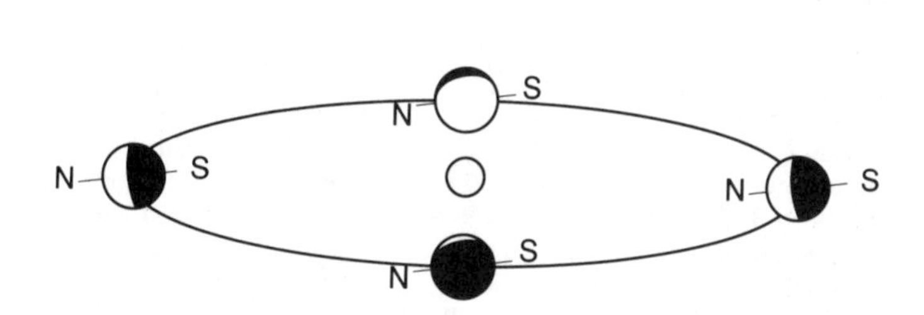


**13.5** Uranus is tilted with respect to its orbit more than any other planet, 98°. This means that for part of each orbit the poles of the planet are pointed nearly directly towards the Sun. An observer near to the poles or Uranus would experience a day 42 Earth years long, followed by a night the same length. Even observers as close as 8° to the equator would experience some extended period of light and dark throughout their year due to the tilt. The cause of the tilt is not yet understood.

Titania (magnitude 13.98) and Oberon (14.25) will need a 200 mm telescope under a good, dark sky to detect them. Since the satellites lie in the plane of the equator which in turn intersects the plane of the orbit at 98°, we see the satellites from different aspects during the Uranian year. In 1986 the moons orbited the planet in a circular fashion; their motion has now changed into an ellipse and by the year 2007 we'll see them edge on, with the moons shuttling back and forth in a straight line.

Only every 42 years do we see the satellites' orbits edge on, like the Galilean satellites of Jupiter. At any other time, their apparent random position will make it hard to find them against the stellar background. With short orbital periods of 8.71 days for Titania, and 13.46 for Oberon, it will take only a few nights' observations and rough sketches to identify the two main satellites. Alternatively an almanac will provide the times of elongation, and a quick calculation should soon prove the identity of a suspected satellite.

The discovery of the planet was confirmed when an orbit for the planet was calculated in 1781, though it wasn't until 1783 that an elliptical orbit was finally worked out. That orbit, the result of calculations by Laplace and others, placed Uranus at an average distance of 2870 million km from the Sun, with a period of revolution of 84.01 Earth years.

When in 1787 Herschel discovered the moons Titania and Oberon the first suggestion that something was wrong with Uranus was made. We have already seen with Jupiter and Saturn that their moons orbit the planet like a miniature solar system, travelling in elliptical orbits close to the plane of the planet's equator. At Uranus, though, the moons appeared to be orbiting about the poles of the planet. An explanation was quickly forthcoming; the moons were in fact orbiting around the planet's equator, but the equator had been tipped almost 90° away from vertical; indeed later measurements would show that the pole had been tipped through 97.9°

Our earthly seasons result from the Earth's modest tilt of 23.5°. What must it be like on Uranus with such a large tilt? Given that the Uranian year is 84 times longer than ours, each season would correspond to 21 of our years. Imagine that you were living near the north pole of Uranus. As spring started, the Sun would be near the horizon, travelling once around the sky every seventeen and a half hours. Over the next 21 years the Sun would move higher in the sky until it was standing almost overhead. As the next 21 years passed the Sun would sink lower in the sky, until in autumn it disappeared below the horizon, not to reappear for 42 years.

The first close-up views of Uranus did not come until January 1986 when Voyager 2 made a close approach to the planet. Up until that time astronomers had used every opportunity to gain information about this distant world. One thing that was known before Voyager 2 arrived was that Uranus had a small system of rings encircling the equator. The discovery was unexpected and showed that luck still plays a part in astronomical discoveries.

On 10 March 1977 a number of groups of astronomers were in the southern hemisphere to observe the occultation of a faint star by Uranus; by observing this event a more exact measurement of the planet's size would be obtained. A group of United States astronomers were aboard NASA's Kuiper Airborne Observatory. The KAO is a converted C-141 Starlifter

**Table 13.1 Uranus' atmosphere**

| Gas | | Fraction |
|---|---|---|
| Hydrohen | $H_2$ | 85.0% |
| Helium | He | 13.0% |
| Methane | $CH_4$ | 2.0% |

aircraft housing a 0.9 m telescope which views through a hole in the roof of the plane. The astronomers were flying at 12500 m above the southern Indian Ocean to observe the event. Because they did not have a precise measurement of the position of the star to be occulted, the astronomers started their observations 47 minutes early to be sure to miss nothing. Almost immediately they noticed the light from the star flicker at least five times. The occultation was duly observed and then the astronomers waited. The flickers repeated themselves, in the reverse order.

The scientists suggested the possibility of thin rings, but at this time the only rings known were those of Saturn's wide system and they were unable to explain how thin rings could exist. Careful examination of the record of the occultation showed that there were nine thin rings encircling the planet. As is common, the new discovery created more questions than it answered. The collisions between particles in the rings should spread them out into broad sheets like Saturn's rings in no time. Clever explanations using the gravitational effects of small moons were proposed, but it was necessary to wait for Voyager 2 to verify the predictions.

Voyager 2 arrived at Uranus on 24 January 1986, $2.8 \times 10^9$ km from Earth and just under 205 years since William Herschel first sighted the planet. In the days before the closest approach, the cameras on Voyager 2 had begun to give the world its first close-up views of this distant planet. Unlike the gaudy colours of Jupiter or the more subtle shadings of Saturn, pictures from the days leading up to contact with Uranus showed little more than a uniformly light blue planet. The atmosphere of the planet is made of exactly the same chemicals as the two inner gas giants, methane with traces of ethyne and ethane, but the lower temperature of the planet gives rise to a very different appearance.

There are clouds, storms and features in the atmosphere of Uranus, just as in the atmospheres of the other planets, but these are hidden from view by a layer of smog. Indeed, using special filters on the cameras and computers to enhance the contrast of the pictures scientists are able to see cloud patterns in the Uranian atmosphere, but these are invisible in visible light. A similar layer of smog exists on both Jupiter and Saturn, but the extra heat they receive from the Sun allows their clouds to rise higher in their skies and rise above the smog layer. On Uranus, though, the clouds are below the haze and so are shrouded from view.

When Voyager 2 arrived at Uranus it was mid summer at the south pole, and even though the north pole of the planet had been in night for over twenty years and the south pole in continual sunlight for the same period, Voyager found that the temperature was constant to within 2°C of −221°C everywhere on the planet. The meagre amount of heat received from the Sun is quickly spread over the planet by strong winds. Atmospheric scientists had been waiting to see what the weather would be like on a planet which received heating almost directly above its poles for part of each year. Theories suggested that the prevailing winds should blow from the south pole towards the equator, due to the Sun heating there, but they were wrong. The prevailing winds, travelling at over 300 km per hour, blew from east to west, the same direction as the planet's rotation.

Voyager 2 was able to measure the rotation period of Uranus to high accuracy by observing the rotation time of the planet's magnetic field. Prior to Voyager's arrival, the best estimates were between 10 and 24 hours, but Voyager measured it to be 17.24 hours. The magnetic field of Uranus was something of a mystery to Earth-based scientists, as they had been unable to detect any evidence of one. Voyager found the field stronger than the Earth's but tilted at 60° to the poles. Also unlike the Earth's field, the magnetic field of Uranus does not pass through the centre of the planet; it misses by 7700 km, about a third of the planet's radius. Observations of Uranus by Voyager have given scientists a good idea of the planet's composition. It was thought prior to Voyager that Uranus had a rocky core surrounded by a liquid mantle and a dense hydrocarbon atmosphere. Voyager's measurements of the planetary bulge did not support this theory. If the planet has a liquid mantle, the equatorial bulge, like that of Jupiter and Saturn, should be quite noticeable, but in fact it was quite minor. To explain this, the mantle is replaced with a super-dense atmosphere extending from the surface we see to the

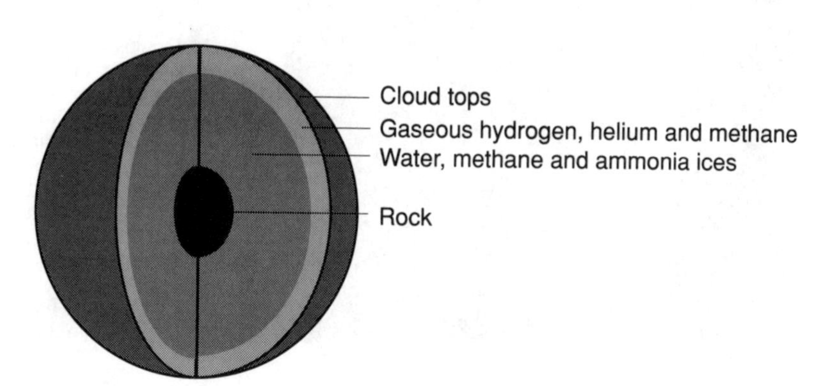


**13.6** The centre of Uranus is a rocky ball surrounded by a layer of water, methane and ammonia ices. There is no layer of metallic hydrogen as there is for Jupiter and Saturn, as the pressures inside Uranus are not high enough. Above the layer of ices is a layer of gaseous hydrogen, helium and methane. The clouds at the top of the atmosphere are similar to those of Saturn and Jupiter, but are much less visible due to a thick layer of obscuring haze high in the atmosphere.

rocky core, with no liquid components. Hydrogen is of course the dominant gas in the atmosphere with 2–3 per cent of methane and other hydrocarbons giving the planet its colour by absorbing red light and reflecting blue.

The one problem with Uranus that Voyager did not solve was that of the tilt. Nothing in the observations was able to shed light on this mystery. The prevailing theory is that if the planets indeed formed by a process of accretion, then their rotations must have been affected by the impacts of meteorites. On some planets these impacts were balanced over the surface so the tilt of the planet was disturbed only a little (Mercury), on others there was a greater imbalance leaving a moderate tilt (Earth), but on Uranus there must have been a large impact near one pole, which remained unbalanced, causing the 98° tilt we now see.

If the planetary disc of Uranus gave a disappointing view, the moons made up for it. Five moons were known, but from Earth they were no more than points of light. It was known that they were large, but smaller than the Galilean satellites of Jupiter. In addition, Voyager discovered ten more moons and two more rings, making fifteen and eleven respectively. The moons all rotate around the planet's equator in the direction of the planet's rotation and with synchronous revolutions so that the same face always points towards the planet. In that respect, the moons do exactly what scientists expected.

Because they are small and cold, the moons of Uranus were expected to be dead worlds, with

**Table 13.2 The Uranian moons**

| Name | Orbital distance | Orbital period | Inclination | Radius |
|---|---|---|---|---|
| 1986U7 | 49 700 km | 8 hours | 0° | ≈20 km |
| 1986U8 | 53 800 km | 9 hours | 0° | ≈25 km |
| 1986U9 | 59 200 km | 10.5 hours | 0° | ≈25 km |
| 1986U3 | 61 800 km | 11 hours | 0° | ≈30 km |
| 1986U6 | 62 700 km | 11.3 hours | 0° | ≈30 km |
| 1986U2 | 64 600 km | 11.75 hours | 0° | ≈40 km |
| 1986U1 | 66 100 km | 12.25 hours | 0° | ≈40 km |
| 1986U4 | 69 900 km | 13.5 hours | 0° | ≈30 km |
| 1986U5 | 75 300 km | 15 hours | 0° | ≈30 km |
| 1985U1 (Puck) | 86 000 km | 18.25 hours | 0° | 85 km |
| Miranda | 129 783 km | 33.9 hours | 3.4° | 242 km |
| Ariel | 191 239 km | 60.5 hours | 0° | 580 km |
| Umbriel | 265 969 km | 99.5 hours | 0° | 595 km |
| Titania | 435 844 km | 209 hours | 0° | 805 km |
| Oberon | 582 596 km | 323 hours | 0° | 775 km |

no internal heating or other sources of energy to change their surfaces. The moons were expected to show cratering dating from their formation and little else. Fortunately, the moons were far from that boring.

Titania is the largest of the Uranian satellites and the second most distant. The surface of Titania was found to be covered by thousands of small craters, but few large ones. As large craters are evidence of early cratering, something must have erased them from the surface. Other surface features are long cracks over 1000 km in length and 5 km deep. These cracks expose lighter coloured material near the bottom. Here was the explanation for the lack of large craters. When Titania suffered large impacts, soft ice from beneath the surface welled up, filling many

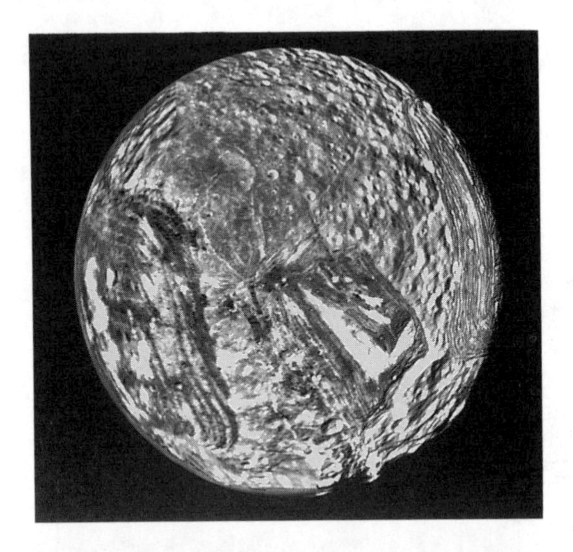

**13.7** Uranus' moon Miranda proved a complete surprise to everyone. This image is a montage of photographs taken by Voyager 2 on 24 January 1986. It shows plateaux, cliffs, canyons, craters, faults and fractures, the most complex landscape yet seen in the solar system. Just to the right below centre is the huge V-shaped feature called the chevron. Two major types of landscape can be distinguished, the older more cratered regions here seen towards the top, and the younger less cratered but more complex regions covering the rest of the surface. Note also the lower right of the planet's limb where mountains can be seen in profile sticking out into space.

Ariel shows a young surface with few craters, features which must have been destroyed by later resurfacing of the moon. Ariel shows evidence of tectonic activity in the chasms which cross its surface and from material which appears to have leaked from beneath the surface creating light patches on the otherwise dark surface.

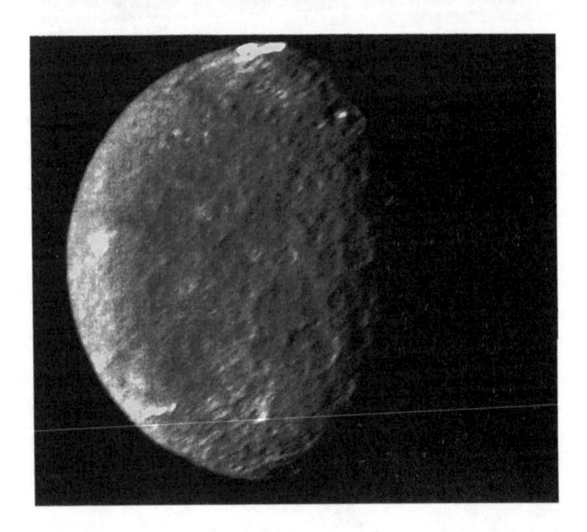

Umbriel is much darker than Uranus' other satellites. It is uniformly covered with craters with the exception of a bright ring, 40 km in diameter, here seen at the top of the picture.

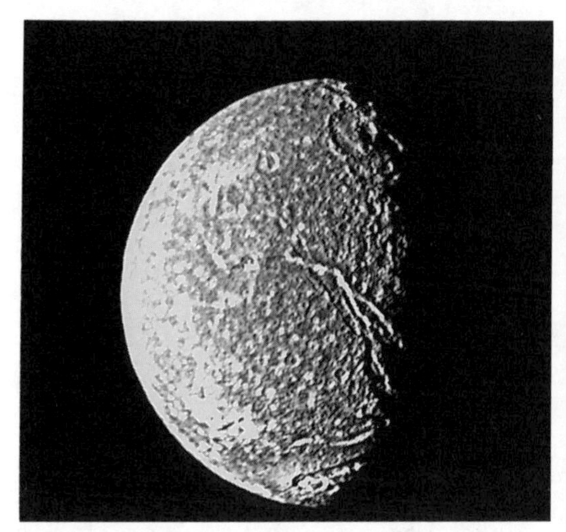

Titania is Uranus' largest satellite. Numerous valleys and faults on the moon's surface, together with the lack of many large craters, indicate that the surface of this moon is relatively recent.

Oberon is the furthest out of Uranus' moons. The surface of Oberon is covered by a large number of impact craters surrounded by bright material thrown out by the impacts. On the lower left a mountain 20 km high can be seen jutting out into space. PHOTO CREDIT: JPL/NASA

of the larger features. Later, as the subsurface ice cooled further, it expanded breaking the surface and leaving the long cracks.

Oberon was thought to be the twin of Titania. Almost the same size, it is the outermost satellite. This moon is covered with an entire history of cratering from the very large old features to the smaller, more recent results of impacts. One photograph shows a mountain jutting from the edge of the planet, a peak over 20 km high. Although the large craters have survived, Oberon too shows some signs of internal melting, with flat light-coloured floors in some craters.

The densities of these two moons provide some explanation for the observed features. The average density of around 1.5 g.cm$^{-3}$ suggests that the moons contain a larger amount of rock than previously expected. This rock contains radioactive minerals, the decay of which can heat the moon. Even so, with an average temperature of −193°C*, a lot of energy is needed to melt water to liquid; perhaps the ice below the surface is not pure water but a mixture of water and methane known as methane clathrate. This melts at a much lower temperature and could be the light-coloured material observed.

Umbriel is the next largest moon of Uranus. Its surface shows many old craters, but unlike the outer two moons, little evidence of lighter material from the interior oozing out, except for two small features, one on a slope and the other the floor of a crater. Umbriel is the darkest of the moons, with an albedo of 0.19 and little variation in surface brightness. It has been suggested that the dark colour may be the result of some event which caused a uniform coating of some dark substance on this moon alone. On the other hand, the dark colour could be left from the materials which formed the moon.

Ariel is the next moon, almost the same size as Umbriel, but in every other respect quite different. Whereas Umbriel is one of the most cratered moons, Ariel is the least cratered. Nearly all evidence of older craters has been wiped away by later catastrophic activity. Titania has some valleys and faults, but on Ariel there is a huge network of cracks and scarps, some up to 30 km deep. Also, where Titania shows some resurfacing, virtually all the surface of Ariel has been recovered at some time by material from below the surface. Where the energy for this resurfacing came from is unknown, but it died away around $3 \times 10^9$ years ago leaving a surface fresh for the light cratering we see now. Perhaps the moon was locked in a

* It is interesting to note that the temperatures of the satellites are higher than that of Uranus. This is caused by their darker surfaces absorbing more energy from the Sun.

**13.8** This region of Miranda shows two different types of terrain. The older cratered surface is interrupted by two lowland regions of parallel ridges which contain a few craters, indicating that, although more recent than the highlands, these features are still ancient. The resolution of this picture is around 600 m and is one of the most detailed images returned by Voyager. PHOTO CREDIT: JPL/NASA

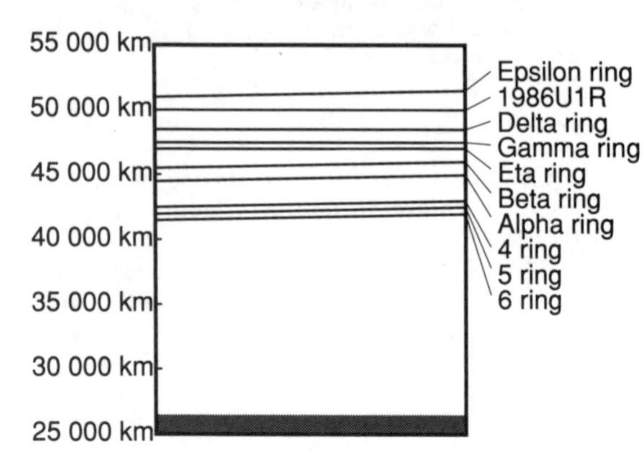


**13.9** Uranus' system of rings was discovered by accident in March 1977 when the planet passed in front of a star. The rings are here shown with their distance from the centre of the planet. Those rings shown as tilted lines are eccentric rings; only the eta, gamma, delta and 1986U1R rings are circular. Unlike Saturn's rings, all the rings of Uranus are thin bands, not broad sheets.

resonant orbit with Umbriel for a time, similar to the situation on Jupiter's Io.

It was the innermost of the known moons which proved to be the most fascinating of Uranus' family. On the surface of Miranda, almost every type of terrain could be seen side by side: craters, valleys, rifts and grooves all were in abundance. Over the old cratered surface were three huge regions called *ovoids* because of their shape. These regions consist of ridges and valleys like those produced when ploughing a field; on the smallest ovoid a large V-shaped feature, the *chevron*, is easily seen. Around the moon runs a huge system of fractures, making steep-sided valleys with terraced walls. Some of these canyons were 20 km deep, ten times the depth of the Earth's Grand Canyon.

The energy for this amount of activity on such a small, cold world is still a mystery. Perhaps Miranda was locked into a synchronous orbit with Ariel at some stage, the tidal heating giving enough energy for some of the rock to begin to fall towards the moon's centre, forming the ovoids. This might explain some of the features on Ariel as well.

The discovery of ten new moons at Uranus disappointed some scientists. They had hoped for at least eighteen, two for each of the thin rings. The first of the moons was found on 31 December 1985 and so was called 1985U1, but the team controlling the spacecraft named it Puck, after another of Shakespeare's *Midsummer Night's Dream* characters. Puck turned out to be 170 km across, roughly spherical and fairly

**13.10** This photograph of the rings of Uranus was taken with the sunlight passing through them rather than being reflected by them. The scattered sunlight shows up the presence of many small particles, revealing the rings as a continuous band rather than as thin discrete rings. Only the ε ring is separate from the main band. The tenuousness of the rings is clear from the ease with which stars, here small streaks caused by the movement of the spacecraft, can be seen through the rings. PHOTO CREDIT: JPL/NASA

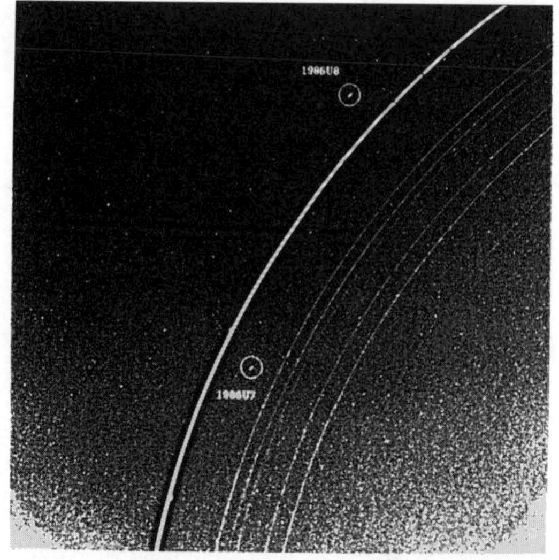


**13.11** The outermost ring in this picture is Uranus' ε ring, kept in place by the two small shepherd moons 1986U7 and 1986U8. The gravitational interaction of these two moons with the ring particles keeps the ring neat and narrow, not allowing any particles to escape from the narrow region between the satellites. PHOTO CREDIT: JPL/NASA

dark. The nine other new moons 1986U1 to 1986U9 are simply smaller versions of Puck, small dark-coloured worlds covered with broken-down methane ices. This black material was an important discovery; later in 1986 scientists discovered a similar black material covering the nucleus of Halley's comet.

The rings of Uranus were known, but they were thin. By the time Voyager arrived, thin rings had been seen at both Jupiter and Saturn, with their accompanying shepherd moons. Uranus' rings have widths of only one to twelve kilometres and contain very little dust. This presented a problem: a ring is a dynamic structure with collisions between the particles reducing the ring particles to smaller and smaller pieces, and keeping a fairly high amount of dust. Something must be removing the dust from the rings.

The answer came from Voyager's ultraviolet spectrometer. It found that the atmosphere of Uranus extended much further from the planet than expected. At the distance of the rings there were still traces of hydrogen gas which was slowly escaping into space. This gas slows the minute particles of dust, causing them to fall towards Uranus and out of the ring system. The entire system of rings surrounding the planet is eroding. Within a short time (a few million years) there will be nothing left.

It is easy to track down Uranus—you may even wonder why it remained undiscovered until 1781. If your telescope is of large enough aperture, an assault on the two brighter satellites is a rewarding challenge.

# 14

# Neptune

If the discovery of Uranus was the result of an amateur astronomer's luck and persistence, the discovery of Neptune was not. The search for the planet further from the Sun than Uranus occupied some of the greatest minds of the nineteenth century, and some of the greatest egos too.

Uranus had shown scientists and the public alike that the limits of knowledge had not been reached, but that theories which had been developed many years earlier could and would hold for as-yet-undiscovered objects. Uranus had doubled the size of the solar system, and shown that Newton's universal law of gravity was still valid. Laplace and Méchain had calculated an elliptical orbit for Uranus in 1783, an orbit which allowed astronomers to predict the motion of the planet to considerable accuracy; the only trouble was that Uranus' movements didn't quite stick to the calculations. It wandered.

The discrepancies between the calculated position and the observed position of the planet were small, but much larger than astronomers expected; still they assumed that as more and better observations were made, better predictions would result. They didn't. Despite the best endeavours of all, Uranus drifted still further from its allotted path. Studies of all available observations of Uranus led some astronomers to search their own early observations to find if they had unwittingly observed Uranus before its discovery and not noticed. One, Pierre Charles Le Monnier, discovered seventeen sightings before Herschel, including four on consecutive nights, but his record-keeping had been so sloppy that he hadn't noticed the change in position. These earlier observations didn't help either.

From 1781 until 1821 Uranus appeared to be running too fast, consistently ahead of its predicted position. Then in 1822 it ran to schedule. Just as everyone was collectively sighing with relief in 1823, it began to run too slow. The effort to explain the errors shifted from simple observation and calculation mistakes to more 'scientific' ideas: perhaps a cosmic fluid was slowing it, but there was no other evidence of such a fluid; perhaps there was a large moon disturbing the planet's motion, but such a moon would have been too large to remain undiscovered; perhaps a comet had struck the planet just before it was discovered, but the probability of that was minuscule; perhaps there was another planet affecting it.

The idea of other undiscovered planets was not really new. In 1758 astronomers working on the return of Halley's comet had suggested that another planet might be responsible for errors in their calculations. In 1834 Alexis Bouvard, who was working on the problem of Uranus' orbit, mentioned in passing to an English amateur astronomer, Thomas Hussey, that if the positions of Uranus were investigated it might be possible to calculate the position of the unseen planet. Hussey in turn wrote to George Airy, the man who was next in line to be Astronomer Royal; expounding the idea. In reply Airy wrote:

> *It is a puzzling subject, but I give it as my opinion, without hesitation, that it is not yet in such a state as to give the smallest hope of making out the nature of any external action on the planet... I am sure it could not be done till the nature of the irregularity was well determined from several successive revolutions.*

In simpler terms, as the orbital period of Uranus is 84 years, Hussey was being told to wait 2–300 years. This opinion of Airy's, and his high

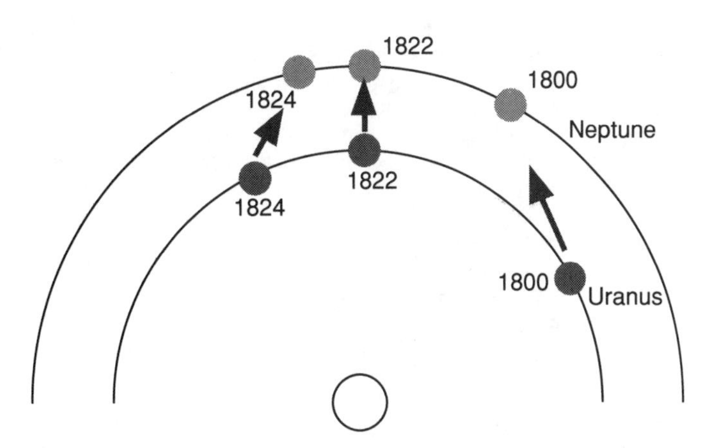


**14.1** The discovery of Neptune depended upon calculating the effect of the planet on the motion of Uranus. From its discovery until 1822 Uranus appeared to be moving too quickly in its orbit. Throughout 1822 the planet moved at the expected rate, but after 1823 the planet appeared to be moving slowly. Many deduced that this speeding up and slowing down was due to the gravitational influence of another planet further from the Sun. Using this hypothesis, John Adams and Urbain Le Verrier were able to calculate the expected position of the outer body, Neptune.

opinion of himself, would be most important in the hunt for Neptune.

Many other astronomers began discussing the possibility of finding another planet. Halley's comet returning one day late in 1835 caused much talk. By 1837 Bouvard had his nephew, Eugène, collecting new observations of Uranus' position. Eugène wrote to Airy explaining his uncle's theory. Airy replied that he thought the problem lay in the distance to Uranus being underestimated, and asked Eugéne to concentrate on that possibility instead. By 1838 the existence, or not, of yet another planet was a cause for major debate and research.

Friedrich Bessel, the famous mathematician, told John Herschel in 1842 that he and a colleague were about to undertake a new analysis of the data to predict the position and orbit of the unknown planet, but ill-health prevented him from ever making a start. In 1841, John Couch Adams, a Cambridge mathematics student, saw Airy's 1832 paper describing the situation and made up his mind to look into the problem when he finished his degree.

Born in 1819, John Couch Adams experienced early schooling which was typically provincial. Yet even with the low grade of teacher found in the English countryside in the early nineteenth century, Adams' talent for mathematics flourished. Adams became interested in astronomy with the 1835 arrival of Halley's comet and by 1837 he had written his first scientific paper, for the local newspaper, on a lunar eclipse. In 1839 Adams won a scholarship to Cambridge University.

Adams' time at Cambridge was far from wasted. He won every award in maths and Greek while he was there. During the summer holidays he tutored other students to earn money and ease the burden on his parents. In his diaries and letters, Adams often worried that his interest in astronomy was distracting him from his real studies in mathematics, yet in 1840 he walked from Cambridge to London to visit the Royal Observatory.

In 1843 Adams completed his degree and took part in the Mathematical Tripos, a series of twelve three-hour exams. Adams scored 4000 points. Francis Bashforth, who was later to become a professor of mathematics, came second with just under 2000. Adams was then elected a fellow of St John's College.

True to his 1832 plan, Adams began to investigate the Uranus problem and in 1843 outlined his plan to James Challis, director of the Cambridge Observatory, who, although supporting Adams, said privately that there was no hope of finding the planet by maths alone.

The problem could be expressed simply. It was possible, knowing the existence of Uranus, to calculate the effect that planet would have on the orbit of Saturn. The relative distances were known and the masses had been calculated. The calculations were long and repetitious, but thanks to Newton's law of universal gravity they were fairly straightforward. What Adams had assigned himself was the problem in reverse: was it possible, knowing the resulting disturbances, to calculate the orbit, position and size of an unknown planet?

Consider some of the effects Adams had to take into account. Both Uranus and the unknown planet were both in orbit around the Sun, so the distance between them and their relative positions would be constantly changing. The strength of the disturbing force would change depending upon the exact orbital distance of the unknown planet and its mass, neither of which was known or could be guessed at accurately. Add to this that the mass and distance were both related and the problem became a good candidate for a guessing game. And that's what Adams did: he guessed.

To give himself a starting point Adams guessed that the planet would be twice the distance from the Sun to Uranus, and so picked 38.4 AU. This number was also in agreement with the unproven Bode's law. By October 1843 Adams had produced his first solution which confirmed the existence of the unknown planet. He shared his results with Challis and asked for more data on Uranus so that he could refine his calculations and predict a position for the planet. Challis wrote to Airy, as Astronomer Royal, requesting the data, which Airy sent along with the suggestion that Adams correspond with him directly in future. Adams continued his investigation, but with increasing lecturing commitments was unable to devote all the time he wanted to the work.

Through the summer of 1844, Adams worked on his second analysis, calculating an elliptical orbit for the unknown planet. By September 1845 Adams completed his new solution, complete with predictions for the observed position of the planet. Adams sent his results to Challis with the request for a search.

Rather than undertake the search himself, Challis suggested that George Airy was the man for the job. He gave Adams a letter of introduction so that Adams could present his work in person whilst on holidays. Upon arriving at Greenwich Observatory, Adams found that Airy was at a conference in Paris, so Adams continued on his way to Cornwall. On his way back to Cambridge, Adams again visited Airy, but he was at another meeting, so Adams left a note of his results and said he would return later. Returning later, Adams found Airy at dinner and was turned away.

Although Airy hadn't met with Adams personally, he did have the results and predicted positions, and a letter from Challis recommending the young man's results, yet still no search was begun. The reason was simple: Airy considered that proper astronomy should be based upon considerable observations, using maths only to describe the results, not to predict new observations. He saw such a theoretical approach as just playing with numbers with no basis in reality—a waste of time.

Despite his views, however, Airy showed Adams' work to William Dawes, a clergyman and amateur astronomer. Dawes was impressed, but as he was still finishing his own observatory did not undertake a search himself, but passed the information along to William Lassell, another amateur and builder of what was at one time, England's largest telescope, a 60 cm reflector. Lassell was at the time in bed with an ankle injury and before he could resume observing he lost the letter.

On 5 November 1845 Airy replied by letter to Adams. Referring to Adams' work as assumptions, he asked why Uranus was slightly farther from the Sun than had been previously expected. Adams' calculations showed that, as it was being slowed by the unseen planet, the lower speed caused it to be further from the Sun, according to Kepler's third law. The exact position of Uranus was, to Adams, irrelevant; his work was on the planet causing the effect, not the effect itself. It is clear that Airy must have only glanced at Adams' work, for he would surely have understood the implications himself upon a full reading.

Adams, after three thwarted meetings, no offer to undertake a search and now a superficial reply, decided to rework his calculations once more to obtain results so sound they could not be ignored again.

It was not only the English who were concerned about the motion of Uranus. The Germans, French and Italians all had people working on the problem. In France Jean Dominique François Arago was director of the Paris Observatory. Popular with the public and astronomers alike for his simple and enthusiastic explanations of science, he was well known for helping young scientists in their careers. So it was that in 1845 he suggested to Urbain Jean Joseph Le Verrier that he tackle the problem of the unseen planet.

Le Verrier's education had much in common with that of Adams. He was raised in provincial France, attending a local high school. In 1831 after winning the annual mathematics prize, Le Verrier was admitted to the École Polytechnique in Paris. He graduated with honours in a wide range of scientific subjects. Le Verrier's first love was chemistry and he published a number of papers in that field, but Joseph Gay–Lussac, a professor of chemistry at the École Polytechnique, recognised an even greater talent for maths, and when a professorship in astronomy became vacant at the École Polytechnique he arranged for Le Verrier to be appointed.

The first problem Arago gave Le Verrier was that of the motion of Mercury, which was always slightly ahead of its predicted position. For three years Le Verrier worked at his calculations explaining most of the effect by the influence of other planets, but with a small unexplained influence remaining. Marking the problem as unsolvable* Le Verrier turned his attentions to comets and their orbits, with great success. It was then that Arago presented the Uranus problem.

In July 1845 Le Verrier began his analysis with Bouvard's tables. On 10 November he presented his first paper to the Paris Academy of Sciences, demonstrating that the problem was indeed real, rather than the result of poor-quality observations. He announced that his next paper would analyse the causes of the discrepancy. It was 1 June 1846 when Le Verrier presented that second paper. Meanwhile Airy had visited Challis at Cambridge, and neither man invited Adams to a meeting.

Le Verrier's second paper concluded with a predicted position for the new planet and an expressed hope that a search would be undertaken at once, but to Le Verrier's amazement no one at the academy offered.

Airy received a copy of Le Verrier's second paper towards the end of June and noticed that Le Verrier's prediction and Adams' agreed to within 1°. Although he had doubted Adams' work alone, there was no way he could ignore both sets of results. Airy immediately wrote to Le Verrier congratulating him on his work. He asked Le Verrier about the calculated distance of Uranus from the Sun, the same question he had posed to Adams, and still little related to the work at hand. Airy omitted to mention that the same work had reached him eight months earlier from a young Cambridge mathematician; he also omitted to congratulate Adams.

Three days after receiving Le Verrier's work Airy announced that he was sure that the new planet would be discovered in a short time, but still made no effort to notify Adams or commence a search. Le Verrier received Airy's letter on 28 June but instead of being encouraged he replied coldly that the Astronomer Royal's question was irrelevant. He offered to send specially calculated positions to Airy if he had 'enough confidence in my work...' Airy replied to Le Verrier declining the offer as he was leaving on a trip; the trip was not until the middle of August, however.

On 29 June Airy visited Cambridge University, in the company of Peter Hansen, director of the Seeberg Observatory in Germany. During their evening stroll, Airy and Hansen ran into Adams on St John's Bridge. After a brief introduction, during which no mention was made of Adams' work, the pair moved on.

On 6 July Airy visited his old professor, George Ely, and mentioned the subject of Uranus. Ely, surprised by his student's actions, urged him to act and begin a search immediately. Finally on 9 July Airy wrote to Challis asking for an urgent start to the search. Writing that the situation was most desperate, Airy neglected to mention his own nine-month delay.

On 29 July 1846 Challis began the search, resigned to the fact that it would take many months and would probably be fruitless. Challis agreed with Airy that it was unwise to place much reliance on the work of Adams and Le Verrier, so instead of beginning with their predicted positions, a section of sky 30° long and 10° wide would be searched systematically. Such a search would involve 3000 stars and use at least 300 hours of telescope time, requiring a month of good weather. Another important point he dismissed was the likelihood that the planet would show a disc rather than the point image of a star. If it showed a disc, then instead of having to view the object over a few days to ascertain its motion, the object should be immediately recognisable. Challis instead decided to laboriously record the position of each object and return to it a few nights later to check for movement.

Le Verrier did not sit back either. On 31 August 1846 he presented his third paper on Uranus' orbit to the Paris Academy. He said the the new planet was 5° east of δ Capricorni and that, as it was just past opposition the planet would be visible all night long, and big enough to show a disc.

* This was one case in which Newton's universal law of gravity really did fail. It wasn't until Einstein produced his general theory of relativity in 1915 that the problem was finally explained.

Adams too was still working. On 2 September he sent his sixth paper to Airy refining further his prediction of the planet's position. He included explicitly a discourse on the distance of Uranus in answer to the astronomer's earlier question, but by this time Airy was away on his trip. Adams received a useless reply from Airy's chief assistant offering more observational data on Uranus.

Adams resolved to present his work to the wider scientific community. On 15 September he arrived at the conference of the British Association for the Advancement of Science, only to find that he had missed the astronomy sessions.

Le Verrier too was having troubles. Although he was widely praised for his work, no one was willing to undertake a search for his planet. Calling in favours, Le Verrier wrote to Johann Gottfried Galle, an assistant at Berlin Observatory asking him to search. Galle received the letter on 23 September 1846 and rushed into the office of Johann Encke asking permission to use the observatory's 23 cm telescope for the search. As it was Encke's birthday and he did not intend to use the telescope that night, he agreed. Heinrich d'Arrest, a young student at the observatory, heard the discussion and asked to be allowed to assist.

Galle and d'Arrest began looking for the new planet as soon as it was dark. Starting with the position given by Le Verrier, RA 21h46m dec -13°24′, they began looking for the planet's disc, but without success. D'Arrest suggested using a star map. These were not very reliable, but with no other choice they began. With a new chart of the region in Aquarius, Galle observed through the telescope and d'Arrest matched stars on the chart. As Galle called the positions, d'Arrest called 'on the chart' as he found them. Then, at RA 21h53m25.84s Galle saw a star d'Arrest couldn't find on the map. They had been observing for under an hour, but the object was less than a degree from Le Verrier's position.

Breaking up his birthday party, they took Encke to the observatory and observed through the night, but they couldn't be sure if the object had moved, or if it showed a disc. The following night Encke, d'Arrest and Galle observed once more. In a few moments they were able to confirm that the object had moved and that indeed it was a disc. On the morning of 25 September Galle notified Le Verrier that the planet indeed existed. Meanwhile Encke began notifying the scientific community, crediting the discovery to himself and Galle and omitting d'Arrest. News of the discovery reached Airy on 29 September while still on holiday. No one recorded his response.

On 29 September the news had not yet reached England and Challis had just received Le Verrier's third paper giving the position. That night while observing, he saw an object at the predicted position showing a disc, but didn't follow the observation up. At dinner the next day Challis mentioned to the Reverend William Kingsley his observation. Kingsley was excited by the observation and asked to see the object for himself. Challis agreed to show him. They reached the observatory, also the Challis' home, with clear skies above. Mrs Challis insisted upon serving tea before they went observing and by the time they had finished the skies had again clouded over.

On 1 October 1846 the *Times* in London announced the discovery of the new planet, a discovery confirmed by a London amateur astronomer, John Hind, despite the cloud and bright Moon.

On the same day Le Verrier wrote to Galle thanking him for his help. Galle had suggested the name Janus for the planet, but Le Verrier was offended as he thought it was his right to name the new world. Le Verrier told Galle that the French Bureau of Longitudes had named the planet Neptune. This wasn't quite true: Le Verrier had suggested the name to them but it had not been accepted. A few days later Le Verrier changed his mind. He decided that the new planet should be named after himself, but as he had already mentioned the name Neptune to a number of people, how could he undo the damage?

Le Verrier went to see François Arago who was himself in favour of Neptune, but Arago agreed to help if Le Verrier urged that Uranus be renamed Herschel. Le Verrier's work on the motion of Uranus was about to be published. At the last minute he had the title changed to *Recherches sur le mouvment de la planéte Herschel (dite Uranus)*, but the text had already been set so it could not be altered and it used 'Uranus' exclusively.

On 5 October Arago told the Paris Academy of Sciences that Le Verrier had asked him to name the planet and that he had chosen Le Verrier. He made the case that comets were named after their discoverers, so why not planets? He also formally proposed Herschel for

Uranus and Olbers for the minor planet Juno, to satisfy the English and German delegates. The idea was rejected as most people had already adopted Le Verrier's first suggestion of Neptune.

While the French were celebrating their discovery of the latest planet, they did not suspect that a challenger was in the wings. On 3 October an article by John Herschel appeared in the *Athenæum*, a London weekly magazine congratulating Le Verrier and reporting his own comments three weeks earlier at the British Association of the Advancement of Science conference at which he predicted the new planet would soon be found. He also mentioned that similar calculations had been undertaken by Adams independently.

On 5 October Challis wrote to Arago saying that in his observations of 29 September he had observed the object and noted it as a disc, this was before he had heard of the Berlin discovery; he was trying to stake a claim for himself. What Challis omitted to say was that his observations which began on 29 July were made using Adams', not Le Verrier's calculations. In fact he made no mention of Adams at all. He also neglected to mention that he hadn't bothered to verify his observation of 29 September.

On 14 October it was Adams' turn to state his position. He wrote to Le Verrier congratulating him and saying: 'I do not know whether you are aware that collateral researches had been going on in England and that they led to precisely the same results as yours.'

Of course Le Verrier did not know of Adams' work as during his correspondence with Airy, Airy had never mentioned Adams' name. In writing to Le Verrier after the discovery, Airy admitted that he had seen calculations prior to Le Verrier's but said that the calculations 'were not so extensive as yours'. Even in this letter Airy did not mention Adams by name.

Having written to Le Verrier, Airy continued to position himself for some glory. He wrote to Challis suggesting that he put forward the name Oceanus for the planet, even though he knew Le Verrier's feelings on the matter. Airy also proposed that he write a report on the search and discovery of Neptune for the Royal Astronomical Society as 'I know nearly all the history and yet have taken no part in the theory or observations.' He sent this proposal to Rev. W.J. Adams, though Adams' initials were J.C. and he was not a reverend. On 17 October Challis published his account of the search and investigation in the *Athenæum*, mentioning Adams' 1843 solution and proposing the name Oceanus.

The French were furious. They attacked Adams, Airy, Challis and Herschel at the Academy of Science and the French press took up the issue with a series of vicious articles. In fact the papers became so abusive that Arago and Le Verrier made moves to distance themselves from the reports. While the dispute was still raging the scientists not involved decided to use Le Verrier's first suggestion for the name and formally adopted Neptune.

On 13 November 1846 the Royal Astronomical Society held an inquiry into the events surrounding Airy and Challis. Airy presented a paper giving his side of the story, Challis a paper giving his side, and Adams presented his paper *An Explanation of the Observed Irregularities in the Motion of Uranus*, the paper he had been trying to bring to people's attention for more than a year. In addition, Adams had calculated the first orbit for Neptune, sending it to Airy on 15 October. Adams did not express his feelings in the matter, but had nothing but praise for Le Verrier:

> *I mention these dates merely to show that my results were arrived at independently and previously to the publication of M. Le Verrier, and not with the intention of interfering with his just claims to the honours of the discovery . . . the facts as stated above cannot detract, in the slightest degree, from the credit due to M. Le Verrier.*

When it came awards time in 1846, the Royal Society gave its Copley Medal to Le Verrier while the Royal Astronomical Society, unable to make up its mind, did not make an award to anyone. At a party given by John Herschel in 1847, Adams and Le Verrier met each other for the first time. Far from showing animosity, the two became good friends for life. In 1848 Adams was finally given the recognition he deserved, with the Royal Society's award of the Copley Medal.

After the greatest demonstration of Newton's gravitation since the first predicted return of Halley's comet, what became of the chief participants?

Galle became director of the observatory of Breslau, dying at age 98 in 1910. He had the distinction of seeing Halley's comet twice as a professional astronomer.

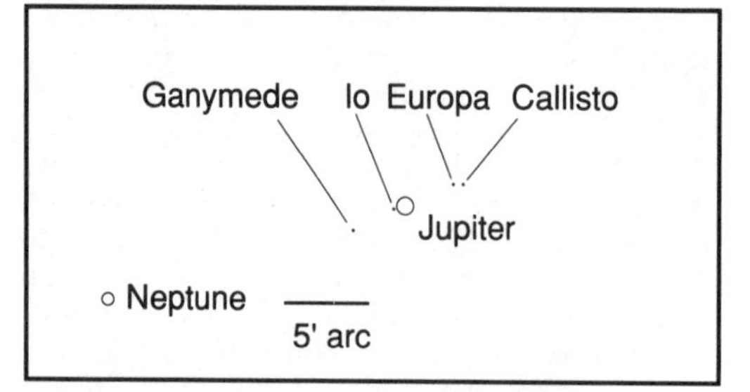


**14.2** On the evening of 28 December 1612, Galileo Galilei was observing the moons of Jupiter when he noticed a star in the field, just 17 arc minutes from Jupiter. On subsequent nights Galileo again observed the star, even commenting that he thought it had changed position. Research in the 1980s showed that the star which Galileo saw and noted was Neptune, 234 years before its discovery.

D'Arrest became director of the University of Copenhagen's observatory. He died at age 52 in 1877 after contributing much to the study of nebulæ and spectroscopy.

Challis continued as director of the Cambridge Observatory during which time his scientific work became more and more unorthodox. It was only though the support of friends that he was able to continue in the position, though eventually he was forced to step down. He died in 1882 just before his 79th birthday.

Airy continued on his way as Astronomer Royal, disliked by most. He retired from the post in 1881 having made no significant contributions to science while there. He died ten years later still organising his papers in order to publish an autobiography.

Le Verrier went from strength to strength. In 1854 he took over as director of the Paris Observatory following the death of Arago. He was dismissed in 1870 for devoting too little money to astronomy, but was reinstated in 1873. He died in 1877 at the age of 66.

Adams had a rough time. His position at Cambridge was not renewed in 1852 because he had not joined the clergy. Time after time he applied for positions but was turned down. Finally in 1861 he was appointed to replace Challis as the Cambridge Observatory director. He was offered a knighthood in 1847 and the position of Astronomer Royal when Airy retired, but he refused both. John Couch Adams died at the age of 72 in 1892.

While all the fuss was going on in 1846 about who could claim the glory of Neptune's discovery, other astronomers were getting down to the business of studying the planet. William Lassell, the amateur astronomer who may have discovered Neptune had it not been for a sprained ankle and losing the letter giving the positions, was observing the new planet with his 61 cm reflector. On 10 October, just two and a half weeks after the planet's discovery, Lassell discovered that Neptune had a moon. The Sun then got in the way and precluded viewing, so the moon's existence wasn't confirmed until July the following year.

Much was made at the time of certain people's observations of Neptune being made but not recognised until after the discovery by Galle. So who was it who really saw Neptune first? The answer to that question wasn't realised until 1980 when Charles Kowal and Stillman Drake were examining the orbit of Neptune with a view to more accurate calculations of its position. They had calculated that Neptune and Jupiter were very close to each other in the skies of December 1613. In 1613 there wasn't much telescopic astronomy going on; in fact Galileo was probably the only person in the world who was making observations.

Galileo was fascinated by Jupiter, as we have seen. He spent many hours plotting the positions of the planet's moons and other observations. When checking a notebook he kept exclusively for the Jovian satellites, Kowal and Drake found a star marked at exactly the right position for Neptune for 28 December 1612. Thus, Galileo was the first person to see Neptune, mistaking it for a star.

Kowal and Drake continued their search through his observations finding another dated 28 January 1613. In this observation Galileo noted the position not only of Neptune, but also of a star (SAO 119234). Along with the drawing he made, Galileo wrote, 'post stella fixa a. alia in eadem linea sequebat, ita ut est b que etia precedet nocte observata fuit; sed videbat remotiones inter se.' (past the fixed star a [SAO 119234] in a straight line, this is b [Neptune] which was observed last night; *but it seemed further from the other.*)

So not only did Galileo observe Neptune once more; he noted that it had moved. Galileo also made a note in his book to follow up the observations over succeeding nights, but never did. What problems would this have caused in seventeenth century Europe had it been known?

Already there was concern over Galileo's work, his dispute with the Church over the structure of the universe; the existence of another planet, one hidden from view, would have really rocked the foundations.

At its great distance from the Sun, very little can be deduced about Neptune from the Earth. Following the discovery of the first Neptunian moon, Triton, in 1846, 103 years were to pass until Gerard P. Kuiper discovered a second, Nereid, in 1949. The names of the Neptunian moons, like nearly all other astronomical bodies come from mythology. Triton was the son of Neptune; he was a merman. Nereid was named after the sea nymphs which attended to Neptune. A further 40 years were to pass until more moons were found, along with much more.

With an opposition magnitude of 7.65, the outermost gas giant is below naked eye visibility, but it is easily found with binoculars or finderscope once its position is known. Simply plot Neptune's position on a star atlas, and 'star-hop' from brighter stars.

With an orbital period of 164.8 years, Neptune has yet to complete a single orbit since its discovery in 1846. When found, the planet was situated about 11′ of arc from magnitude 5, μ Capricorni, and just over 1° from Saturn. By the year 2010 the planet will again reside in the same field with the red giant star.

Neptune's angular motion across the sky is only about 2° per year, half that of Uranus, or about four Moon diameters. Uranus and Neptune remain close together in the sky until about 1993, when the separation between the two begins to increase. They will not be seen together again until 2164 when they approach within 1° of each other. Currently Neptune is in the constellation of Sagittarius; following the ecliptic closely it moves into Capricornus during 1997, where it will reside for the next thirteen years.

A telescope of 150 mm aperture and high magnification will be needed to resolve the planet into a disc. At 2.5″ in diameter, small telescopes will have difficulty in distinguishing it from a star. No detail can be expected on the tiny disc with any amateur telescope. Like its sister planet Uranus, Neptune appears as a greenish-blue planet. Methane present in the atmosphere is also responsible for its colour, a large percentage of the short blue wavelengths are reflected while the longer red is mostly absorbed.

The only satellite of Neptune available to amateur instruments is Triton. At 13.69 magnitude, Triton is slightly brighter than the two observable moons of Uranus. Although closer to the planet than the Uranian pair, Triton is easier to see as there is not as much glare from the parent body, but an instrument in the 200 mm range will still be required to catch a glimpse of it. Triton's rotation period is short, only 5.88 days; therefore observations over two or three nights will easily identify it from the background stars. The times of greatest elongation can also be obtained in an almanac for confirmation of a suspected sighting.

**14.3** Neptune and its moon Triton are seen here at the centre of the photograph as a dot with a bump on it. The bump is Triton. This photograph was taken on 12 July 1985 using a 250 mm reflecting telescope using a two minute exposure at prime focus on 400 ISO film.
PHOTO CREDIT: STEVEN QUIRK

The tiny disc does not lend itself to photography through the telescope, and the same method used for Uranus with a standard SLR camera is recommended. A good star atlas will help identify Neptune and any other nearby planets or minor planets from the background stars. Photographs taken a week or two apart will show a shift in the planet's position.

There is no work of real value that the amateur can do with either Neptune or Uranus. Passing spacecraft have added more to our knowledge than could ever be hoped for by terrestrial means on the lesser gas giants. They are both easy targets to add to your list of observable planets, and will prepare you for the real challenge: Pluto.

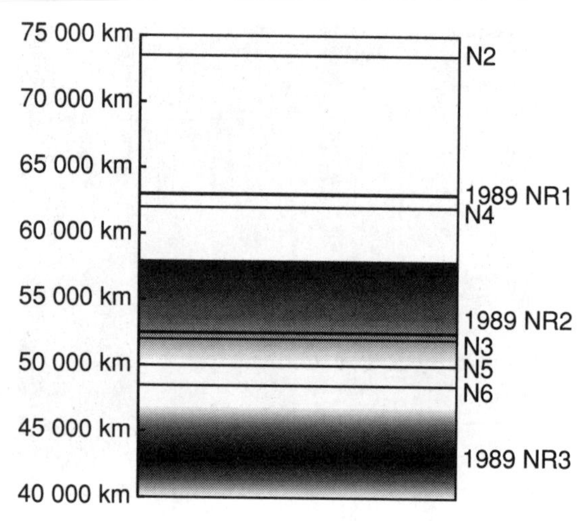


**14.4** Here Neptune's three rings are shown with their distance from the centre of the planet and with the orbits of five of Neptune's moons. Rings 2 and 3 are broad bands with some central thickening while Ring 1 is a narrow ring. All the rings are distinctly clumpy, unlike the rings of the other gas giant planets. This clumpiness gave rise to suggestions that the rings might only be arcs rather than full circles, but photographs taken after Voyager 2 passed Neptune confirmed that the rings are complete circles.

It was just after the twelfth anniversary of its launch that Voyager 2 had its closest approach to any planet. On the evening of 25 August 1989 Voyager 2 passed within 4900 km of the cloud-tops of Neptune's north pole, sending back views of that planet and its satellite system that had never been seen before. Once more the space-craft Voyager 2 has done much to extend our knowledge of this planet; things which scientists had had tantalising hints of were fully revealed to a waiting world. Before Voyager's visit only a little was known about Neptune: scientists had some idea of the size of the planet from the occultations of stars, they knew of two moons and they thought that there may be rings.

Knowing that all the other gas giants had rings it would not seem to be too much out of the ordinary for Neptune to have rings also, but finding them had been a problem. The best way to find rings would be to watch the occultation of a star by the planet, in the same way that Uranus' rings were discovered, so astronomers began looking. On 7 April 1968 two astronomers observed an occultation by Neptune from New Zealand. They noticed a dip in the brightness of the star just before occultation, but being interested in Neptune's atmosphere, did not pay much attention to the event. Ten years later after the discovery of Uranus' rings, the two remembered their observations and resurrected the recording. They found that it was consistent with a broad ring 10 000 km above the clouds.

A systematic series of observations was commenced in both Australia and Chile on 10 May 1981. On that night nothing was observed, but ten days later in Arizona while observing a fainter occultation, a dip was independently recorded by two telescopes, though oddly Flagstaff Observatory, also in Arizona, did not observe anything. The next major occultation was on 15 June 1983 when observers in Australia, the United States and China all made observations. Despite good weather at all sites, no one saw any evidence of rings. Further observations were made on 22 July 1984, 7 June 1985 and 20 August 1985. At all these occultations some observers saw evidence of rings, while others saw none at all; at other occultations no one could find anything.

The solution the scientists came up with was that Neptune did have rings, but only partial ones, or else very thin, clumpy ones, so it was with great expectations that the results of Voyager's visit were awaited. While it was still quite a distance from the planet, Voyager found what was thought to be the solution: two ring arcs. The first was 62 000 km from Neptune's centre and stretched 50 000 km either side of a small moon, the other was 52 000 km from the planet's centre, but only formed 3 per cent of a complete ring; it too associated with a small moon. It seemed that the hypothesis about the partial rings was correct.

A fuller explanation emerged after the encounter as Voyager moved away from the planet. From that aspect any rings would be much easier to see as they would be backlit by the Sun. The effect can be seen in a cinema. When looking at the screen you can't see the beam of light from the projector, but when you look back up the theatre the particles of dust in the air are backlit and the beam of light is quite obvious. Photographs taken from behind Neptune showed three rings, complete rings which were quite clumpy but which completely encircled the planet. What had earlier appeared to be partial ring arcs were simply the thickest bits of the complete rings.

The three rings are quite distinct. The innermost ring is quite broad and very faint, being just 17 000 km above the cloudtops. It is broad because of the lack of moons to help keep the particles in a narrow region. The two outer rings of the planet are similar to the thin rings seen around the other gas giants, with a retinue of smaller moons to keep particles in well-defined regions. Both the outer rings are decidedly clumpy, which explains the different results of the stellar occultation experiments. At some of the occultations the star would have passed behind a thick section of the ring, leading to a drop in brightness which could be measured; at other times a thin part which could not be detected from Earth intervened.

Scientists did not know what to expect as Voyager neared Neptune. Uranus had proved disappointing from a pictorial point of view, its lack of surface features leaving many dissatisfied. The lack of features at Uranus was due to its low temperature, so what of Neptune which was even colder? Pictures of the planet taken in late January 1989 made many breathe easier. They showed markings on the surface visible even though the spacecraft was still more than 300 million km distant.

As Voyager approached the planet the true nature of Neptune's clouds became clear. The atmosphere, while not as active as Jupiter's or Saturn's, certainly did show activity. To match Jupiter's red spot, Neptune has a dark blue spot at 22° south. The spot, almost the same size as Earth, is an anticyclone. It is not known if the feature is long-lived like Jupiter's spot, but its appearance changed little over the few months Voyager was close enough to view it.

Around the edge of the Great Dark Spot white clouds, similar to cirrus clouds on Earth, stretched away. Photographs taken when the spot was near the terminator showed that these white clouds were 60–70 km higher in the atmosphere than the clouds making up the visible surface. Near the planet's south pole is a smaller dark spot with white cirrus clouds at its centre. Between the two, another patch of white cirrus cloud, called the Scooter, travels around the planet much faster than the other cloud patterns.

**14.5** Neptune's rings are clumpy, as can be seen by this combination of images taken on each side of the planet. The central part of the centre ring, 1989NR2, in particular, is much more prominent in the left-hand image than the right. The broad band of diffuse material which makes up 1989NR2 can be seen stretching either side of the central bright band. The diffuse nature of the inner ring, 1989NR3, is also clear. PHOTO CREDIT: JPL/NASA

No one expected much weather on Neptune. With the little amount of heat received from the Sun, there would be nothing to drive it. The fact that patterns in the atmosphere can be seen shows that much of the atmosphere is moving without friction between adjoining layers. Such motion had not been seen elsewhere in the solar system, the patterns on Jupiter and Saturn resulting from turbulent flow between layers. The clouds which give the planet its blue colour are methane, so are the white clouds, but in that case the methane has been forced up into a higher and colder level and has condensed into white crystals.

Apart from observations of the atmosphere Voyager made measurements of other quantities to determine more about the planet. One experiment looked at the radio emission from the planet. It detected a cycle of 16h03m, revealing the rotation period of the planet's core, the

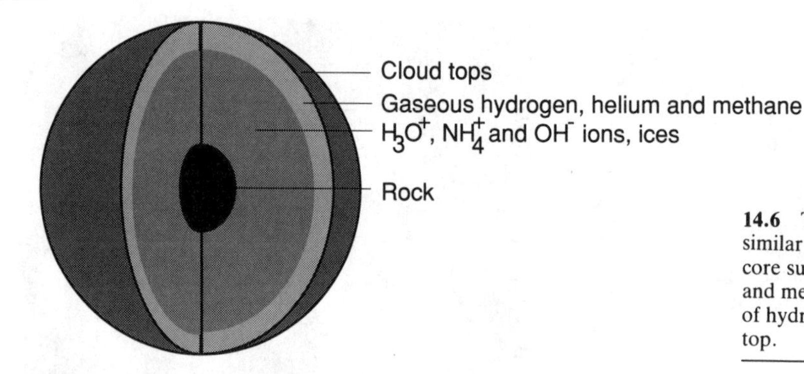


**14.6** The structure of Neptune is similar to that of Uranus: a rocky core surrounded by water, ammonia and methane ices and a gaseous layer of hydrogen, helium and methane on top.

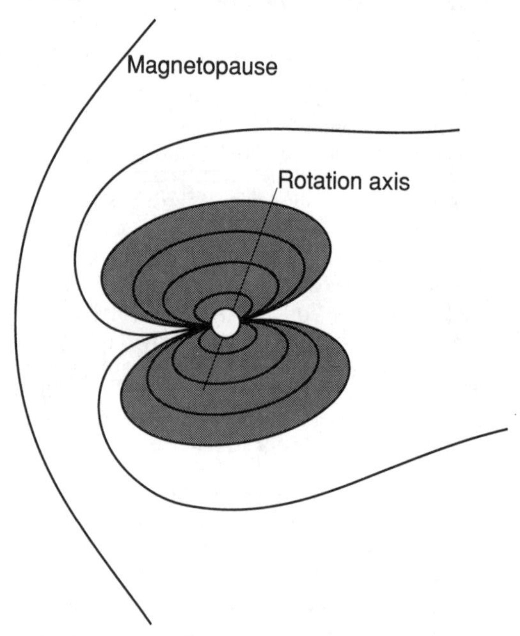


**14.7** The magnetic field of Neptune is tilted at an angle of 50° to the axis of rotation of the planet, and is located 80 per cent of the way from the centre of the planet to the surface.

cloud tops rotating about the planet more quickly.

Another thing it investigated was the planet's magnetic field. Initially scientists were puzzled: as Voyager got closer and closer to Neptune it was unable to detect a magnetic field at all. This turned out to be because the spacecraft was approaching the field from close to the south pole where it is closest to the planet. Neptune's magnetic field turned out, like Uranus', to be tilted with respect to the planet's axis of rotation, by 50°. Not only that, but the field is well away from the planet's centre, 80 per cent of the way towards the planet's south pole.

As it approached Neptune, Voyager 2 began to discover moons, six in all. The first found, 1989N1, turned out to be Neptune's second biggest at 400 km, bigger than the previously known Nereid which was measured at only 170 km. One close-up photograph of 1989N1 was taken. It is a dark world roughly triangular in shape. The other five moons also turned out to be irregular in shape and dark coloured.

The last body Voyager was to encounter in the solar system was Neptune's largest moon, Triton. At 2720 km in diameter, Triton is only slightly smaller than our Moon. True to form, Voyager's last object was a remarkable body, and unlike anything the scientists had predicted. Triton turned out to have a tenuous atmosphere stretching 800 km from the surface. The atmosphere is almost completely nitrogen, with methane present as a tiny trace. The pressure is low, 1 Pa, but enough to support thin clouds and haze layers.

It is the surface of the moon which is remarkable. Long double ridges and groves stretch across the moon, but never more than a few hundred metres high as the frozen nitrogen and methane which make up the surface are unable to support tall features. It had been said that Triton would have pools of liquid nitrogen on its surface, but it doesn't; it is even too cold for

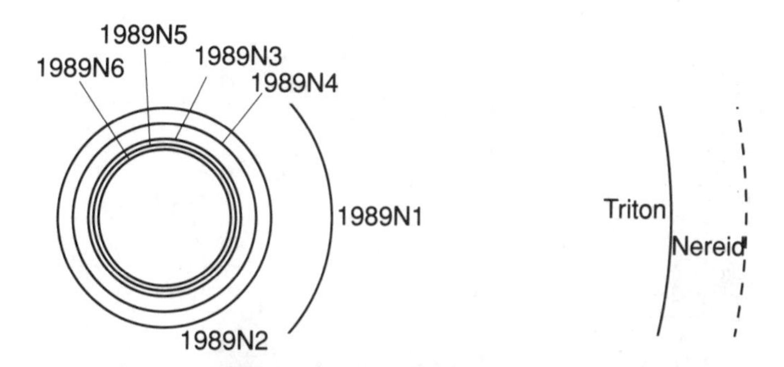


**14.8** Prior to the arrival of Voyager 2 the only moons Neptune was known to have were Triton and Nereid. Here the orbits of Neptune's eight moons are shown to scale, except for Nereid's orbit which is highly eccentric and titled to the plane of the planet's orbit. The orbit of Triton is also unusual as it is retrograde, the only large moon in the solar system to travel in that direction.

that, at −236°C, nitrogen is solid, and Triton is the coldest place in the solar system. The planet does have a polar cap in the south, stretching almost all the way to the equator. The 'snow' in the polar cap seems to be a mixture of frozen methane and nitrogen. Organic molecules, formed from the interaction of the solar wind and the methane, give the ice a pink colour.

Though there is no liquid nitrogen on the planet's surface, there is certainly some beneath it. On one image of the moon, Voyager captured a geyser throwing dark particles 8 km above the surface of the moon, forming a cloud which stretched 150 km westwards in the winds of Triton's atmosphere. The eruption seems to be caused by gaseous or liquid nitrogen below the surface expanding and blowing dark dust particles into the upper atmosphere where they form clouds which drift westwards. Why the coldest moon in the solar system should have such active features is not completely understood, but is probably related to tidal forces due to the moon's eccentric orbit.

Crater counts on the surface of Triton show that its surface is very young, the youngest parts being only 500 million years old. This suggests that the surface of the moon was completely erased at some time in the past. An explanation for this might also explain the orbit of the moon. Triton is the only large moon in the solar system which orbits its planet in a retrograde sense; the orbit is also tilted with respect to the equator of Neptune by 70°. It is proposed that Triton was not an original satellite of Neptune, but was captured at some time in the past. If this is the case, then Triton would have lost a lot of energy when it was caught and some of this would have caused the moon to melt. This melting erased the surface and what we now see is the result of later cooling.

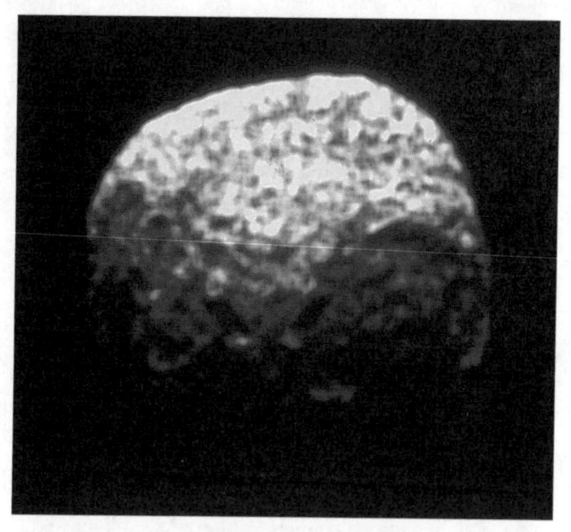

**14.9** The satellite 1989N1 was the first of the new Neptunian moons discovered by Voyager 2. It is a small, bumpy world 400 km in diameter, orbiting 117 500 km from Neptune. Its pockmarked surface is typical of other small moons throughout the solar system. PHOTO CREDIT: JPL/NASA

As the last body Voyager 2 explored, Triton certainly lived up to expectations, showing us that what we think we know about the solar system and what we really do know are often quite different.

**Table 14.1 The Neptunian moons**

| Name | Orbital distance | Orbital period | Inclination | Radius |
|---|---|---|---|---|
| 1989N6 | 48 000 km | 7.0 hours | 5° | 50 km |
| 1989N5 | 50 000 km | 7.5 hours | | 90 km |
| 1989N3 | 52 000 km | 8 hours | | ≈150 km |
| 1989N4 | 62 000 km | 10 hours | | ≈150 km |
| 1989N2 | 73 000 km | 13 hours | | ≈150 km |
| 1989N1 | 117 500 km | 27 hours | | 400 km |
| Triton | 354 000 km | 141 hours | 160° | 2720 km |
| Nereid | 5 510 000 km | 8642 hours | 28° | 170 km |

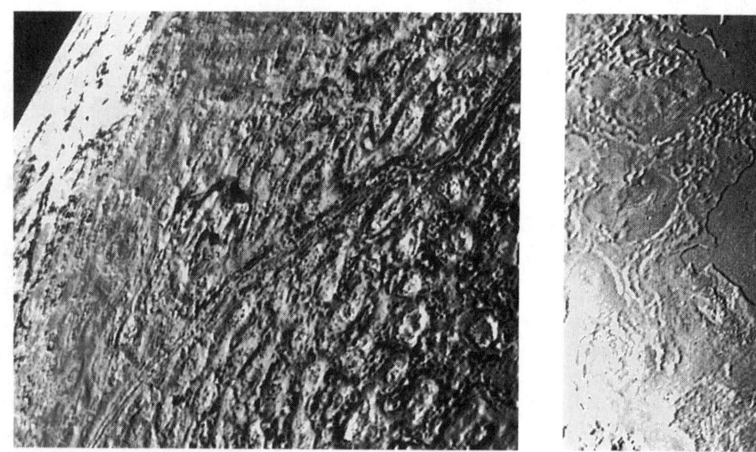

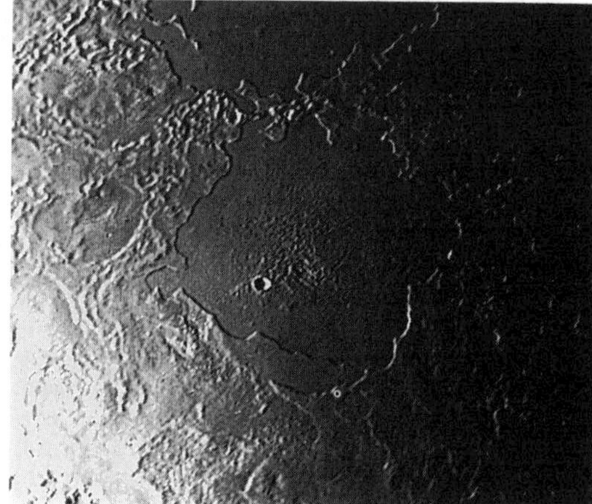

**14.10** These two images of Triton show two different types of terrain. The image on the left shows cratering and strange depressions which may have been caused by the melting and collapsing of the moon's icy surface. The image on the right shows a younger region of flat plains and low terraces, showing that some of the moon's surface has been reworked since its formation. PHOTO CREDIT: JPL/NASA

*Voyager 2's encounter with Neptune*

25 August 1989 was a date which was noted on many calendars around the world. It marked the last encounter of Voyager 2 within the solar system, the day on which we would get our best views of the last of the large planets, Neptune. Voyager 2 is tracked by the NASA deep space network (DSN), three 70 m parabolic antennae: at Tidbinbilla near Canberra, Australia, Madrid, Spain and Goldstone, United States. These antennae are used to transmit and receive signals from Voyager 2 and other United States spacecraft. Because of the immense distance to Neptune, and the amount of information to be obtained and relayed to Earth, during the encounter the DSN was supplemented by the Very Large Array in New Mexico, the Parkes radio telescope in Australia and another radio telescope in Usada, Japan. Using antennae spread across the world allows 24-hour contact with the spacecraft for the receipt of data.

Because Neptune was at declination −22° on the day of the encounter and because it had all been done before for Uranus, NASA decided that the Neptune and Triton closest approaches would take place with the spacecraft in Australian skies and that Tid-

**14.11** This is the telescope owned by the Australian National Radio Astronomy Observatory near Parkes, New South Wales used to receive the signals from Voyager 2 during its flyby of Neptune. The dish of the telescope is 64 m in diameter and is normally used for radio astronomy observations. As this photograph was being taken, the telescope had been tracking Voyager 2 for about an hour on the day of the closest approach. Beneath the telescope, partially hidden by the bushes, is the small portable building housing the equipment for the radio occultation experiment. In the distance are the offices of the observatory. PHOTO CREIDT: DAVID REIDY

binbilla and Parkes would be the antennae receiving the signals at the crucial time. It was to Parkes that I journeyed for the encounter.

The Parkes radio telescope is run by the Australian Commonwealth Scientific and Industrial Research Organisation (CSIRO). It was commissioned in 1961 and has been used since that time for radio astronomy. Refurbished and improved, the telescope has had some notable achievements including pioneering work on quasars and observations of the galactic centre. Its use for spacecraft tracking is unusual but not rare. The telescope was used during the Apollo Moon landings, and for the Voyager encounter with Uranus in 1986 and later that year as the primary link during the Giotto spacecraft's encounter with Halley's comet. By agreement between NASA and the CSIRO, the telescope was once more made available for Voyager at Neptune.

The preparations at Parkes began in February 1989 when the vanloads of equipment needed for the encounter began to arrive to be installed in and around the telescope. Chief among these were the hydrogen maser amplifier and the receiver equipment and the recording equipment to keep backup copies of all the data received and to record

**14.12** This is the main control desk of the Australian National Radio Astronomy Observatory's radio telescope at Parkes in New South Wales, the primary telescope used during the Voyager 2 flyby of Neptune in 1989. On the monitors above the control desk, the pointing and status information from the telescope can be seen on the left next to a monitor showing the latest picture from Voyager 2. The graphs below these monitors show the results of the last check on the telescope's pointing accuracy. The pinned-up sheet shows the latest computer solutions for pointing the telescope accurately, solutions which proved to be much more accurate than anyone had predicted. To the right of the panel a duplicate set of monitors display the information in case of failure. PHOTO CREDIT: DAVID REIDY

the vital radio science occultation experiment data.

Although the use of the telescope was handed over to NASA, the astronomers at the telescope remained very involved taking advantage of the state-of-the-art equipment which NASA supplied to do other useful astronomical work with the telescope. At the same time the scientists whose jobs it would be to run the telescope for NASA began to devise new and better ways to perform their tasks so that this encounter would be the best, in spite of it being the most difficult, ever.

It might be thought that pointing a large, computerised radio telescope at Neptune and Voyager would be a simple operation, and indeed if you're not too concerned about the level of signal you receive, it is; it could even be done manually. But when you have a very low-powered spacecraft over four hours away at the speed of light, every little bit of signal is important; even a few per cent off maximum would mean a loss of data, so the pointing had to be accurate, to within 15″ of arc. Consider some of the effects which the astronomers had to allow for: the exact position of the planet, the position of the moving spacecraft near the planet, 15″ is only seven planetary diameters, the refraction of the signal through Earth's atmosphere and the mechanical movements of the telescope, which varied from moment to moment with temperature and wind and the direction in which the telescope was pointing. Taking all this into account, sophisticated computer software was required to achieve the accuracy NASA required.

In addition, a method was needed to check the pointing of the telescope to ensure the best possible reception. Normal astronomical practice allows the signal to drop off and be reacquired to check the pointing, but with every bit of data important this method was not practical, so another technique was developed by which the telescope drew a tiny circle in the sky while the signal strength was monitored, never allowing it to drop by more than a tiny fraction. Using the results of this small circle, the signal strength could be checked and the telescope's position adjusted. By combining these two computer programs, and with a lot of practice, calculations and mathematical modelling, the astronomers hoped to achieve results better than ever before.

As the day drew near, public interest in the events that would be taking place at Neptune increased and provision was made to allow people to view the results as soon as they were available. When signals from Voyager were received at Parkes, they were sent by microwave link to the deep space network site at Tidbinbilla where they were combined with the signals received by the 70 m antenna there. From there the signals were relayed via satellite to the Jet Propulsion Laboratory (JPL) centre in Pasadena, California. At JPL the signals were assembled by computer into the images and other data the spacecraft had sent. These results, which were then distributed by normal communications channels to monitors throughout the world, including the visitors centre, library and control rooms at the Parkes telescope, could be viewed just six minutes after their receipt on Earth.

On the morning of 25 August there was expectation in the air at Parkes. Although the telescope had been tracking the spacecraft for a few months now, today was to be the big day. The reports of the previous night's work were excellent, better than what was required. Throughout the day meetings were held to make sure everyone was ready, and standby crews for everything from the air conditioning systems to the backup power supplies were on duty to guard against anything going wrong. Throughout the day visitors arrived at the visitors centre, and were rewarded by views of Neptune, its rings and satellites which were just minutes old.

The work at the telescope began at three o'clock with the routine testing of the dish, receivers and control systems. A few minutes before four o'clock the telescope tilted towards the east, waiting for the planet and spacecraft to come within range; acquisition of signal was predicted for 16:06. In the visitors centre a chart recorder showed the signal level from the telescope, at 16:06 nothing happened, then a few minutes later the pen sprung across the page indicating that the signal from the spacecraft was being received. Throughout the afternoon visitors to

**14.13** The aerial cabin high above the main dish of the Parkes radio telescope. In the centre of the base of the cabin you can see the opening of the receiver mounted inside the cabin. The signals reach this receiver having bounced from the surface of the 64 m diameter dish below. No one is in the cabin while observations are being made. All adjustments are made from the control room in the base of the telescope. PHOTO CREDIT: DAVID REIDY

the telescope arrived at the centre, while in the library astronomers and their families watched each new image arrive. Each hour NASA provided a live briefing over the link to inform all of the new discoveries, and brave scientists gave their interpretation of data which had arrived just minutes before.

The most important part of the very close approach to the planet was not to be the images it could send back, but the radio science occultation experiment. In this experiment, the radio signal from the spacecraft would be allowed to pass through the atmosphere of Neptune as the spacecraft passed behind the planet. By observing the strength of this signal, and comparing results with the readings of the Sun's brightness made by another instrument of the spacecraft, scientists hoped to find out much about the composition and dynamics of the atmosphere of Neptune. The only problem was that the spacecraft would be out of contact with Earth for 46 minutes. Three quarters of an hour before the spacecraft was due to travel behind the planet, all data transmission was stopped so that the signal level of the spacecraft's transmitter could be increased and so that data transmission didn't interfere with the signal passing through the atmosphere. Although not transmitting, Voyager was still recording images and other data, but storing them on an internal tape recorder for transmission the following day.

With the chart recorder showing the telescope signal just millionths of a second after its receipt from Voyager, the signal started to drop at 18:08, wobbling considerably as the strength fell due to the signal penetrating deeper and deeper into Neptune's atmosphere. These wobbles would be the data for many scientists to work on over the following years, probing the structure of Neptune. After two minutes of varying, the signal suddenly disappeared, Voyager was behind the planet. For the next forty minutes we waited for the signal to reappear, at 18:53, as the pen returned to its former full strength position, a cheer went up: Voyager had survived. For another forty-five minutes no data was transmitted to allow for further measurements and to allow calibration for the next occultation, the moon Triton at 23:45.

In the fifteen minutes before the expected Triton occultation, the spacecraft again ceased data transmission and increased signal level. This occultation was to be much shorter, just three minutes, giving astronomers an accurate measurement of the size of the moon, and information about the moon's atmosphere, if one existed. At 23:45, right on schedule, the signal level from the spacecraft dropped quickly, taking only five seconds to reach zero. The signal did not wobble. It decreased smoothly, but slowly enough to show some atmosphere. Three minutes later the signal bounced up to its former level, taking just half a second, shorter than would be expected with an atmosphere. What exact-

**14.14** The top of the receiver mounted in the aerial cabin of the Parkes radio telescope. This device received the signals from Voyager and amplified them so that they could be decoded for their information. Much of the piping seen around the receiver is carrying liquid helium for cooling the receiver to eliminate electronic noise from the weak signal. PHOTO CREDIT: DAVID REIDY

ly this meant had to wait for the experts who had more detailed recordings and other data as well.

For the next hour and a quarter the telescope continued to receive signals, now containing pictures of Triton, showing its ice-covered surface. At 1:01 on 26 August the spacecraft dropped too low in the sky for the telescope to point at it and the signal was lost, but by then the DSN antenna in Spain had been following the craft for a short while and took over tracking. With calls of congratulations on a job well done coming in from the United States and other centres, the results of the evening's work were seen. The computer solutions were so accurate; the telescope had been four times more precise than had been needed, much better than anyone had hoped with the signal dropping only 1 per cent below maximum over the entire nine hours.

The Parkes telescope continued to track Voyager daily for a few more weeks. The NASA equipment was removed, and the telescope finally was returned for radio astronomy on 13 September, ending another successful chapter in the history of the Australian National Radio Astronomy Observatory.

Voyager's encounter with Neptune officially ended on 2 October when the final photographs and readings were taken, but the probe's mission has not ended. It is estimated that there is sufficient fuel to keep the antenna pointed towards Earth until 2015, maybe longer. In recognition of the ongoing nature of the job to be done, the project was, on 2 October 1989, renamed by NASA the Voyager Interstellar Mission.

# 15

# Pluto

The discovery of Neptune was a triumph of mathematics, leaving no doubt in the minds of astronomers of the ability of their science. Many of them were convinced that if Neptune could be found using a piece of paper, so could other planets. Why should Neptune be the last? When names were being suggested for Neptune, someone suggested Janus* but Le Verrier, discoverer of the planet said: 'The name Janus would imply that this planet is the last one in the solar system, and there is no reason to believe that this is so.' Other scientists agreed. Neptune had been discovered by observing the motion of Uranus and noting any irregularities in the position of that planet. Astronomers were certain that after a few decades of observing Neptune, such anomalies would lead them to the discovery of the next planet.

On 13 March 1855 Percival Lowell was born, the eldest of five children. In 1877 Lowell graduated with honours in mathematics from Harvard University. For the next six years he worked in his grandfather's businesses making a sizeable fortune for himself in the process. In 1883 he began travelling throughout Asia as a travel writer and served as foreign secretary and counsellor for the first United States diplomatic mission to Korea.

In 1893 Lowell returned to the United States to start an observatory at Flagstaff, Arizona. Initially his interest was in Mars and the canals he saw crossing the surface, but all aspects of planetary astronomy fascinated him. His interest in a trans-Neptunian planet went back to his college days and his teacher Benjamin Pierce. Now with the facilities to conduct a search and the interest and skills in mathematics to enable calculations, he was ready to begin work. Many said that Lowell's attempt to find the ninth planet using conventional astronomy was to give credence and support to his theories of life on Mars, theories which had, to some extent, put him at the edge of scientific circles.

Lowell's first search for the planet began in 1905 but it proceeded intermittently for four years. The search was not mathematically based. Instead it began as a series of photographs along the ecliptic in the hope that the motion of the planet would give itself away. It was only later that Lowell began systematic theoretical calculations to support his program, but these were not integrated with the observational program. Each time Lowell made a new prediction he would send his results to the staff at the Flagstaff Observatory, and they would be expected to search that region.

When the series of photographs was completed in 1907, Lowell was worried that someone else would steal his ideas, so uncharacteristically he remained silent on the matter, referring to it only by code names. But by 1908 he had determined that the task was so complex no one could simply steal his work and he publicly announced his search for 'Planet X'.

In November 1908 Lowell had some troubling news when he attended a lecture given by William Pickering of Harvard. Lowell and Pickering knew each other, as Pickering had helped Lowell start his Mars observing program in 1894. Since that time, however, Pickering had been publicly critical of Lowell's theories regarding Mars. During the lecture Pickering presented a graphical plot of the anomalies of Nep-

* Janus was the gatekeeper of the gods. With two faces he was able to look both forwards and backwards; he guarded the gate of heaven. It is after him that January is named.

tune's orbit and predicted the existence and position of 'Planet O', the trans-Neptunian planet.

Pickering's prediction for the planet gave it an orbital distance of 51.9 AU, a mass roughly double that of Earth's, and a period of 373.5 years. Using these figures, Pickering predicted that the planet would be magnitude 11.5, if like Neptune, or magnitude 13 if dull like Mars. He said that two quick searches he had made failed to find anything. Pickering asked Lowell's help in searching for the planet, but Lowell declined, forgetting to mention that he had been working on the problem himself for more than three years. Lowell worried that he had a serious rival, one whose calculations seemed more advanced than his own. The worry made Lowell pursue the ninth planet with all the energy he could spare from his Mars work.

Lowell hired William Carrigan of the United States Naval Observatory in 1905 to help with the extraction and calculation of the perturbations in Uranus' and Neptune's orbits. In 1909, after four years of part-time work, Lowell dismissed him saying that he had reached his own conclusion without Carrigan's results. Planet X was 47.5 AU from the Sun, had a period of 327 years, was fainter than mag. 13 and had a mass five times the Earth's. Although Lowell wrote all this to Carrigan, he never published it, nor did he use it in any search program.

After a break of a year, Lowell began a new search for Planet X in July 1910, attacking the problem with mathematics. He was assisted by Elizabeth Langdon Williams who had been working in the publications section for the previous five years. Assuming that the planet was in the same plane as Uranus, and that it was 47.5 AU from the Sun, he began work. Using perturbations to the orbit of Uranus, rather than Neptune, he sought to calculate the mass of Planet X, its orbital eccentricity and the longitude of perihelion. On 13 March 1911, Lowell's 56th birthday, he sent a telegram to Carl Lampland, assistant director at Flagstaff, asking him to begin a new search along the ecliptic for the planet. New calculations of the position, to narrow the search region, were to follow in a few days. Lampland proposed that the observatory buy a blink comparator, an instrument which would greatly enhance the probability of finding anything on the photographic plates being used for the search. Lowell bought one and shipped it between Boston and Flagstaff as needed.

The first of Lowell's new calculations was ready in late April, but the photographic search got little telescope time as Mars was nearing opposition and Lowell and everyone else was making plans for observations around that time. Around the same time Pickering published his new calculations, this time for three more planets, Planets P, Q and R. Planet Q was huge; it had a mass 6 per cent that of the Sun, almost enough for a star, but with a highly elliptical polar orbit.

Pickering's new predictions didn't alter Lowell's opinions; instead they spurred him on to work even harder. From 1911 to 1914 Lowell put the observatory's largest instrument, a 102 cm reflector, to work on the project, but as it could see only a small region of the sky it wasn't much help. Lowell then borrowed a 23 cm refractor from Swarthmore College, without telling them exactly what he wanted to do with it. The computations finished in 1914 when the same results were continually recurring. Lowell stepped up the photographic search once more. Lowell left for a European holiday in May 1914 telling the observatory not to mind startling him by sending a telegram 'Found!'.

No telegram arrived. On 13 January 1915 Lowell was to present his work to the American Academy of Arts and Sciences. He told staff at the observatory that it would be nice if they could announce the discovery before that date, but they didn't. Lowell's paper gained little attention; the Academy even declined to publish it. Lowell finally did at his own expense, but he turned from the project and did little about it again.

The photographic search continued into 1916. On 2 July the final plates were made as the borrowed telescope was needed back at the college. Over the five years of the search more than 1000 plates were made. The plates provided a wealth of new objects: there were 700 variable stars, 515 minor planets and two images of the ninth planet. Lowell never examined these plates himself—that was left to a junior assistant, and as the images of Planet X were at magnitudes 15 and 16, five times fainter than the mag. 13 predicted by Lowell, they were not discovered until much later. Lowell was never to know how close he came. On 12 November 1916 Percival Lowell suffered a massive stroke and died at the observatory, aged 61.

Lowell did not want his work to die with him. He left three million dollars to the observatory

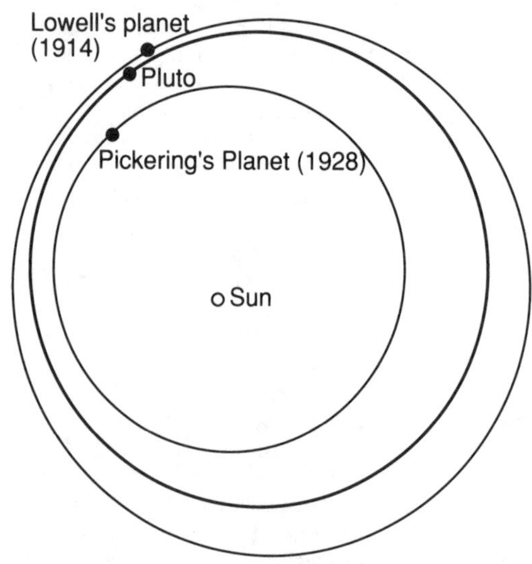


**15.1** A comparison of the orbit of Pluto with the predictions made by Percival Lowell and William Pickering. The positions of the predicted planets are shown for January 1930, along with Pluto's discovery position.

to continue his work, both with Mars and the search for Planet X. In 1927, after a legal battle over who should get what from Lowell's estate, Flagstaff Observatory commissioned a new 33 cm refracting telescope specially made for photographic work. The money actually came from Percival's brother, Abbott Lowell, president of Harvard University. The telescope was ready for use in 1929 and proved to have extremely good optics.

There was only one problem. With all the other work being done at the observatory there was no one to work the telescope. Vesto Slipher, now director of the observatory, wanted to employ someone who was not a research astronomer for the job, as research astronomers usually wanted time for their own projects. In 1928 Slipher had received a letter from Clyde Tombaugh, a student and farm boy from Kansas. Tombaugh was an amateur astronomer who had built his own telescopes, grinding mirrors from sheets of glass. Despite Tombaugh's lack of formal astronomical training, Slipher invited him to join the staff and conduct the search for Planet X, on a trial basis.

Tombaugh, then 22, arrived at Flagstaff in January 1929. The third photographic search for Planet X began on 6 April 1929 with Tombaugh working the 33 cm refractor. Planet X was expected to be fainter than magnitude 13, otherwise it would have been found in the earlier searches, so Tombaugh exposed plates for one hour in the telescope, allowing stars to magnitude 17 to be seen. The use of the blink comparator meant that two plates of the same region of sky had to be made within a week of each other and developed under as nearly the same conditions as possible. Lowell planned to make three plates of each region, two for blinking and the third for verification.

Tombaugh started his search in the constellation of Cancer to the east of Lowell's prediction. Each plate Tombaugh took showed between 50 000 and 400 000 stars—a possible 400 000 Planet Xs. On the nights of 11 and 30 April 1929 Tombaugh photographed Planet X, but the plate of the 11th had cracked due to the cold, and the plate of the 30th was poorly matched in development, so the two plates were never blinked.

Tombaugh was originally only meant to do the photography, but by June 1929 so few plates had been compared that he decided to do it himself, spending half a month at the telescope photographing and the other half comparing the plates. Some of the plates near the Milky Way contained a million star images and Tombaugh had to look at each pair of star images to decide if there was movement. On the plates were images of variable stars, which may be bright on one plate and dim on another, or a minor planet might appear, so that it would cause a false alarm until the amount of movement gave it away. More discouraging than the three days it took to check each pair of plates was the scepticism of other astronomers, many of whom doubted Lowell's theories since doubt had been heaped upon his Mars work.

In January 1930 Tombaugh was once more in the constellation of Gemini, his trip around the ecliptic complete. On 21 January he photographed the region around δ Geminorum, on 23 January he again photographed it, and finally on 29 January took his third shot of the area. The plates of 23 and 29 January were the most similar so it was these he decided to compare. Tombaugh began the comparison on 15 February. Three days later he had compared only a quarter of the images on the plates when he noticed one moving back and forth. Checking his watch he noted that at 4 pm on 18 February 1930 he

**Table 15.1 Pluto—predictions and reality**

| Orbital Elements | Lowell 1914 | Error | Pickering 1919 | Error | Pluto 1930 |
|---|---|---|---|---|---|
| Mean distance | 43.0 AU | 9% | 55.1 AU | 39% | 93.5 AU |
| Eccentricity | 0.202 | 19% | 0.31 | 25% | 0.248 |
| Inclination | ≈ 10° | 42% | ≈15° | 1% | 17.1° |
| Longitude of node | | | ≈100° | 3% | 109.4° |
| Longitude of perihelion | 204.9° | 5% | 280.1° | 16% | 223.4° |
| Period | 282 y | 14% | 409.1 y | 65% | 248 y |
| Perihelion date | February 1991 | | January 2129 | | September 1989 |
| Longitude (1930) | 102.7° | 2% | 102.6° | 2% | 108.5; |
| Mean and annual motion | 1.2411° | 14% | 0.880° | 39% | 1.451° |
| Mass | 6.6 Earths | 843% | 2.0 Earths | 186% | <0.7 Earths |
| Magnitude | 12–13 | 17% | 15 | 0% | 15 |

had found Planet X. Forty-five minutes later, having checked the plate of 21 January and another shot of the region, he was sure he was right; he called to Carl Lampland that he had found it.

Slipher wanted to keep the news quiet until the observations had been thoroughly checked, so he made Tombaugh promise not to tell anyone. That night, as weather conditions made observation impossible, Tombaugh went to the pictures. A month later Tombaugh's parents learnt of the discovery when, following the formal announcement, the local newspaper editor called for their opinion.

Tombaugh was able to photograph Planet X again on the nights of 19, 21 and 22 February. On the night of the 20th Tombaugh, Slipher and Lampland examined the object visually with the 61 cm telescope. They were unable to make out a disc, or any satellites which would confirm it as a planet, but it didn't have any of the characteristics of a comet and was moving much too slowly for a minor planet.

Slipher still did not want to announce the discovery, partly because he wanted no mistakes, and partly because he wanted Flagstaff to be the first with a computed orbit. He invited John Miller of Sproul Observatory to come to Flagstaff to make the calculations. Miller had taught both Slipher and Lampland when they had been at college. On 12 March 1930, at 10 pm Slipher sent a telegram to Harvard College Observatory announcing the discovery. The next day on what would have been Percival Lowell's 75th birthday and the 149th anniversary of Herschel's discovery of Uranus, it was made public. The planet was within 6° of the spot predicted by Lowell in 1915.

The interest in the new planet was incredible. It was news around the world. Lowell and his observatory received greater interest and adulation than it ever had in Lowell's lifetime. Not everyone was happy, though; in order to stop others calculating the orbit, Slipher released only one position for the planet. Even so, two calculated orbits appeared before Flagstaff's. One on 7 April by Armin Leushner, Ernest Bower and Fred Whipple of the University of California, Berkley, gave an orbital distance of 41 AU at an angle of 17° to the ecliptic.

On 9 May British astronomer Andrew Crommelin showed that the planet appeared on a plate taken on 27 January 1927, thus extending the known records by three years and allowing much more accurate calculations to be made. These gave an orbit with a period of 265.3 years and an eccentricity of 0.287. These values showed that the object was definitely a planet and part of the solar system. The values were close to those now known and also close to those predicted by Lowell.

As soon as more observations were made it was found that this new planet was a strange one indeed. Its greatest distance from the Sun was 7375 million km, while its closest approach was just 4425 million km, closer than the eighth planet, Neptune. In fact, for twenty years out of its orbital period of 247.7 Pluto would be the eighth planet from the Sun. That period commenced in 1979 and will continue until 1999; perihelion occurred in September 1989. Along with the orbital eccentricity (0.25), the tilt of the orbit also was impressive: 17.2°, much more than any other planet. This means that the planet often wanders into constellations which no other planet visits.

For the third time in astronomy the problem arose of what to name the new planet. Mrs Lowell proposed Zeus, but this was unsuitable as Jupiter is another name for Zeus. She then decided that the old gods were worn out and proposed Percival. She later changed her mind again and proposed Constance, her own name. Finally she urged Planet X.

Slipher refused to be pressured into a decision. He personally favoured Minerva, goddess of wisdom, but the name was already used for a minor planet. Acceptable alternatives were Cronos and Pluto. Cronos was rejected eventually because it was another name for Saturn and because no one liked the astronomer who suggested it. Pluto was suggested by an eleven-year-old school girl from Oxford, Venetia Burney. She thought the name of the god of the underworld was suitable for a planet so far from the Sun. Pluto was officially proposed by Slipher on 1 May 1930. As a symbol for the planet Slipher proposed the letters PL, the first letters of Pluto and the initials of Percival Lowell.

Once Pluto was shown to have a planetary orbit, debate arose as to who, Lowell or Pickering, had correctly predicted it. Lowell's orbital elements for Pluto were fairly close, except for the actual position in the sky, better overall than Le Verrier's predictions for Neptune. The predicted position of the planet wasn't too far wrong: 6° from the actual location. Pickering's work too was 6° in error. When comparing Pickering's 1919 effort and Lowell's work, Pickering was closer in inclination, ascending node, mass and magnitude while Lowell was closer in distance, eccentricity and longitude of the ascending node. It appeared that the roles of Adams and Le Verrier had been replayed.

As more data on the planet was collected, the question changed from, 'Who was closest?' to 'Is Pluto the planet that either predicted?', and the biggest problem was mass. No matter what telescope was used, Pluto failed to show a disc. Even the 2.5 m Mount Wilson telescope, the largest in the world, failed to find a disc. Pluto had to be small, very small, and there was no way to reconcile the mass required to either Pickering's or Lowell's work. The problem was: how could such a small planet, one which must have less mass than Earth, noticeably disturb Uranus when the closest they ever got was $1.6 \times 10^9$ km?

Was it possible that Pluto really was very dense, and for its small size had a large mass? The matter was not settled until 1978 when Pluto's moon was found. Work on the orbit of Pluto had been continuing at the United States Naval Observatory since 1930 in an effort to refine still further the perturbations in the orbits of Uranus and Neptune and to find, if possible, a tenth planet from its effect on the orbits of the outer planets. To this end, a set of high-resolution plates of Pluto were made using the USNO's 155 cm reflecting telescope at Flagstaff over three nights in April and May 1978.

The plates were transferred to the Washington headquarters of the USNO to be measured. James Christy put one of the plates into an automatic measuring machine to obtain an accurate position for Pluto and noticed something strange. The image of Pluto had a bulge. Checking each of the other five plates made on the same night he found the bulge still there, so it wasn't a fault on the plate. Could it be a star image behind the planet? Checking plates from a sky survey he found that there was no star in that position. Examining the plates made on the other nights showed the bulge changed position, a clear indication that it was indeed a moon.

Christy went back over all the plates of Pluto taken with the 155 cm telescope. Two more instances of the bulge were found on 1965 plates and five more on 1970 plates. Using the data from the plates, Robert Harrington of the USNO attempted to calculate an orbit for the moon. With confirming observations being made with the 4 m Cerro Tololo telescope, Harrington calculated that the moon had a period of 6 days 9 hours and was 20 000 km from the surface of Pluto.

Provisionally named 1978P1, Christy was given the honour of naming the moon. He called it Charon after the mythical boatman who ferried souls across the river Styx into the underworld, a suitable companion for Pluto.

Knowing the orbit of the moon, it was possible for astronomers to calculate the mass of Pluto and the moon. Pluto was found to have a mass of 0.002 Earths and Charon of 0.0006 Earths—tiny masses, much less than our Moon. Observations of Pluto's spectrum let astronomers know that the surface of the planet is covered by methane ice; this gave an indication of the planet's size. Pluto must have a diameter of less than 3000 km and Charon 1400 km.

What is important about these latest results is that they settle once and for all who was right, Lowell or Pickering. Neither! With the tiny mass

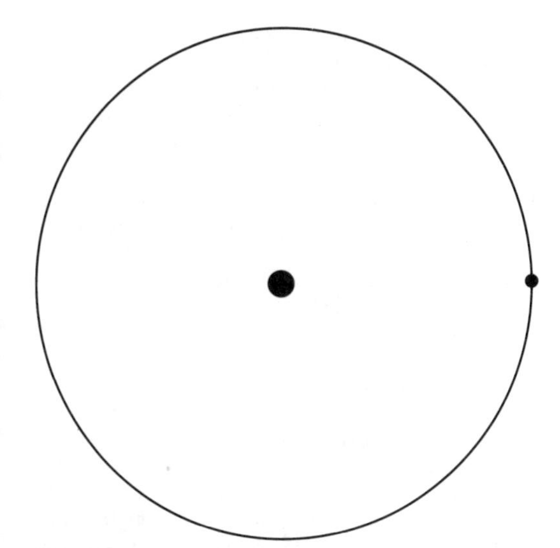

**15.2** Here, to the same scale, are Pluto, Charon and the orbit of Charon about Pluto. This shows the relative sizes of the bodies and the closeness of one to the other. Seen from Pluto Charon would subtend an angle of 3°24′, the largest of any satellite in the solar system.

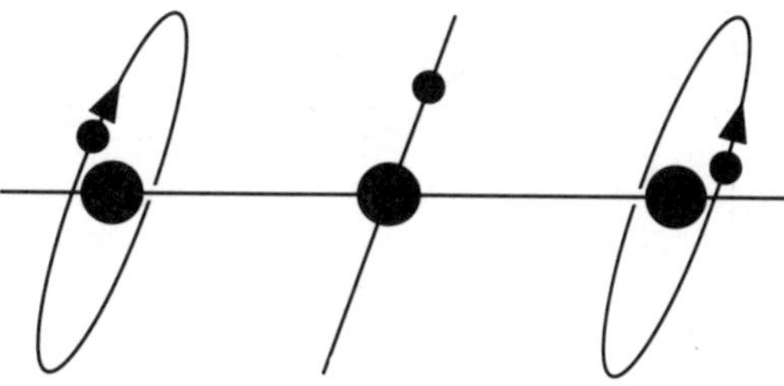

**15.3** For only a brief period during each revolution around the Sun an observer on Earth can see Pluto and Charon eclipsing each other twice every six days. The most recent series of eclipses (right) began during 1985 with Pluto and Charon just clipping each other. In 1987 and 1988 (centre) Pluto and Charon were completely in front of each other, until in late 1990 (left) the series of eclipses finished until the next set begins in the twenty-second century.

it has, Pluto could not possibly be responsible for the disturbances of the orbits of Uranus and Neptune, the measurements Lowell and Pickering used in their works. No, Pluto was discovered by the systematic hard work of a Kansas farm boy, Clyde Tombaugh.

Once Pluto was found to have a moon, research into this remote world began. The orbit of Charon turned out to be tilted to the ecliptic by 118° with Charon orbiting the equator of Pluto. Further observations of the two worlds revealed another interesting feature; not only does Charon keep one face pointed always towards Pluto, but Pluto also keeps one face permanently turned towards Charon: both bodies are in synchronous orbits.

As Pluto orbits the Sun in its 248-year cycle there will be times, as there is with Uranus, when the plane of the orbit is directed towards Earth, during spring and autumn on Pluto. For a few years in each orbit it would be possible to see eclipses of Pluto by Charon and Charon by Pluto. Remembering that the moon was only found in 1978, it was extremely lucky that these eclipses began just over six years later in March 1985 and continued only until October 1990. The next set of eclipses won't begin until the twenty-second century.

These eclipses have given more information about Pluto than just about anything else. By observing the amount of light coming from the planet and its moon as they eclipse it is possible to build up a map of the surface of the two worlds. It had been known for a long time that the brightness of Pluto varied over a small range with a period of 6.4 days; this was what had led to the calculation of the planet's period. By carefully monitoring the light during each eclipse, the brightness of the part of the planet hidden by the moon could be calculated. As the eclipses progressed, the moon gradually passed in front of all the planet's moon-facing hemisphere, repeating the sequence as the cycle of eclipses neared its end.

The exact timing of the eclipses has allowed very precise measurements of the sizes of the two bodies to be made. Pluto is 2284 km in diameter and Charon 1192, almost exactly half the size, by far the largest ratio of planet to moon in the solar system; the Earth-Moon system is next at 1.23 per cent.

Three astronomers, Marc Buie, David Tholen and Keith Horne, working with a telescope in Hawaii have developed maps of the surfaces of Pluto and Charon from the data obtained during the eclipse cycle. The maps produced have, of course, very low resolution, but they do show that both bodies have some bright and dark markings. Pluto has bright polar caps which are three times as bright as the darker markings

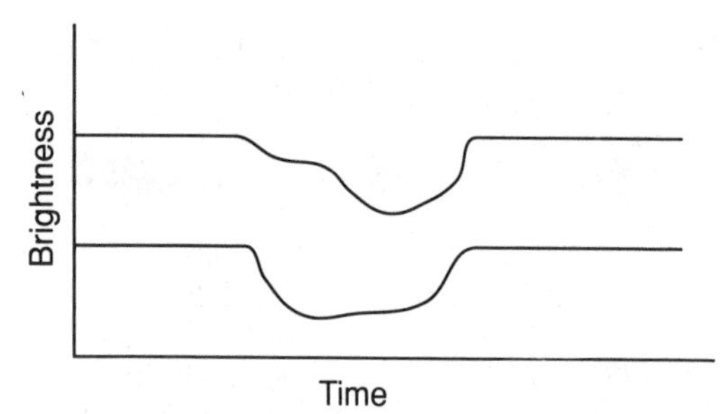


**15.4** When Pluto and Charon pass in front of each other, careful measurements using a telescope and photometer yield light curves such as those shown in this picture. The differing shapes of the light curves are due to different parts of the surface being obscured by the changing portion of the body being covered. By using a computer and a large number of occultation observations, astronomers Marc Buie, David Tholen and Keith Horne have been able to produce maps of the surfaces of the two bodies.

around the equator which appear reddish in colour. The planet's overall albedo is 0.5. Charon is much more uniform in brightness. It doesn't have polar caps or an equatorial band, just striping at the mid-latitudes, a dark band in one hemisphere and a light band in the other, with an albedo of 0.3.

The bright markings on Pluto are thought to be fresh methane ice, recently deposited over older and dirtier water and methane ices. If these are relatively fresh falls then there must be some sort of atmosphere. Discovery of an atmosphere involved finding an occultation of a star by Pluto. Because of its small size, such occultations are rare, but one occurred over the Pacific in June 1988. Astronomers from around the world gathered in Australia, New Zealand and Pacific islands to observe the event. Again the Kuiper Airborne Observatory was used to observe the event from above most of the Earth's atmosphere.

The occultation was observed successfully by many groups. When the data were analysed it was found that the star started to fade when 1500 km from the centre of Pluto, and that the light of the star dimmed gradually before being completely hidden. According to Jim Elliot, the scientist who organised the KAO flight, the results show that Pluto's atmosphere is around 300 km thick with a thicker 46 km layer of haze above the planet's surface.

The orbit of Pluto has long puzzled astronomers. If the planet can come closer to the Sun than Neptune, why hasn't it at some time come close enough to that planet to be captured or to be flung from the solar system? The reason is that the orbital period of Pluto is exactly 1.5 times that of Neptune. Because of this, whenever Pluto is at perihelion, Neptune is safely in another part of its orbit and unable to greatly influence the planet. It is for this reason that after Voyager 2 reached Neptune it couldn't be directed to Pluto, which was then only a month from perihelion.

The viewing of Pluto may seem a daunting task for the beginner, and most amateurs have never seen nor even bothered to try to find the most distant known planet. This is surprising, although it must be admitted that Pluto isn't exactly the most exciting of the planets to view, being no more than a faint star at the limit of large amateur telescopes, but to find this world at the extremes of our solar system is a challenge that one cannot ignore if a complete tour of the planets is planned.

Pluto reached perihelion in 1989, and is heading towards aphelion in 2113, so the next few years are the best time to observe Pluto as its perihelion magnitude is slightly brighter than 14. Once outside Neptune's orbit it will become fainter than 14 and reach about 14.5 at aphelion.

The minimum telescope aperture required to view Pluto is at least 250–300 mm. You also need a good dark sky, and knowledge of the star field in which the planet lies. Assuming access to a suitably large telescope and an observing site free from light pollution, all that is left is the ability to separate the planet from the stellar background. This is not easy, unless you have access to an expensive photographic atlas.

The limiting magnitude of popular star atlases varies from around 6 down to 9.5. If you consider there are about 3000 stars to magnitude 6, some 325 000 to magnitude 10, and a staggering 32 000 000 to magnitude 15, you will appreciate that there is little chance of finding a magnitude 14 planet situated among a field loaded with stars of similar brightness, especially since no matter how much magnification is used you will not be able to see a disc.

There are several ways to locate Pluto. Firstly, if you have an equatorial mounted telescope with setting circles, you can 'dial up' the co-ordinates as published in an almanac. Secondly, you can take the position given in an almanac, access a

**15.5** Two photographs of Pluto taken five days apart. The first photograph, on the left, was taken on 12 July 1985, the second on 17 July. The motion of the planet is obvious over the five days, but without this movement the planet would be indistinguishable from the stars in the field. PHOTO CREDIT: STEVEN QUIRK

photographic atlas in a university or college library and sketch a map of the surrounding area. Or thirdly, use a camera mounted at the prime focus of a telescope.

All methods will work equally well, but they need two sessions at the telescope. It is best to find a bright star near Pluto, then 'star hop' until you feel that the planet is within the centre of the field. With visual observation you will need to draw all field stars, being attentive with their positioning; follow up after a week or so with another drawing, then compare the two. Pluto is of course, the one that moved. As long-exposure photographs will reach fainter magnitudes than the naked eye, you will be able to use a smaller aperture telescope than is required to see the planet visually. Exposure time, film speed and telescope aperture are all factors that will determine your success. Once you find Pluto, you will have the satisfaction of knowing that you have joined a small band of amateurs who can boast that they have observed all the known planets.

With the size of the solar system now set at 12 000 million km from side to side, once more, the question arises: is *this* the last planet? Can anyone hope to emulate Herschel, Galle and Tombaugh and again discover a planet? Tombaugh did not cease his work. Following a break after the announcement of Pluto, he was back at work comparing stars from where he left off with the search. When he finally ceased comparing plates in 1943, Tombaugh had examined the entire sky visible from Flagstaff, forty-five million stars in total. In fourteen years, Tombaugh found a comet, nearly 800 minor planets, a cluster of 1800 galaxies and one planet.

From his work it is certain that no planet brighter than magnitude 16.5 exists in the regions he searched. A planet the size and colour of Jupiter would have been found if it were closer than 500 AU, a planet the size of Pluto would have been found to a distance of 60 AU. Remembering that both Lowell and Pickering based their work on perturbations to the orbits of Uranus and Neptune, and that Pluto turned out not to be the planet causing these wobbles, could these disturbances lead to the discovery of another planet? Many astronomers believe that they will.

Astronomers from around the world have begun the task of looking at past records of the motions of the outer planets and attempting to calculate the position of the object causing the effects, assuming that the effects are real. Some astronomers attribute the apparent wanderings in the orbits to poor observations in earlier times; indeed more recent observations do lead to reasonably good orbits. Others, though, consider that the effects are real. Planets which have been predicted by astronomers range from those with orbits taking over a thousand years to smaller bodies much closer. One set of figures predicts a planet with a period of 251 years and an orbit almost matching Pluto's, but as this planet would have a magnitude of 12.2 it is hard to see how it wouldn't have already been found.

It is not difficult to believe that if another planet exists it should have been found already. It took Joseph Adams just two years, part-time, working with pencil and paper, to calculate the position of Neptune with perhaps a million calculations being needed. Given the power of computers routinely available today, such a calculation would take seconds, or only minutes on some of the more powerful home computers now available, once a suitable method had been arrived at.

The search for an even more distant planet has led many scientists to adopt the brute force method: simply check everything, and if it's there it will show up. Such a program has been undertaken using the 600 000 objects recorded by the United States' Infrared Astronomical Satellite (IRAS). The astronomers in the search are looking for faint, cool objects; they then see if the object appears on the Palomar sky survey plates; if it doesn't, it is investigated further. In this way they hope to find the planet.

Other scientists do not believe a tenth planet exists, or at least that there is no data to suggest that it does. The astronomers at the Jet Propulsion Laboratory whose jobs it was to calculate extremely accurate positions of Neptune and Uranus for the Voyager 2 mission achieved remarkable success using only observations of the planets since 1910. By ignoring the earlier measurements they removed the problems which may have been caused by inaccurate observations or recording. The discrepancies in the positions of the planet based on earlier sightings are small enough to be observational errors.

The situation with regards to a tenth planet is, at present, still open. The evidence on either side of the argument is shaky, but it seems clear that if there is a tenth planet, it is unlikely to be found in the near future, and if it is, it probably won't fit any of the predictions.

# 16

# Comets and meteors

The planets and their moons are the staid, normal part of the solar system, but there is one part which fascinates astronomers and non-astronomers alike. Infrequent visitors to the region around the Earth and the Sun, comets are an important part of the Sun's family, perhaps the last traces of the material which first came together to form the planets.

Throughout history, the passage of comets through the solar system has been observed with interest. In the dark skies of the pre-industrialised world, one comet a year would have been seen, and once a decade a really bright comet would come along. The arrival of a comet was always unannounced and its appearance variable, perhaps it was this unpredictability, unlike everything else in the skies, that led to comets being associated with disaster.

Comets were a worry for early philosophers. Unable to deny their existence, yet sticking to their concept that the heavens were fixed and unchanging, they decided that comets were a phenomenon of the atmosphere. It was the Great Comet of 1577 that turned the tide. This comet was so bright that its discovery is credited to no one; too many people found it simultaneously. Observers at that time paid much attention to the comet in an effort to determine once and for all what it was. At his observatory on Hven, Denmark, the great astronomer Tycho Brahe observed the comet and made very accurate measurements of its position. Using these measurements Tycho was able to show that not only was the comet not within the Earth's atmosphere, it was further away than the Moon. This was a major blow to the fashionable theory at the time that the planets orbited fixed to crystal spheres, pushed by angels. This comet was clearly passing through the spheres, and without the sound of breaking glass.

Though Tycho's work was important, in the investigation of comets one name is foremost. Edmund Halley was born in London in 1656. By the time he reached Oxford University he was an accomplished scholar in both mathematics and music. In 1676 Halley published his first astronomical work, a short paper on the orbits of the planets. When he was 26, Halley observed the Great Comet of 1682, but his interest was purely curiosity. In 1704 Halley was appointed professor of Geometry at Oxford University and it was then that he started his work on comets. Halley was a great supporter of Isaac Newton and his theories; indeed, it was Halley who first persuaded Newton to publish his works, so when he began to investigate comets he naturally applied Newton's laws. Halley found that Newton's gravitational theory meant that a comet could have one of only four paths around the Sun: a circle, ellipse, parabola or hyperbola. Clearly observations showed that the orbits of comets were not circular, so which of the other shapes could they be?

Halley concluded that the exact shape of the orbit would depend upon the amount of energy the comet had; a large amount of energy and the path would be a hyperbola, a little and it would be an ellipse. The path would only be a parabola if it had a very specific energy and if it were not disturbed on its journey around the Sun, events so unlikely that parabolic orbits could be ignored. The differences between the orbits were obvious. An elliptical orbit is a closed path around the Sun; any comet following such a path would return time after time, in the same way that the planets follow their elliptical orbits. A hyperbolic orbit, on the other hand, is an open path; a comet on such a trajectory would visit the Sun once and disappear from the solar system.

**16.1** Comet Halley during its 1986 return was a spectacular sight from country locations. This photograph was taken on 23 March 1986 and shows Halley's comet in the constellation of Cetus. Note the amount of structure that can be seen in the comet's tail and the distance over which it stretches. PHOTO CREDIT: KEN WALLACE

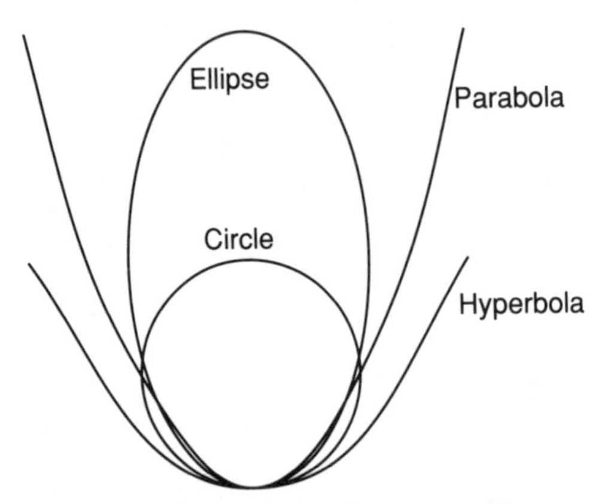


**16.2** All orbits must be based upon one of four simple figures known as conic sections. The circle, ellipse, parabola and hyperbola can all be made by slicing a cone in a certain manner. The circle and parabola are too unstable to survive long in any real system, so the only shapes observed in nature are ellipses and hyperbolas. When applied to comets this means that comets following an elliptical path will return to the Sun again and again, while hyperbolic orbit comets will pass the Sun once and never return. The problem faced by astronomers is that near to the Sun all the conic sections have much the same shape, making it difficult to distinguish between orbits.

The problem Halley faced was the similarity between elliptical and hyperbolic orbits in the region near the Sun, the only region in which comets could be observed. Halley began exacting calculations on the orbits of comets, beginning with the assumption that the orbits were indeed parabolic, and then seeing whether an ellipse or hyperbola fitted better. During these calculations, which involved looking at observations of many comets seen over hundreds of years, Halley noticed that the orbits of the comets of 1531, 1607 and the comet he had observed in 1682 were remarkably similar. He concluded that this was the comet he had been seeking, one which followed an elliptical path and which had returned to the Sun.

Many astronomers were sceptical, saying that it was just a coincidence, but Halley predicted that the comet would return in 1758. Halley did not live to see his prediction verified, however. He died in 1742. Throughout 1758 amateur and professional astronomers alike scoured the skies for the comet. On Christmas Day an English amateur, Johann Palitzsch, found the comet. Observations of the comet's orbit verified that this was indeed Halley's comet, a name by which the object is still known.

Halley's comet was the first to be recognised as a short period comet, one which returns to

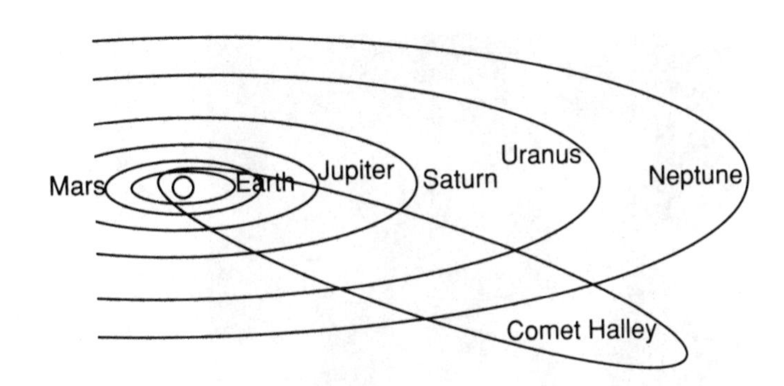


**16.3** The orbit of Comet Halley around the Sun takes it from outside the orbit of Uranus to within the orbit of the Earth. The orbit is tilted to the plane of the solar system by 162° so the comet travels through the sky in the opposite sense to the planets, taking 76.1 years to complete one orbit.

the Sun with a period of less than 200 years. It wasn't until 1820 that another comet joined Halley's, Comet Encke was found to have a period of just 3.3 years and was much fainter than Halley's. Many other comets have been seen since Halley made his prediction, some of them also orbit the Sun in elliptical orbits while others are one-time visitors following either very long elliptical orbits or hyperbolic orbits so that no-one can hope to ever see them again.

Each year a few comets are visible from Earth, although some require a telescope as they are only faint objects far from the Earth and Sun, or small like Comet Encke. About once each year a comet appears which is bright enough to be seen with the naked eye from a dark observing site. Such comets usually go unnoticed in the city because of the bright skies caused by city lighting. Once each decade or so, though, a comet bright enough to be seen from the city will appear, such as Comet West in 1976, or Halley in 1986 (from the southern hemisphere at least).

It is in the discovery of comets that amateur astronomers make some of their most valuable contributions to professional astronomy. The study of comets is an important part of modern planetary physics, but astronomers need to observe each comet for as long as possible to gain the most information, which means discovering the comet as early as possible.

Professional astronomers rarely have sufficient time or equipment to scan the whole sky continuously searching for a new comet. Amateur astronomers watching the sky each night are much better placed. Throughout the world dedicated amateur astronomers search each night for a tiny fuzzy patch which might signal the appearance of a new comet. Some amateur astronomers have had amazing success. At the time of writing, William Bradfield of South Australia has discovered fourteen new comets, mostly from his home. Bill's chief competition comes from United States' husband and wife team Eugene and Caroline Shoemaker; they use a stereomicroscope to compare Schmidt camera plates against earlier plates of the same region of sky and they often come neck and neck with Bill.

The differences between these comet hunters, aside from the obvious visual versus photographic technique, is the area of sky searched and the discovery magnitude of each comet. The Schmidt camera is used under dark-sky conditions and typically comets are picked up at magnitudes 14 to 16. Visually, Bill Bradfield searches in the vicinity of the Sun, just after evening or before morning twilight, where comets are brighter and more easily visible in modest amateur instruments. The two methods in effect complement one another; when a comet is a long distance from the Sun it is faint and requires professional equipment, while Bill's technique covers an area of sky unsuited to photographic techniques by its proximity to the Sun.

The work of amateurs is recognised by the International Astronomical Union by naming comets after their discoverers. The IAU will allow up to three discoverers to be associated with a comet, provided that all discovered the comet within a few hours of each other. Thus comets end up with names such as Comet Bradfield 1986, or Comet IRAS-Iraki-Alcock named after one satellite and two astronomers. Discovering a comet is the only way of being sure that an astronomical body will be named after you.

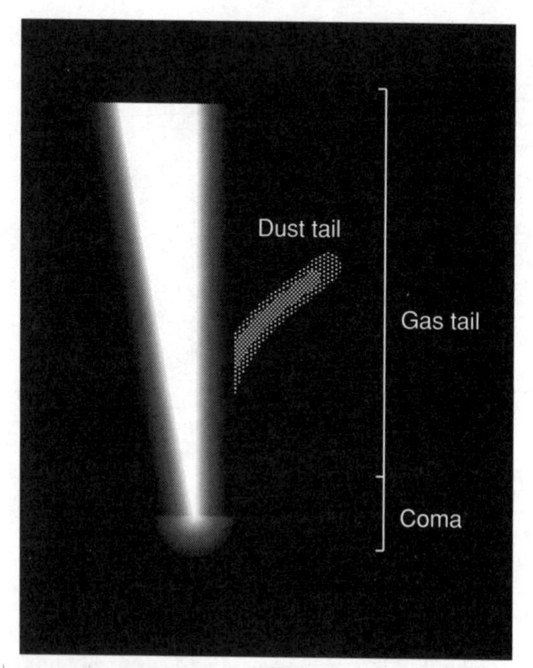


**16.4** The three main divisions of a comet visible from Earth are the *coma*, the region surrounding the nucleus, the *gas tail* made up of gases escaping from the nucleus and pushed by the solar wind so that it always points away from the Sun regardless of the direction of travel of the comet, and the *dust tail*, a tail of heavier particles which trails behind the comet.

Comets are also designated by the year of discovery, followed by a letter to show the order in which they were found in that year; therefore 1990b was the second comet to be discovered or recovered in 1990. To complicate matters, once the orbit has been calculated, the comet is assigned a Roman numeral to show the order of perihelion passage; this new tag may not necessarily coincide with the year of discovery. Therefore Comet 1990 IV would have been the fourth comet to pass perihelion in 1990.

Very few comets are found by accident; most are the result of systematic searching. If you intend to get serious about comet hunting you will need to be prepared for many fruitless hours at the eyepiece. It has been estimated that the average comet hunter spends about 200 to 300 hours before a discovery; sometimes this figure may be as high as 1000 hours.

Any instrument can be used for comet hunting. Binoculars, a small refractor or a Newtonian all work equally well. The telescope should have a low magnification—as a rough guide, choose an eyepiece that gives a power equal to the objective diameter (in millimetres) divided by five. For example, for a 150 mm aperture telescope the best power would be 30x.

The telescope mount is an important consideration, and although an equatorial is preferred for most astronomical work, the altazimuth is ideally suited to the sweep technique used in comet hunting. It is handy if the altitude axis can be locked while sweeping, but the prime requirement is a sturdy mount; the popular and economical home-made Dobsonian mount is well suited.

The sweep technique is simple and does not take long to master. The search should start just after evening twilight. The observing site needs a good, unobstructed horizon, as it is in the area close to the Sun that the search should begin. Start by positioning the telescope about 60° north or south of the sunset or sunrise point and slowly sweep the telescope in azimuth until a full 120° has been covered. The telescope is then raised in altitude by half the field diameter, and slowly moved in the reverse direction. The process is repeated for the next hour or two. Each sweep should be as slow as possible; if the motion is too rapid a faint nebulous haze can easily be missed. In the morning the procedure is reversed, the sweeps starting at an altitude of about 60° and finishing at the horizon just before twilight.

A good star atlas is an essential tool for the comet hunter, to assist in the identification of faint nebulae, galaxies and clusters. Most new comets will not have developed tails, and if a nebulous haze appears in the field, the position should be noted and checked against the atlas; the sweeps become faster as you become familiar with the common comet 'lookalikes'. If a suspect comet is found, it should move in relation to the background stars in an hour or so.

Knowing what comets do and what they look like, the natural question is: what are they? When observing a comet there are two main parts to note: the *head* or *coma* region, which is the brightest part of the comet, and the *tail*, which stretches through space over many millions of kilometres. It was realised very early on that the tail of a comet does not stream out behind it, but always points away from the Sun, no matter in what direction the comet is travell-

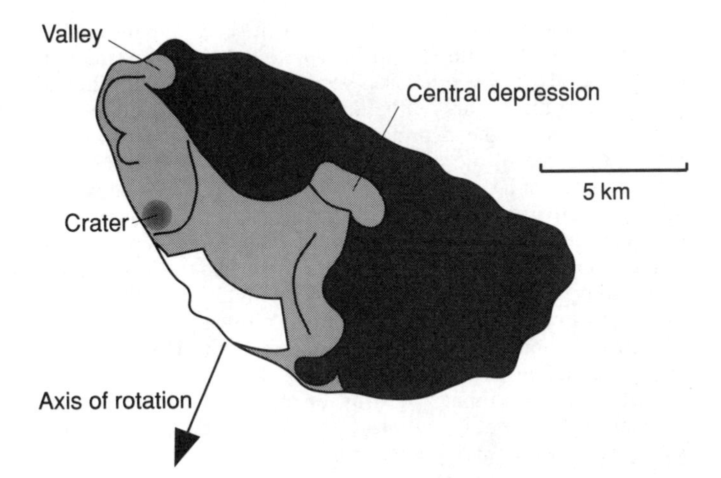


**16.5** The European Space Agency's spacecraft Giotto travelled close to the nucleus of Comet Halley during its 1986 apparition. Using data obtained during the flyby, astronomers were able to construct this view of the comet's nucleus.

ing. This means that whatever it is making up the comet's tail, it must be tenuous enough to be blown by the solar wind.

The substantial bit of the comet is the *nucleus*, that small object shrouded within the coma which provides the material for the rest of the show. When it is in the outer reaches of the solar system the nucleus is a small, hard lump which is undetectable from Earth. It is only as it nears the inner solar system, about the distance of the minor planets, that the heat from the Sun is sufficient to begin melting and vaporising the materials of the nucleus.

As the materials of the nucleus begin to vaporise, the pressure of the solar wind blows them away from the comet and out into space, forming the distinctive cometary tail. This tail, although it stretches for many millions of kilometres, is so tenuous that it would be indistinguishable from a vacuum here on Earth. It is only the action of the charged particles in the solar wind, causing it to glow, which makes it visible at all.

It was Fred Whipple of Harvard University who, in 1950, invented the model of a comet's nucleus we use today. Called the *dirty snowball model*, Whipple hypothesised that the nucleus of a comet would contain roughly equal amounts of ice and dirt. It is the ices which vaporise as the comet approaches the Sun; the dust is then released to make a secondary dust tail. All of this was conjecture as the nucleus of a comet is hidden in amongst the glare of the coma gases.

The nuclei of comets remained invisible until 1986 when Halley's comet was visited by spacecraft from Japan, the Soviet Union and the European Space Agency. The Soviets used two spacecraft bound for Venus to fly within 8000 km of Halley's nucleus. VEGA 1 and 2 photographed the envelope of dust and gases surrounding the nucleus, and were able to view several bright plumes of gas erupting from it. Using the data obtained by the VEGA craft, the next spacecraft was able to approach much closer.

The ESA's craft was named Giotto, after an Italian painter who had included Halley's comet in a painting as the star of Bethlehem*. Unlike the two Soviet craft which had to continue on to Venus, Giotto was dedicated to the task of observing Halley's comet and so could travel as close to the nucleus as possible without fear of its destruction ruining the meeting. So sure were scientists that Giotto would probably be destroyed by its encounter that the craft did not have recording apparatus; all data had to be transmitted directly back to Earth, to the Parkes radio telescope in Australia.

Giotto wasn't destroyed, though its encounter with the comet did leave it damaged and pointing in the wrong direction. Photographs returned from the spacecraft showed Halley's nucleus immersed in a thick cloud of mist and dust. The nucleus was much darker than expected, having an albedo of only 0.03, and is an

* Comet Halley could not have been the star of Bethlehem, it passed by the Earth in 12 BC.

irregularly shaped body 16 × 8 × 8 km in size. The dark coating is thought to be dust left behind as the water and other ices evaporate into space, or it could be tars left behind by chemical reactions between the hydrocarbon compounds making up some of the comet's ices.

The nucleus of a comet turns out to be a place of violent activity, erupting jets of gas streaming out from the core to form the cloud of gases which will eventually be the tail. Each trip around the Sun causes some more of the material of the nucleus to evaporate and be lost to space, so a comet must get smaller each time it passes perihelion. The calculated mass lost is around ten million tonnes per trip, or about 0.1 per cent of the comet's total mass. At this rate a comet would only survive a thousand perihelion passages before being reduced to nothing.

This loss of material explains why the regular comets are much less bright than the unexpected visitor. Regular comets have been around the Sun many times, so most of their material will have already been lost, leaving little to form spectacular tails. Only Halley's comet still has enough material to put on a spectacular show. New comets, however, will have a full load of material to vaporise, producing long bright tails as they pass the Sun.

The jets of escaping gas are violent enough to disturb the motion of a comet, acting like a rocket engine to push the comet away from its predicted orbit. These effects can only be guessed at, and so they make the exact prediction of the position of a comet uncertain. The disruption is not enough to be noticed by a telescopic observer, but it can make it difficult to aim a spacecraft precisely.

The tail of the comet is its most intriguing feature. Observing the tail of a comet as it passes the Sun will show many fine structures which change from day to day in unpredictable ways. Near the surface of a comet's nucleus, the density of gas is tiny, around one-millionth as much as the density of our atmosphere; further down the tail, the density is even less. About 100 000 km from the nucleus the effects of the solar wind are such that the atoms making up the tail have their electrons stripped off and become ionised. It is this ionisation that makes the tail of a comet glow. The tail is so tenuous, that over its entire 100 million km length there is about as much material as in a couple of large office buildings.

On numerous occasions the Earth has passed through the tail of a comet. The most publicised passage was in 1910 when the Earth passed through Halley's comet's tail. Throughout much of the 'educated' world, people were panicked by reports of poison gases in the tail of the comet and many people made a lot of money selling remedies for the expected catastrophe. It is true, there are traces of cyanide and other poisons in cometary tails, but there is certainly nowhere near enough to cause any sort of problems as the Earth passes through. Naturally enough, the incident passed by without any problems at all.

Along with the plasma tail comes the comet's other tail, the dust tail. Dust tails are easily distinguished from gas tails: they aren't as long, they are a different colour as they reflect sunlight rather than glow themselves, and they are curved. The particles of dust released from the nucleus are much heavier than the atoms of gas so the solar wind has less effect on them and the tail tends to trace out the path of the comet through the sky. Measurements of the dust tail suggest that it usually has as much material in it as there is in the gas tail, indicating that the comet is roughly a 50–50 mixture of ices and dust, as the dirty snowball theory suggests.

Cometary photography is a rewarding, though sometimes a difficult, pastime. For bright comets, satisfactory photographs can be taken using a standard 50 mm lens working at its fastest *f* stop. With the camera mounted on a tripod and loaded with fast black and white or colour film, an exposure of a few minutes or less will record most of the comet. For longer exposures, the camera should be mounted 'piggy-back' on a telescope with motor drive and guided on a suitably bright star, to prevent trailing caused by the Earth's rotation.

For detailed photographs of a comet's fine structure, a telephoto lens mounted piggy-back on a telescope will work extremely well. As a comet's movement is independent of the apparent motion of the stars, the telescope drive is practically useless in tracking for detailed photographs. If the comet is not moving too fast it is possible to use the fast and slow controls on the telescope drive to keep the nucleus centred; if this proves difficult, manual tracking may be necessary. Close-up photographs of the coma region may be taken using a camera mounted at the prime focus of a telescope, a separate guide scope being used to track the comet. The real difficulty with any form of narrow-field photo-

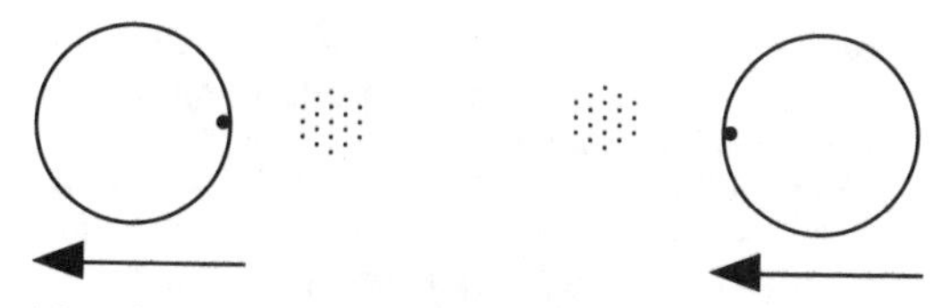

**16.6** Near the time of sunset (left) an observer is travelling in the trailing side of the Earth so to enter the atmosphere a meteor must first catch up with the Earth. At sunrise (right) an observer is on the leading side of the Earth and so will run into meteors much more readily. This explains why more meteors are seen in the hours leading up to dawn than after sunset and why meteors seen in the morning are usually faster than those seen during evening.

graphy is that the comet's nucleus or central condensation is not always well defined, and accurate guiding is often near-impossible.

Observing and photographing comets is one of the more interesting areas of planetary astronomy. As a comet races towards perihelion and then back to the outer solar system, the changes and development in the tail and coma are fascinating to observe. Your name may never be immortalised unless you work hard and discover a comet, but most amateurs are perfectly willing to enjoy other people's finds.

The dust left behind by the comet doesn't just disappear; it hangs around in the solar system following roughly the same orbit as the comet which left it. Just as the Earth occasionally passes through the tail of a comet, so it occasionally passes though the stream of dust left behind by a comet. As the dust is spread along the comet's orbit, such passages are often a yearly event. When the Earth passes through a stream of this dust, grains hit the Earth's atmosphere, heating up and ionising the air around them until it glows. Such events leave a bright streak across the sky, a 'shooting star'. Meteors, as they are more properly known, can be seen on any clear night, and from a dark-sky location an average rate of five to ten per hour can be expected. These are known as sporadic meteors; they follow no set pattern and orginate from random points in the sky. After midnight the rate of sporadic meteors increases, sometimes to two to three times the evening average. The rate changes because of the Earth's motion; when evening falls, an observer is looking in the opposite direction from that in which the Earth is travelling, and any meteoroids must catch up to the Earth before entering the atmosphere. After

**Table 16.1 Meteor showers throughout the year**

| Name | Best date | Notes |
|---|---|---|
| Quadrantids | 3 January | Medium speed, blue |
| Corona Australids | 16 March | |
| Lyrids | 21 April | |
| η Aquariids | 4 May | Very fast moving objects associated with Comet Halley |
| Lyrids | 15 June | Blue |
| Ophiuchids | 20 June | |
| Capricornids | 25 July | Very slow, yellow |
| δ Aquariids | 30 July | Slow, long paths, two radiants |
| Piscis Australids | 30 July | |
| α Capricornids | 1 August | Yellow |
| ι Aquariids | 6 August | Two radiants |
| Perseids | 11 August | |
| κ Cygnids | 20 October | Bright, exploding |
| Draconids | 9 October | |
| Orionids | 20 October | Very fast, associated with Comet Halley |
| Taurids | 31 October | Slow, very bright, associated with Comet Encke |
| Andromedids | 14 November | |
| Leonids | 16 November | |
| Phoenicids | 4 December | |
| Gemenids | 13 December | White, associated with the minor planet 3200 Phaeton |
| Ursids | 22 December | |

It is impossible to predict how spectacular a meteor shower will be as they change greatly from year to year. The date given is close to the centre of the period, but nights on either side of this date may also give good views.

midnight the observer's hemisphere faces the direction of Earth's travel, and therefore even a stationary meteoroid may hit the atmosphere at a greater speed than one that must chase the Earth. This is shown in figure 16.6.

This effect can be compared to a collision in a car travelling at 60 km per hour. For another vehicle to strike the rear of yours it must travel faster than you, and the impact, if the other driver was moving only 5 km faster than you, would be slight. If you were to a strike a stationary vehicle in front of you at the same speed, the damage would be much worse. The analogy does not always hold true, however, as some meteoroids may strike the receding atmosphere faster than some from the preceding side.

The connection between meteors and comets is clear when meteor showers are considered. These are annual events where the number of meteors seen by the naked eye is much higher

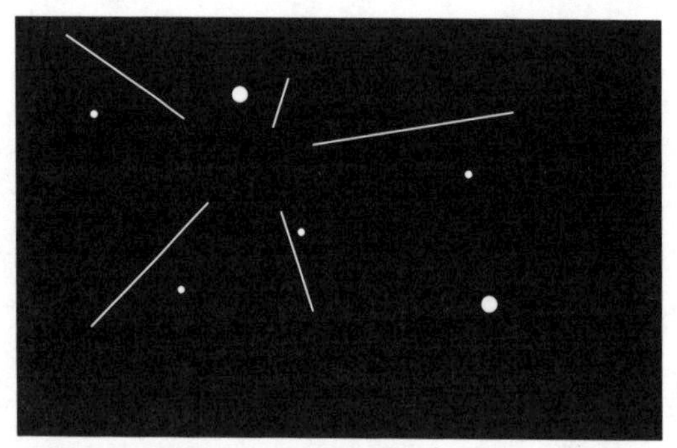

**16.7** Meteors which are part of a shower travel along parallel paths. When these strike the upper atmosphere the effect of parallax makes it look like all the meteors are radiating from one point in the sky, the radiant. If the paths of a number of meteors in a shower are plotted on a star map, all their tracks will point towards one small region indicating that all the meteors have the same origin. By doing just this, amateur astronomers in the southern hemisphere have discovered many previously unknown meteor showers.

than the usual number of sporadic meteors. The annual meteor showers are listed in Table 16.1. These showers mark the annual passage of Earth through a particularly dense cloud of dust particles left by a comet.

A meteor shower can be a spectacular event, lasting from a few hours to days, depending on the width of the stream. In November 1833, an estimated 240 000 meteors fell over a nine-hour period, and in 1966 a peak hourly rate of 100 000 was averaged in a shower over Arizona. The Draconid or Giacobinid meteor shower is associated with the short period comet Giacobini-Zinner. In the years 1933 and 1946, splendid displays were seen at a rate of 5000 meteors per hour. The short duration of the shower, typically about one hour, shows that the width of the stream is only in the order of 150 000 km. Imagine meteors falling like rain—surely the highlight of any amateur's observing career.

It was the 1833 event that caused scientific interest and changed science's attitude to meteors; up until that time they were considered a purely atmospheric event. It was noted that the meteors seemd to radiate from a common point, the constellation of Leo. As the Earth's rotation carried Leo across the sky the meteors still appeared to radiate from the same point. The only explanation for this was that the meteors were not of an atmospheric but an interplanetary nature.

Astronomers term the apparent focal point where showers orginate the *radiant*. The radiants are named after the constellation in which they appear, or the comet that is associated with the shower. The showers associated with the 1833 and 1966 events are known as the Leonids and Perseids. These occur annually, as do most showers, but with varying degrees of intensity. The Leonids are best every 33 years, showing that the dust is not distributed evenly along the comet's orbit. The Perseids, as a rule, maintain a constant rate each year.

Members of meteoroid streams travel through space in parallel paths. The apparent divergence from the radiant is only an illusion, due simply to the effect of perspective. When you look down a long straight road, the trees or buildings on either side of the road converge in the distance; the effect is the same for a shower happening 100 km above the observer.

The relative speed with which a stream contacts the atmosphere will determine the appearance of each shower. Some are swift and others are slow moving, depending on the direction of travel of the stream in relation to the Earth. A stream does not necessarily follow the planetary convention of moving around the Sun in an anticlockwise direction, and consequently a stream moving in the opposite direction will contact the Earth with an impact speed equal to the Earth's plus that of the stream, creating very fast meteors. A stream moving in the same general direction will need to catch up to the Earth and its relative speed will be equal to its own less that of the Earth, resulting in slow-moving meteors.

As mentioned earlier, observing meteors is fun, but for your work to be of any real value it is essential that you be familiar with at least the brighter stars. Observation of apparently sporadic meteors has led to the discovery of new shower radiants. The method involves plotting every meteor seen on a star chart. After the observing session it should be clear if the meteors were sporadic or originating from a radiant. Amateur astronomers in Australia have been active in this area during recent years and have found many previously unknown southern hemisphere showers.

In the early days of the Apollo missions, amateurs kept a watch on metoer showers for

NASA, as it was feared that a severe storm could jeopardise the launch of a spacecraft and the lives of astronauts. More recently, data collected by an Australian group, the National Association of Planetary Observers (NAPO), on the $\pi$ Puppid meteors were used to help redirect the spacecraft Giotto from Halley's comet to observe comet Grigg-Skjellerup.

Very bright meteors that reach negative magnitudes are called *fireballs or bolides*. These usually leave a trail that lingers long after the meteor has vanished from sight. The trail itself is known as a *train* and exceptionally bright fireballs can leave a train that may last for several hours. While observing with friends in the country during January 1982, we were surprised to see the entire countryside light up. Drawn from the eyepiece, our attention was focused on a fireball that approached from behind and in a few short seconds exploded overhead. We were stunned. Before looking skyward, our first thoughts were that a vehicle was approaching from behind with headlights on high beam, so bright were the surrounds. The train from that stunning event remained for fifteen to twenty minutes as a sliver as bright as the Milky Way. It then began to distort into a serpentine shape, and faded below naked eye visibility after a further ten minutes.

Pointing a fixed camera to the sky is a simple method of recording meteors, and good results may be obtained with a standard or wide angle lens and fast film. It is futile trying to photograph sporadic meteors as the chances of capturing them with the camera pointed in a random direction is small. To show the futility of this type of work, a fellow amateur specialising in galactic novae search, has over a three-year period taken several thousand photographs with a standard 50 mm lens. Altogether, on these photographs, he has only recorded two meteors!

Obviously, the best way to increase the odds is to point a camera on a tripod in the direction of a known radiant when a shower is due. The camera should be offset from the radiant point, as most meteors will be seen outside this area. The lens should be set at its fastest speed, *f* 2 or faster, and exposures should not exceed the sky fog limit of the observing location. The beauty of this type of work is that the camera can be left unattended for the duration of the exposure, leaving you free to relax, make a cup of coffee, or observe with a telescope. Even if you get a negative result from your evening's work, you will still have some good pictures of star trails.

The recording of a fireball on film can be of value if another amateur has photographed the same event from a different location. Some amateurs carry out this regular patrol work using wide-angle lenses stopped down around *f* 16 to enable very long exposures. Only very bright fireballs are recorded in this way, and by comparing the two photographs it can be determined by triangulation where possible remnants or meteorites may have landed. A ground search can then be instigated in the hope of recovering any fragments.

Meteor observing, like most amateur activities, needs the co-operation of many individuals. It is fine to work alone, but for your work to be of value it should be reduced and recorded by a group specialising in this field. Contact with such bodies can be made through your local amateur group.

A quick study of the orbits of meteors and comets shows that they originate in many different parts of the solar system. Looking at the 113 known short-period comets, many of them have periods of between five and eight years, indicating that their greatest distances from the Sun are near to the orbit of Jupiter. Among the short period comets, Halley, with its aphelion near Neptune, is the exception rather than the rule. Clearly Jupiter is not the source of these comets. Furthermore, a source within the solar system fails to explain the long-period comets which have nothing to do with Jupiter or any other planet.

Another problem with cometary orbits is that they are inherently unstable. The orbits of the planets are stable: they follow staid paths around the Sun and their orbits don't cross—even Neptune and Pluto never come close to each other. The orbits of comets, on the other hand, cross many planetary orbits, so the perturbations caused by the close passage of a comet near a planet would be enough to alter its orbit.

From where then do comets originate? The first satisfactory theory was proposed in 1950 by Dutch astronomer Jan Oort. Oort noticed that for many new comets a careful study of their orbits showed aphelia at around 50 000 astronomical units. He proposed that at this distance there might be a large cloud of comets, subsequently called the *Oort cloud*. At this distance the comets would be undetectable, but every so often the gravity of a nearby star might perturb

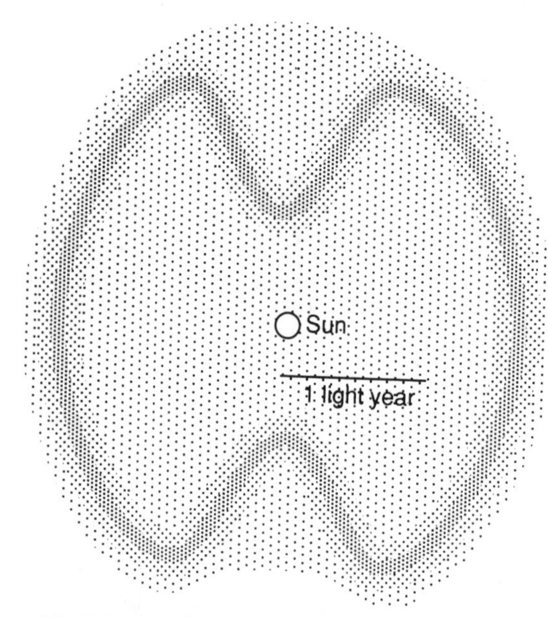


**16.8** It has been hypothesised that the comets which approach the Sun come originally from a large hollow cloud of comets surrounding the Sun at a distance of about one light year. The comets in the cloud, called the Oort cloud, occasionally fall towards the Sun following a collision or because of the gravitational influence of another star.

them, or a collision might occur. knocking one comet on its way towards the Sun. Only when the object neared the Sun would it appear obvious to us. Once past the Sun the comet would continue on its path back to the Oort cloud, only to return again to the Sun in three or four million years. Oort calculated that for the number of comets we see (around ten each year) to be accounted for, the Oort cloud must contain about a billion ($10^{12}$) comets.

Comets represent perhaps the oldest material in the solar system. About half of the comet is made of ices of one form or another, the most abundant being water ice, but carbon dioxide ice, solid ammonia and some hydrocarbons have also been observed. The dust making up the other half of the comet is probably similar to sand. If comets have a rocky core, and observations as yet do not make this certain, it will probably be like the meteorites we see throughout the solar system.

Comets formed in a region of space where the temperature was between $-220$K and $-170$K. This corresponds to the region of space near Uranus and Neptune now. It is supposed that the comet material condensed in this region and then, when the Sun began to shine, the material pushed further from the Sun, some to form the Oort cloud and the rest to be dispersed through the solar system.

If all the comets we see now come from the Oort cloud, then why are there periodic comets? The answer lies with Jupiter and the gas planets. Their gravitational fields are large enough to slow a comet as it enters the solar system, robbing it of energy so that it is unable to leave the region around the Sun. This explains the association of so many comets with Jupiter: its field is so strong that comets are easily captured by it. Saturn too has a collection of comets with aphelia near its orbit. Comet Halley, as we have seen, is associated with Neptune.

Encounters with planets also explain why there aren't more comets in the inner solar system. Not only can the gravity of a planet rob a comet of energy, it can also give energy to the comet. This extra energy can return a previously captured comet to the Oort cloud, or even give it a hyperbolic orbit so that it never returns to the Sun. Another way for planets to clean the solar system of comets is to run into them. With so many passing through the inner solar system, the likelihood of a collision is fairly high.

Just such a collision took place on 30 June 1908 in the wilderness of the Tunguska River in Russian Siberia. Though no-one saw it, the impact of the comet with the atmosphere sent shock waves around the world. A bright fireball accompanied the impact and over a thousand square kilometres of forest were flattened, the pattern of the tree falls radiating away from the blast point. In spite of the force of the explosion, no crater was formed, indicating that the object which caused the impact must have vaporised completely in the atmosphere. The nucleus of a comet is the only explanation. It has been suggested that the object may have been a small part of Comet Encke. The only known human casualty of the impact was a man in a trading settlement 60 km away who was knocked from his chair.

The Tunguska event less than 100 years ago, together with the numerous impacts of meteorites on the Earth and other planets, reminds us of how vulnerable we are here from objects in space. Yet even with that hanging over our heads there is still much of wonder to explore in the region of space we call home.

# Glossary

*Albedo* A measure of the amount of light which reflects from a surface. The Moon, for example, has an albedo of 7 per cent, meaning 7 per cent of the light which reaches it from the Sun is reflected back into space.

*Altazimuth co-ordinates* A system used mainly by amateur astronomers to describe the position of an object in the sky. Altazimuth co-ordinates have two parts, *altitude* and *azimuth*, which specify the height of the object above the horizon and its direction from north respectively.

*Altazimuth mount* A telescope mount that allows you to turn the telescope in two directions, from side to side, (azimuth), and up and down, (altitude). An altazimuth mount is the easiest to set up, but to follow a star through the sky you must constantly adjust the mount's position in both axes.

*Altitude* In astronomy this means the measurement in degrees of the position of an object above the horizon. It is part of one system for specifying the position of an object in the sky.

*Aperture* The diameter of the lens or mirror of a telescope.

*Aphelion* The point in the orbit of a planet or comet at which it is furthest from the Sun.

*Asteroid* see *Minor planet*

*Astronomical unit* An astronomical unit is the average distance of the Earth from the Sun. It is equal to $1.50 \times 10^8$ km.

*Atom* An atom is the smallest particle of matter which has chemical properties. Everything we see around us is made from atoms. There are only 91 naturally occurring atoms, but with nuclear reactors man has made another fifteen.

*Azimuth* This is the measurement of the direction an object appears to be in. To calculate it you draw an imaginary line from the object directly to the horizon below it. You then measure the compass direction of this calculated point starting at north and working towards east. This angular measurement is the azimuth of the object.

*Billion* In this book we use the English form of billion; that is, a one followed by twelve zeros. Some United States books use billion to represent a one with nine zeros; in this book that is called a thousand million.

*Black hole* A black hole is a region of space which is distorted by an incredibly strong gravitational field. The field may be caused by a huge collapsed star which can no longer shine as its fuel is used up. Black holes, if they really exist at all, will be some of the most fascinating objects in the universe.

*Celestial equator* An imaginary line drawn in the sky above the Earth's equator.

*Celestial pole* The celestial poles are the points in the sky directly above the Earth's north and south poles. The north and south celestial poles (NCP and SCP) are the points in the sky about which all the stars seem to rotate.

*Celestial sphere* The apparent sphere of sky that surrounds the Earth. It relates to the ancient concept of the Earth being surrounded by a sphere containing the stars and planets.

*Chromatic aberration* An imperfection introduced by lenses in which the light from an object is split into colours, in a similar way to a prism. Chromatic aberration introduces coloured fringes to the images observed through the telescope. It is reduced by the use of achromatic or apochromatic lenses.

*Coma* The region of gas surrounding the nucleus of a comet. It also describes a defect in a telescope where the image of a star appears to be a fuzzy blob with a small tail.

*Comet* An object made from water and other ices which orbits the Sun, usually in highly elliptical or hyperbolic orbits. Comets are thought to have been formed at the same time as the solar system and are congregated in Oort's cloud one to two light years from the Sun.

*Constellation* An area of sky agreed upon by international consent by which objects are named. Constellations are derived from the pictures ancient astronomers thought they could see in the arrangement of the stars.

*Declination* One of the two co-ordinates used in the equatorial co-ordinate system. Declination is equivalent to the latitude system used on Earth. It goes from +90° at the north celestial pole to −90° at the south celestial pole, with 0° at the celestial equator.

*Deep sky* A term used to describe nebulae, clusters and galaxies: objects which are further away than most stars.

*Density* A measure of the relationship of mass to volume. Density is calculated by dividing the mass of an object in kilograms by its volume in cubic metres, or litres. Throughout this book, $kg.m^{-3}$, or $g.cm^{-3}$ have been used.

*Dichotomy* The time when an astronomical object is exactly half illuminated. The only objects in the solar system that can show this to Earth are our Moon at first and last quarter and the planets Mercury and Venus when they are at greatest elongation.

*Eccentricity* The measurement of the flattening of an ellipse.

*Eclipse* An eclipse appears when one object passes into the shadow of another.

*Ecliptic* The ecliptic is defined as the path the Sun appears to take as it passes through the heavens throughout the year. All the planets, except Pluto, appear close to the ecliptic.

*Elevation* Elevation is the same as altitude.

*Ellipse* The shape of the orbits of the planets around the Sun. An ellipse looks like a flattened circle. The axes of an ellipse are called the *minor axis* and the *major axis*.

*Elongation* The angular distance of one object from another. In astronomy the elongation normally refers to the distance an object appears to be from the Sun. For example, if a comet was 46° away from the sun we would say that its elongation was 46°.

*Epoch* A date chosen to be the reference point for observations. The most commonly used one is Epoch 1950.0, which relates observations to the positions of the stars on 1 January 1950.

*Equatorial co-ordinates* A system used by astronomers to measure the position of an object in the sky. Equatorial co-ordinates consist of two parts, *right ascension* and *declination*, similar to the latitude and longitude system used on Earth.

*Equatorial mount* A mount which allows your telescope to follow a star through the sky by turning it along only one axis. Equatorial mounts are necessary for you to do any photography with your telescope.

*Equinox* For an Earth-based observer the equinox is the day on which both night time and daytime are exactly twelve hours long. For an astronomer it marks the time at which the Sun crosses the celestial equator. The *vernal equinox* is when the Sun crosses from south to north, the *autumnal equinox* when it crosses the other way.

*Filter* Any of a variety of devices which are attached to eyepieces or telescope tubes to modify the light passing through them.

*Finderscope* A small refracting telescope attached to a larger telescope to help point the large telescope.

*First point of Aries* The point defined by astronomers as the start of the equatorial right ascension system. The point is the place at which the ecliptic crosses the celestial equator.

*Focal length* The distance from a lens or mirror to its focus.

*Focal ratio* The focal length of a lens or mirror divided by its aperture. In photography, this refers to the speed of a camera lens.

*Focus* The point where all the converging light rays from a lens or mirror meet.

*Galaxy* A large integrated collection of stars. Galaxies are the largest single structures in the universe. They may contain hundreds of millions of stars and huge amounts of gas and dust.

*Gamma* ($\gamma$) *rays* The highest energy type of electromagnetic radiation.

*Gegenschein* A dull glow opposite the Sun in the sky caused by small particles of dust scattered through the solar system.

*Geocentric* A system which has the Earth at the centre. The view of the solar system proposed by Ptolomy was a geocentric system.

*Heliocentric* Any system which has the Sun at the centre. Our present view of the solar system is a heliocentric system.

*Hypersensitisation* A chemical means of increasing the suitability of photographic film for astronomical work.

*Inclination* The angle that the orbit of one astronomical body makes with the orbit of another body.

*Julian date* A system used by astronomers to specify the date on which an event happened. Julian date is the number of days, including decimal parts, since 12:00 1 January 4713 BC.

*Light year* A light year is the distance a beam of light will travel in one year. It is equal to $9.46 \times 10^{12}$ km.

*Magnitude* This is the system astronomers use to measure the brightness of stars. It was originally based on the Babylonian system of having six main classes of stars, but it has since been refined by scientists. The basis of the system is that each magnitude is a certain number of times brighter than the magnitude above it. The exact scale for the magnitude system is the fifth root of one hundred, or about 2.51.

*Mare, (pl. maria)* The large dark patches on the Moon. Mare is the Latin word for sea and is pronounced 'mar-ay'.

*Meridian* The local meridian is the line running directly overhead from north to south.

*Meteor* A tiny piece of material which enters the Earth's atmosphere at high speed. The particle heats up causing the air around it to glow, producing the bright streak which is sometimes called a shooting star.

*Meteorite* A large piece of material probably left from the formation of the solar system which impacts upon a planet or moon. In the case of the Earth, features such as the Barringer meteor crater are the result; on the Moon, craters such as Copernicus formed from such impacts.

*Milky Way* The Milky Way is the name we give to our own galaxy. At night the Milky Way can be seen as a faint cloudy band of light stretching across the sky, the light of many millions of stars too faint to be seen individually. Our Sun is but one of these stars, located two-thirds of the way from the centre to the edge of the galaxy.

*Minor planet* A minor planet is one of the large number of small bodies which orbit the Sun. Most minor planets are located between the orbits of Mars and Jupiter, but some approach the Sun closer than Mercury, while others are almost as distant as Neptune.

*Nadir* The point directly beneath your feet, that is directly opposite the zenith.

*Nebula (pl. nebulæ)* A huge collection of gas and dust particles which may shine with reflected or transmitted light, or which may block the light of stars behind causing a dark patch in the sky. Nebulæ are both the birthplaces and graveyards of stars.

*Neutron star* Neutron stars are the remains of a massive star after it has exhausted all its nuclear fuels. The atoms in the star collapse, forming neutrons. Neutron stars are also known as *pulsars* because of their appearance to Earth-based observers.

*Nuclear fission* The process currently used in nuclear reactors to produce power. Nuclear fission involves the nucleus of a large atom, such as uranium or plutonium, being made to split into smaller pieces along with the release of heat.

*Nuclear fusion* The process used by stars to produce energy. In nuclear fusion four hydrogen atoms are joined to make one helium atom. In the process a small amount of mass is lost. This mass is converted to energy according to Einstein's equation $E = mc^2$.

*Nucleus* A nucleus is the large ball of ices and dust which forms the central part of a comet.

*Occultation* When one astronomical body passes in front of another.

*Oort's cloud* An hypothesised collection of comets in orbit around the Sun at a distance of one to two light years.

*Orbit* The path followed by one body as it moves around another.

*Parallax* As the Earth orbits around the Sun some of the closer stars appear to shift in relation to the background of stars. This apparent yearly motion is known as the parallax of the star.

*Parsec* A parsec is the distance a star would be from Earth to have a parallax of 1 arc second. It is equal to 3.26 light years or $3.09 \times 10^{13}$ km.

*Perihelion* The point in the orbit of a planet or comet at which it is closest to the Sun.

*Plasma* Plasma has been called the fourth state of matter. A substance is a plasma if the atoms of which it is made have had their electrons torn away by heat or some other process. Stars are balls of hydrogen plasma. A household example of a plasma is the inside of a fluorescent lighting tube or a candle flame.

*Precession* The motion of the stars and planets which causes them to slowly change their equatorial co-ordinates. Precession results from the Earth wobbling on its axis every 25 800 years.

*Pulsar* A pulsar is a neutron star which spins very rapidly upon its axis. Because of its spinning, a beam of light emitted from its polar regions sweeps past the Earth at regular intervals (as often as 300 times per second). This beam appears as a flash or pulse of light, hence the name pulsar.

*Radiant* The point from which meteor showers appear to originate. The radiant is purely an effect of perspective.

*Reflèctivity* see *Albedo*.

*Right ascension* One of the two co-ordinates used in the equatorial co-ordinate system. RA is equivalent to the longitude system used on Earth. It is marked into 24 hours.

*Satellite* A body which orbits another. We can call the Moon a satellite of the Earth, or Triton a satellite of Neptune. Technically the Earth and all the other planets are satellites of the Sun.

*Scientific notation* A shorthand method of writing very large or very small numbers used by scientists. Instead of writing all the zeros in a number, scientists will count them up and write them as a power of ten. For example, one billion can be written as 1 000 000 000 000 or as $1 \times 10^{12}$ in scientific notation. A small number such as one trillionth can be written either as 0.000 000 000 000 000 001 or $1 \times 10^{-18}$.

*Seeing* A method used to describe the clarity and stillness of the night air when making observations.

*Sidereal time* The method of time keeping which uses the motion of the stars rather than the Sun. Twenty-four hours of sidereal time is the same as 23h56m4s of normal solar time.

*Solar Wind* The steam of sub-atomic particles that the Sun throws out into space. These particles cause the Aurorae Australis and Borealis when they reach the Earth.

*Solstice* The solstice marks the time when the Sun's passage reaches the furthest point north or south. It coincides with the longest day, in the case of the summer solstice, and the longest night, in the case of the winter solstice, of the year.

*Spectroscope* A device which splits light up into its component colours, producing a spectrum.

*Spectrum* The spread of colours which makes up the

light from distant stars and planets. By studying the make-up of the spectrum, scientists can tell what materials were present when the light was formed or reflected.

*Tail* The gases and dust left behind by a comet in its orbit about the Sun. The tails of comets can stretch over 100 million km.

*Transit* The movement of one astronomical body across another or through some point in space.

*Twilight* The time between sunset or sunrise and when astronomers consider the sky dark enough to begin observing. Astronomical twilight is when the sun is less than 18° below the horizon. Nautical twilight is when the sun is less than 12° below and civil twilight when it is less than 6°.

*Van Allen belts* The belts or radiation caused by particles trapped in the Earth's magnetic field. The belts were discovered in January 1958 by the United States' Explorer 1 satellite.

*Zenith* The point directly overhead, that is 90° altitude.

*Zodiac* The name given to twelve of the thirteen constellations through which the ecliptic passes.

*Zodiacal light* A band of light which stretches along the ecliptic. It is caused by small particles of dust stretching along the Earth's orbit.

# Bibliography


Airy, George B. (1846) 'Account of some Circumstances Historically Connected with the Discovery of the Planet Exterior to Uranus', *Monthly Notices of the Royal Astronomical Society*, London, vol. 7, pp. 124–5

Audouze, Jean and Israël, Guy (1988) *The Cambridge Atlas of Astronomy*, 2nd edn, Cambridge: Cambridge University Press

Badash, Lawrence (1989 'The Age-of-the-Earth Debate' *Scientific American*, vol. 261, no. 2, pp. 78–83

Batrakov, I.V. et. al. (1988) *Ephemrides of minor planets 1989*, USSR Institute for Astronomy, Leningrad

Binzel, Richard P. et al (1985) 'The Detection of Eclipses in the Pluto-Charon System', *Science*, USA, vol. 228, pp. 1193–5

Brahic, André (1989) 'The Baffling Ring Arcs of Neptune', *Sky & Telecope*, USA, vol. 77, no. 6, pp. 606–9

Burke, James (1985), *The Day the Universe Changed* London: British Broadcasting Corporation

Cook, Lt James (1771), *The Journal of HMS Endeavour 1768–1771*

Drake, Stillman and Kowal, Charles T. (1980) 'Galileo's Sighting of Neptune', *Scientific American*, vol. 243, pp. 52–9

Gavaghan, Helen (1989) 'A celestial odyssey for Galileo', *New Scientist*, London, vol. 123, No. 1685, pp. 35–9

Gregory, Richard (1923) *The Vault of Heaven*, London: Methuen

Harrington, Robert S. and Harrington, Betty J. (1979) 'The Discovery of Pluto's Moon' *Mercury*, USA, vol. 8, pp. 1–3, 6, 17

Hecht, Jeff (1989) 'Radar echoes reveal ice continents on Titan', *New Scientist*, London, vol. 123, No. 1674, p. 16

Henbest, Nigel (1989) 'Neptune: Voyager's last picture show', *New Scientist*, London, vol. 123, no. 1681, pp. 19–22

—— (1989) 'Triton steals the show from Neptune', *New Scientist*, London, vol. 123, no. 1680, p. 4

—— (1989) 'Pluto: the planet that came in from the cold', *New Scientist*, London, vol. 122, no. 1662, pp. 21–6

Herschel, William (1781), 'Account of a Comet', *Philosophical Transactions of the Royal Society*, London, vol. 71, pp. 492–501

—— (1787) 'An Account of the Discovery of Two Satellites revolving round the Georgian Planet', *Philosophical Transactions of the Royal Society*, London, vol. 77, pp. 125–9

—— (1787) 'An Account of Three Volcanos in the Moon', *Philosophical Transactions of the Royal Society*, London, vol. 77, pp. 229–32

—— (1788) 'On the Georgian Planet and its Satellites', *Philosophical Transactions of the Royal Society*, London, vol. 78, pp. 364–78

—— (1912) *The Scientific Papers of Sir William Herschel*, London: The Royal Society and the Royal Astronomical Society

Hickey, James (1952) *Introducing the Universe*, London: Eyre & Spottiswoode

Hughes, David W. (1987) 'Pluto, Charon and eclipses', *Nature*, USA, vol. 327, pp. 102–3

Kowal, Charles T. and Drake, Stillman (1980) 'Galileo's observations of Neptune', *Nature*, USA, vol. 287, pp. 311–3

Lane, Albert L. et. al. (1986) 'Photometry from Voyager 2: Initial Results from the Uranian Atmosphere, Satellites, and Rings', *Science*, USA, vol. 233, pp. 65–9

Littmann, Mark (1988), *Planets Beyond*, New York: John Wiley & Sons

Mason, John (1989) 'Neptune's new moon baffles astronomers' *New Scientist*, London, vol. 123, No. 1674, p. 15

Meeus, Jean (1982) *Astronomical Formulæ for Calculators*, 2nd edn., Richmond: Willmann–Bell

Moore, Patrick (1976) *Guide to the Planets*, London: Lutterworth Press

—— (1978) *The Amateur Astronomer*, London: Lutterworth Press

—— (ed.) (1963) *Practical Amateur Astronomy*, London: Lutterworth Press

Morrison, David and Owen, Tobias C. (1988) *The Planetary System*, New York: Addison–Wesley
Muirden, James (1968) *Astronomy for Amateurs*, London: Cassell
NASA (1984) Viking — *The Exploration of Mars*, US Government Printing Office
Newcomb, (1898) *Popular Astronomy*, New York: MacMillan & Co
Press, Frank and Siever, Raymond (1978) *Earth*, 2nd end, San Francisco: W.H. Freeman and Company
Reidy, David and Wallace, Ken (1987), *The Southern Sky*, Sydney: Allen & Unwin
Roatsch, Thomas (ed.) (1988) *Data of the Planetary System*, Berlin: Akademie–Verlag
Sagan, Carl (1980) *Cosmos*, London: Macdonald
Smith, B.A. et. al. (1986) 'Voyager 2 in the Uranian System: Imaging Science Results', *Nature*, USA, vol. 233, pp. 43–64
Staal, Julius (1963) *Focus on Stars*, New York: Horizon
Tholen, David J. et. al. (1986) 'Improved Obital and Physical Parameters for the Pluto-Charon System', *Science*, USA, vol. 237, pp. 512–4
Tombaugh, Clyde W. (1946) 'The Search for the Ninth Planet, Pluto', *Mercury*, USA, vol. 8, pp. 4–6.
Wilson, Colin (1980) *Starseekers*, London: Hodder & Stoughton

# Index

Entries which refer to figures or tables are in **bold** type. **CP** indicates that the entry is illustrated in the Colour Plates.